Infrastructure Robotics

Infrastructure Robotics

Methodologies, Robotic Systems and Applications

Edited by

Dikai Liu
University of Technology Sydney
Sydney, Australia

Carlos Balaguer
Universidad Carlos III de Madrid
Spain

Gamini Dissanayake
University of Technology Sydney
Sydney, Australia

Mirko Kovac
Imperial College London
London, UK

IEEE Press Series on Systems Science and Engineering
MengChu Zhou, Series Editor

IEEE PRESS
WILEY

Published by John Wiley & Sons, Inc., Hoboken, New Jersey.
Published simultaneously in Canada.

For general information on our other products and services or for technical support, please contact our Customer Care Department within the United States at (800) 762-2974, outside the United States at (317) 572-3993 or fax (317) 572-4002.

Wiley also publishes its books in a variety of electronic formats. Some content that appears in print may not be available in electronic formats. For more information about Wiley products, visit our web site at www.wiley.com.

Library of Congress Cataloging-in-Publication Data

Names: Liu, Dikai, editor.
Title: Infrastructure robotics : methodologies, robotic systems and
    applications / edited by Dikai Liu [and three others].
Description: Hoboken, New Jersey : Wiley, [2024] | Includes bibliographical
    references and index.
Identifiers: LCCN 2023041014 (print) | LCCN 2023041015 (ebook) | ISBN
    9781394162840 (hardback) | ISBN 9781394162857 (adobe pdf) | ISBN
    9781394162864 (epub)
Subjects: LCSH: Robotics. | Infrastructure (Economics)
Classification: LCC TJ211 .I48144 2024 (print) | LCC TJ211 (ebook) | DDC
    629.8/92–dc23/eng/20231101
LC record available at https://lccn.loc.gov/2023041014
LC ebook record available at https://lccn.loc.gov/2023041015

Cover Design: Wiley
Cover Images: Courtesy of Dikai Liu; Courtesy of Prof. Mirko Kovac, Laboratory of Sustainability Robotics, Empa & Aerial Robotics Laboratory, Imperial College London. Video taken by Schwarzpictures.com; Carlos Balaguer

Set in 9.5/12.5pt STIXTwoText by Straive, Chennai, India

Contents

Contents **xiii**

About the Editors

Dikai Liu received his BEng, MEng, and PhD degrees from the Wuhan University of Technology in 1986, 1991, and 1997, respectively. He currently holds the position of distinguished professor at the Robotics Institute of the University of Technology Sydney (UTS), Australia. His primary research interests lie in the field of intelligent robotics, with a specific focus on perception, human–robot collaboration, brain–robot interface, human–robot teaming, robot systems, and design methodology. Besides conducting fundamental robotics research, he has successfully translated his research into practical applications, including infrastructure maintenance, construction automation, manufacturing, and health/aged care. Prof. Liu has led the development of over 10 intelligent robotic systems designed for real-world applications. Examples include autonomous robots for steel bridge maintenance, bio-inspired climbing robots for inspection in confined spaces in steel structures, intelligent robotic co-workers for human–robot collaborative abrasive blasting, smart hoists for patient transfer, and autonomous underwater robots for underwater structure maintenance. Since 2006, his research has received numerous awards, including the 2019 UTS Medal for Research Impact, the 2019 ASME DED Leonardo da Vinci Award, the 2019 BHERT Award for Outstanding Collaboration in Research and Development, and the 2016 Australian Engineering Excellence Awards.

Carlos Balaguer received the BSc, MSc, and PhD degrees from Polytechnic University of Madrid in 1977, 1981, and 1983, respectively. Since 1996, he has been a Full Professor at University Carlos III of Madrid where he is the coordinator of the RoboticsLab, a research group in the field of intelligent robots. His main research topics are intelligent robots design and control, humanoid robots, healthcare and assistive robotics, soft robotics, manipulation, and locomotion planning. He was a member of the Board of Directors of the euRobotics (2015–2021), an association of European robotics with more than 300 affiliated organizations. He was also the President of the International Association for Automation and

Robotics in Construction (IAARC) for the period 2001–2004. He participated in 29 competitive European Union projects and being the coordinator of several of them. He organized numerous important scientific events; among them, he was the General Chair of the IEEE/RSJ International Conference on Intelligent Robots and Systems (IROS'2018), the biggest worldwide scientific conference in the field of Intelligent Robotics with about 5000 attendants. He received several awards, among them for the best book in Robotics by McGraw-Hill (1988); the best paper of the ISARC'2003 in Eindhoven (The Netherlands); IMSERSO's Award 2004 for assistive robots' research; the Industrial Robot journal Innovation Award of the CLAWAR'2005 in London (UK); Tucker-Hasegawa Award 2006 in Tokyo (Japan) for a major contribution in the field of Robotics and Automation in Construction; and FUE's Award 2014 for AIRBUS-UC3M Joint R&D Center.

Gamini Dissanayake is an emeritus professor at the University of Technology Sydney (UTS). He was the James N Kirby Distinguished Professor of Mechanical and Mechatronic Engineering at UTS until his retirement in 2020. He graduated in Mechanical/Production Engineering from the University of Peradeniya, Sri Lanka. He received his MSc in Machine Tool Technology and PhD in Mechanical Engineering (Robotics) from the University of Birmingham, England. He taught at University of Peradeniya, National University of Singapore, and University of Sydney before joining UTS in 2002. At UTS, he founded the UTS Centre for Autonomous Systems that grew to a team of 75 staff and students working in Robotics by 2020. His main contribution to robotics has been in Simultaneous Localization and Mapping (SLAM), which resulted in the most cited journal publication in robotics in the past 20 years. SLAM is the robotic equivalent of a human finding their way around in a city without GPS and maps, thus underpins many robot applications ranging from household vacuum cleaning robots to self-driving cars. He has also been involved in developing robots for a range of industry applications including cargo handling, disaster response, mining, infrastructure maintenance, and aged care.

Mirko Kovac received his BSc and MSc degrees in Mechanical Engineering from the Swiss Federal Institute of Technology in Zurich (ETHZ) in 2005. He obtained his PhD at the Swiss Federal Institute of Technology in Lausanne (EPFL) in 2010 and pursued a Postdoc at Harvard University until 2012. He is now director of the Aerial Robotics Laboratory and full Professor at Imperial College London. He is also heading the Laboratory of Sustainability Robotics at the Swiss Federal Laboratories for Materials Science and Technology (Empa) in Dübendorf in Switzerland. His research group focuses on the development of novel, aerial robots for distributed sensing and autonomous manufacturing in complex natural and man-made environments. Prof. Kovac's specialization is in robot design, hardware development, and multimodal robot mobility. He has received numerous awards, including the ICE Howard Medal in 2021, as well as several fellowships, including the ERC Consolidator Grant in 2022 and the Royal Society Wolfson Fellowship in 2018.

Preface

Maintaining civil infrastructure assets, including bridges, tunnels, water mains, power and telecommunication transmission towers, underwater wharf piles, and sewers, has traditionally been labor intensive and often hazardous for workers. The need for safe, efficient, and effective infrastructure maintenance has led to a desire, across industry sectors, to automate maintenance operations. Intelligent robots that can work either on their own or collaboratively with humans in a structural environment offer a highly promising solution to maintenance operations.

Significant progress has been made in developing robotic systems for civil infrastructure inspection and maintenance. Increasing numbers of students, practitioners, researchers, and professionals are becoming interested in this exciting field. The editors, who have been conducting infrastructure robotics research for over 20 years, think that a book that provides comprehensive, state-of-the-art information about infrastructure robotics will address an important gap in the literature, as there are currently very few books available on this topic. Internationally renowned researchers in infrastructure robotics have authored various chapters of this book.

The book is organized into two parts. Part I discusses the methodologies that enable robots to operate in civil infrastructure environments for inspection and maintenance. It includes an introduction to infrastructure robotics (Chapter 1), the design of infrastructure robotic systems (Chapter 2), robot perception, localization and SLAM (Chapter 3), machine learning and computer vision algorithms (Chapter 4), robotic coverage planning (Chapter 5), and physical human–robot collaboration for infrastructure maintenance (Chapter 6). Part I does not cover the fundamentals of robotics, such as kinematics, dynamics, and control because there are many textbooks that cover these topics

Part II presents 11 case studies that highlight different types of intelligent robotic systems designed for maintaining various civil infrastructures. These include climbing robots, underwater robots, wheeled robots, legged robots, and unmanned aerial vehicles (UAVs). Each case study focuses on a specific type of infrastructure and its corresponding robotic systems. The case studies cover a wide range of civil infrastructures, such as steel bridges (Chapter 7), underwater structures (Chapter 8), tunnels (Chapter 9), underground construction (Chapter 10), underground water mains (Chapter 11), wastewater pipes (Chapter 12), confined spaces in steel structures (Chapter 13), electrical power lines (Chapter 14), oil refineries (Chapter 15), and solar panels (Chapter 16). Additionally, Chapter 17 discusses aerial repair and manufacturing techniques. These case studies provide insights and lessons learned from the development and deployment of robots for practical applications. Readers can gain knowledge and understanding of the challenges, successes, and practical considerations associated with implementing robotic systems for infrastructure maintenance.

The intended readers of this book are researchers, engineers, and graduate students who are interested in design, control, and use of intelligent robotic systems for inspection and maintenance of civil infrastructures. Professionals in civil engineering, asset management, and project management may be also interested in reading this book. We hope that the readers would find this book interesting and useful in their research as well as in practical engineering work.

The editors would sincerely like to express their gratitude to the contributing authors for their professionalism as well as their commitment to the success of the book and to those researchers who contributed to the work included in the book. The editors would like to thank Mr. Dinh Dang Khoa Le for assisting the compilation of the book chapters. The editors would also like to express their appreciation for the support from the IEEE RAS Technical Committee for Robotics Research for Practicality.

Sydney, Australia

Dikai Liu
Carlos Balaguer
Gamini Dissanayake
Mirko Kovac

Acronyms

ASME	American Society of Mechanical Engineers
ASNT	American Society for Nondestructive Testing
AUT	automated ultrasonics, corrosion mapping or C-scan
API	American Petroleum Institute
CML	corrosion monitoring locations
LASER	Light amplification by stimulated emission of radiation
LIDAR	light detection and ranging
MFL	magnetic flux leakage
NDE	nondestructive examinations, or nondestructive evaluation
NDI	nondestructive inspection
RFID	radio frequency identification
UT	ultrasonic testing

Part I

Methodologies

Part I consists of six chapters that discuss the methodologies enabling robots to operate in civil infrastructure environments for inspection and maintenance. Chapter 1 introduces infrastructure robotics and reviews examples of research in several topics. Chapter 2 discusses the design of infrastructure robotic systems, including the design process and examples. Chapter 3 presents the theoretical foundations for perception, localization, mapping, and simultaneous localization and mapping (SLAM). Chapter 4 discusses machine learning, computer vision algorithms and their applications in water utilities and transport. Chapter 5 presents methods for robotic coverage planning in civil infrastructure inspection and maintenance. Chapter 6 presents methodologies for physical human-robot collaboration and their applications in infrastructure maintenance. Part I does not cover the fundamentals of robotics, such as kinematics, dynamics, and control, because there are many textbooks that cover these topics.

Infrastructure Robotics: Methodologies, Robotic Systems and Applications, First Edition.
Edited by Dikai Liu, Carlos Balaguer, Gamini Dissanayake, and Mirko Kovac.
© 2024 The Institute of Electrical and Electronics Engineers, Inc. Published 2024 by John Wiley & Sons, Inc.

1

Infrastructure Robotics: An Introduction
Dikai Liu and Gamini Dissanayake

Robotics Institute, University of Technology Sydney, Sydney, NSW, Australia

1.1   Infrastructure Inspection and Maintenance

Civil infrastructure, such as bridges, tunnels, oil refineries, and pipelines, plays a vital role in both the economy and the community. Due to aging, environmental factors, increased loading, damages caused by human and natural factors, and/or inadequate maintenance, civil infrastructure is progressively deteriorating. Appropriate periodic inspection and maintenance are required to ensure that the designed life of service of civil infrastructure can be achieved or extended. The purpose of inspection, e.g., visual examination or testing, is to assess the condition of infrastructure and detect any defects that may affect the infrastructure's functionality. Maintenance involves activities of cleaning, painting, or fixing problems identified during inspection to keep the infrastructure in good condition. Typical civil infrastructure such as bridges and power transmission towers are large and complex structures. Manual inspection and maintenance require that people to work in confined spaces, at heights, near water, or near live and high-voltage power lines, leading to significant health and safety issues.

Bridges are essential in transport infrastructure worldwide. Bridge maintenance or replacement is one of the biggest expenditure items in transport infrastructure development and maintenance. For example, there are approximately 42,000 steel bridges in Europe, and 210,000 and 270,000 steel bridges, respectively, in the United States and in Japan [Balaguer et al., 2005]. Corrosion is the primary cause of failure in steel bridges [Hare, 1987], and is minimized by painting the steel structure. Inadequate maintenance may result in structural failures such as the Mississippi Bridge incident in Minneapolis that led to 13 fatalities, 145 injuries, and a replacement cost of US $234 million.

Truss joints, rivets, and box girders are common in steel bridges. One example is the Sydney Harbor Bridge (Figure 1.1a), which is a very complex structure. Thus,

Infrastructure Robotics: Methodologies, Robotic Systems and Applications, First Edition.
Edited by Dikai Liu, Carlos Balaguer, Gamini Dissanayake, and Mirko Kovac.

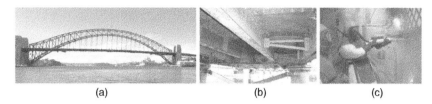

(a)   (b)   (c)

Figure 1.1 (a) Sydney Harbor Bridge, an example of a complex steel bridge structure; (b) A section of a steel bridge structure; (c) Grit-blasting operation.

steel bridge inspection requires special equipment such as special lifts and scaffolds. Some areas may not be easily inspected by humans due to safety concerns (Figure 1.1b). A team is needed to support each inspector, which implies high cost and low productivity.

Steel bridge coating maintenance consists of two procedures: (i) removing rust and old paint, and (ii) repainting. The most effective and efficient method of large-scale paint stripping is grit-blasting (Figure 1.1c), and herein lies the critical problem. Grit-blasting is extremely labor-intensive and hazardous and is arguably the most expensive operation needed in steel bridge maintenance. Workers have to not only spend long periods of time handling forces of 100N or above [Joode, 2004] but also need to take precautionary measures to avoid exposure to dust containing hazardous chemicals. Paints used until the 1980s in most steel bridges in Australia and much of the industrialized world contain red lead. As the long-term health damage due to exposure to lead is now obvious [Information Services, 1998], parts of a bridge being maintained need to be fully enclosed to avoid contamination of the environment and their potential health risk to the general public.

Confined spaces are common in civil infrastructure. Examples of confined spaces include box girders (Figure 1.2a,b) and pipelines. Inspection and maintenance inside a confined space are very dangerous for human workers. In many situations, for example inside the box girder shown in Figure.1.2c, accessing some parts of the confined space is difficult for a human inspector. Furthermore, as rescuing a human inspector in the event of an accident is near impossible, potential safety risk is therefore extremely high.

Inspection and maintenance of underwater infrastructure such as bridge piles, wharf piers, underwater pipelines, and offshore oil/gas platforms are also extremely challenging. These structures vary in shape and size and are normally covered with marine growth. For example, oysters and barnacles can grow up to 20 cm thick, obscuring the structure's geometry and condition (Figure 1.3a). A bridge may have a number of piles, which makes the area cluttered (Figure 1.3b). Deterioration and defects of underwater structures include local scour, deteriorated piles, cracks, and exposed steel reinforcements. In order to comprehensively

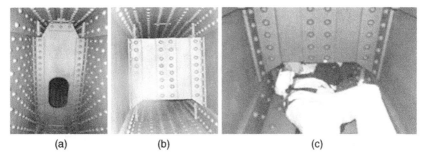

(a) (b) (c)

Figure 1.2 (a) An example of confined space with partition plates each with a standard-sized manhole; (b) A confined space with a diaphragm with spaces above and below; (c) A human inspector has to slide through a space under the diaphragm to access a part of the confined space. Source: [Ward et al., 2014].

(a) (b)

Figure 1.3 (a) Bridge piles covered by marine growth; (b) A bridge with a number of piles.

inspect underwater infrastructure, marine growth must first be removed. This is currently done by human divers using either ultra-high-pressure water jet blasting or mechanical cleaning tools. Methodical planning is required for manual cleaning to ensure the safety of the divers and their support teams, especially because diving in turbid water and in high currents is dangerous. Given the low productivity and expense of having human divers performing such work, complete cleaning is rarely attempted. Rather, marine growth is removed from a strip of a few selected piles to allow partial inspection. Therefore, manual inspection of underwater infrastructure is essentially a sampling process, which is also an extremely dangerous, labor-intensive, and expensive exercise.

Truss structures such as electrical power transmission towers and telecommunication towers are another example of infrastructure. These towers can be about 100 m high. For inspection and maintenance, human workers must climb the towers, carry maintenance tools, and conduct inspection, rust removing, and painting. They often must resort to awkward poses during operation (Figure 1.4a,b).

(a)　　　　　　　　　　　　　　　　(b)

Figure 1.4 (a) A team of human workers painting a transmission tower using brushes; (b) A human worker conducting waterjet cleaning at the top of a tower.

Supplementing manual labor with intelligent robotic aids has the potential to have a significant health, safety, and economic impact on infrastructure inspection and maintenance.

1.2 Infrastructure Robotics

Infrastructure robotics involves developing methodologies that enable robotic systems to construct, maintain, and repair various types of civil infrastructure. These include bridges, buildings, roads, tunnels, underwater structures, and underground water mains. Robotic systems can take various forms, such as legged climbing robots, wheeled robots, underwater robots, drones, and unmanned aerial vehicles (UAVs). The focus of this book is on infrastructure inspection and maintenance, excluding the exploration of robotics for the design and construction of civil infrastructure, which is a separate research topic.

It is reasonable to state that almost all robotics research has the potential to be applied in inspection and maintenance of civil infrastructure. Due to the nature and complexity of civil infrastructure and the specific requirements of inspection and maintenance operations, developing intelligent robotic systems for infrastructure has specific research and development challenges. These challenges include (but are not limited to) [Liu et al., 2014]:

- Sensing technology and sensor networks that can be used in field infrastructural environments for perception and robot navigation, robot interaction with the environment or structural members, and assessment of surface and structural conditions;

- Design of novel (including bio-inspired) mechanisms for robots to move, stay, and support themselves safely in complex environments such as steel bridges and truss structures;
- Methodologies for real-time awareness of environments and operations, including navigation, localization, and map building;
- Efficient algorithms for real-time robot motion planning and collision avoidance in complex 3D environments;
- Design methodologies such as collaborative design and bio-inspired design;
- Multi-robot systems and human-robot teaming;
- Methodologies that facilitate safe and intuitive physical human-robot collaboration (HRC);
- Robot learning and computer vision that allows robots to adapt to various structural and field environments and conduct various tasks safely.

Besides research challenges, there are many engineering challenges in developing robots for civil infrastructure inspection and maintenance. Examples include design of lightweight robots that can move and climb in compact and complex structures, high torque-to-weight ratio actuators that allow robots to have enough payload for performing maintenance tasks, and fail-safe robotic system design.

There is a significant amount of research on infrastructure robotics research and development of robotic systems for infrastructure maintenance. Examples include:

1.2.1 Inspection and Maintenance of Steel Bridges

One of the early works was the robotic system developed to remove corroded paint and rust from the surfaces of steel beams [Lorenc et al., 2000]. This system is composed of a large crane boom, an actuated platform, a robot arm, a vision system, and proximity sensors. It is suitable for small and open bridge structure. A CAD drawing of a bridge structure is assumed to be available and used to reduce the challenges in perception. A manipulator mounted to a crane was also used by Boeing for aircraft maintenance [Schmitz, 2003]. These types of robotic systems can be used for inspecting and maintaining large smooth surfaces but are not suitable for complex, compact, and enclosed areas.

An autonomous grit-blasting robot for surface preparation in steel bridge maintenance was developed and deployed in real applications (Figure 1.5) [Liu et al., 2008, Paul et al., 2010]. This robot consists of a mobile platform, a 6-DoF root arm mounted to the platform, a sensor package, and a grit-blasting tool mounted to

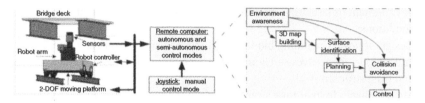

Figure 1.5 A robot for stripping rust and old paint using grit-blasting.

the end-effector of the arm. When this robot is placed on a scaffolding platform under a steel bridge, it autonomously performs sensing and perception to get the geometric information on the environment around it, builds a complete 3D map of the environment [Sehestedt et al., 2013], localizes itself, identifies surface condition, plans its motion and blasting trajectories [Paul et al., 2013], and then starts blasting to remove rust and old paint. This robotic system was extensively tested and verified in a number of field trials on a steel bridge.

The Wall-pushing Autonomous Maintenance roBOT (WAuMBot), as can be seen in Figure 1.6, removes rust and old paint. It vacuums and paints inside a confined space in a steel bridge. This robot has four limbs, each with a scissor lift mechanism to push against the walls of the confined space. The robot body is a 3-DoF manipulator that allows the robot to move forward or backward. Using the sensors mounted on the robot and algorithms for navigation, mapping, planning, and control, the robot can operate in a confined environment fully autonomously.

1.2.2 Climbing and Wheeled Robots for Inspection of Truss Structures

Inspection of truss structures and confined spaces is challenging because it requires a robot that is able to move in all possible directions along truss structure

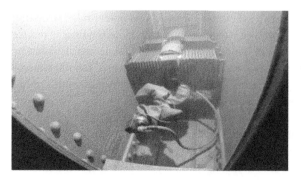

Figure 1.6 A Wall-pushing Autonomous Maintenance roBOT (WAuMBot) inside a still tunnel of the Sydney Harbor Bridge.

members or inside a confined space. At the same time, the robot needs to maneuver sensors and tools on the surfaces of structural members or surfaces inside confined spaces. Due to the challenges of navigation in complex truss structures or inside confined spaces, UAVs may not be an effective solution. One option is to use a wheeled robot, but a wheeled robot needs to be able to safely support itself (against gravity) onto a truss structure or inside a confined space, climb and go through manholes, and negotiate obstacles such as rivet arrays and positive and negative corners between surfaces.

A more feasible strategy is to develop legged climbing robots for this application. Legged locomotion enables operation in difficult and complex terrains where the use of wheels is either inefficient or infeasible. Research on adapting legged locomotion for climbing started in the early 1990s with a strong focus on development of robots for specific applications [Berns et al., 2003]. With advances in robotics research, there has been increasing interest in the use of climbing robots. Examples include the SM2 robot designed to walk along the I-beam structure of a space station [Nechyba and Xu, 1994, 1995], a climbing robot that can move in complex 3-D metal structures [Balaguer et al., 2005, Balaguer, 2000], the Shady3D robot for climbing trusses [Yoon and Rus, 2007], the Stickybot – a gecko-inspired robot that climbs smooth vertical surfaces such as glass [Kim et al., 2008], the RiSE climbing robot [Spenko et al., 2008], and the climbing robots for maintenance and inspection of vertical structures [Schmidt and Berns, 2013].

Most of the climbing robots developed to date are for climbing large flat surfaces or cylindrical surfaces with a large diameter, including smooth walls of glass, brick, steel, or concrete. These robots have difficulties to transfer from one surface to another angled surface (e.g. from wall to ceiling), climb non-flat steel surfaces such as steelwork with rivet arrays, reinforcement plates, or intersections of structural members.

Commonly used adhesion mechanisms for legged or wheeled robots for climbing steel structures include grasping, vacuum, permanent magnet, and electromagnet. Permanent magnets are used in the climbing robotic system developed by Nguyen et al. [2020]. This robot has both a mobile mode and a transforming mode that allows it to adapt to a wide range of surfaces (e.g. flat, curved, or rough).

A novel flying-climbing mobile robot was designed for inspection of steel bridge structure [Pham et al., 2021]. This design combines a drone's maneuverability with the mobile robot's climbing capability, allowing the robot to fly and land to location in a steel structure and conduct inspection. Permanent magnets are used in this robot for adjusting the distance between the robot body and the steel surface, allowing the robot to switch from landing, taking-off, and moving modes.

Many researchers have worked on climbing mechanisms [Webster et al., 2017, Nguyen and Liu, 2017]. Prototype robots have been developed with some of them

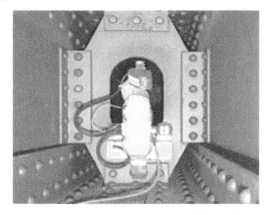

Figure 1.7 A climbing robot inspecting the steel condition inside a confined space tunnel.

extensively tested or deployed in the field. One example is a climbing robot for inspection of steel bridges [Ward et al., 2014]. This robot consists of a 7DoF body and two footpads each with three magnetic toes (Figure 1.7). Once placed inside a steel structure, it autonomously explores the environment [Quin et al., 2016], navigates around, plans safe motions [Yang et al., 2016], localizes itself in the environment, goes through manholes, and collects visual information for condition assessment.

1.2.3 Robots for Underwater Infrastructure Maintenance

Intelligent robots that could clean and inspect underwater infrastructure components, such as bridge piles, wharf piles, underwater pipelines, have significant safety, cost, and health benefits. Developing robots for inspecting underwater structures in harsh shallow underwater environments has many challenges that are caused by unpredictable water current, low visibility, and marine growths that obscure the objects' geometry and condition. Extensive research and development on remotely operated underwater vehicles (ROVs) [Jin et al., 2013] and autonomous underwater vehicles (AUVs) [Caccia et al., 2000] have been conducted globally. Most ROVs and AUVs are designed to operate at depths where water currents are relatively slow and consistent, for example for detecting submerged wrecks and rocks, and mapping underwater environments. Therefore, the aforementioned challenges make the use of ROVs problematic in near-surface and tidal environments. AUVs that have been designed for the ocean environment for survey, monitoring, and oceanographic research may not be suitable because they are often heavy and may not respond fast enough to water current.

Some of the ROVs and AUVs have manipulative abilities. These are typically not designed to operate at relatively shallow depths, nor in the intertidal zone where wave action can be strong and inconsistent [Papadopoulos and Kurniawati, 2014]. The near-surface and tidal environment is hostile. External disturbances can

Figure 1.8 An underwater robot firmly attached to a bridge pile by grasping with its arms uses a waterjet to remove marine growth.

strongly affect the dynamics of a robotic system, particularly near fixed underwater structures. For cleaning and inspection of structures in shallow water environments, the robots must have perceptual capabilities under low visibility and potentially rapid platform motion. The control for robot motion needs to deal with thrusters that have complex nonlinear behavior, particularly in the presence of changing currents. Some seemingly simple operations easily achievable with a ground-based robotic system, such as holding position, can become very difficult for an underwater robot in such environments.

"Crawfish" is a recently developed underwater robot, modified from BlueROV2, for ultrasonic fatigue crack detection and coating of underwater structures. Hardware architecture and software architecture are discussed in Preuß et al. [2022], and Wendt et al. [2022], respectively. A snake eel-inspired multi-joint underwater robot was developed [Lyu et al., 2022] for inspection of undersea infrastructure. Due to its slim design, this robot has high movement flexibility and passing ability inside complex structures. An ROV was also designed [Kizor V et al., 2021] for biofoul cleaning and inspection of underwater structures such as bridge piles and dams.

Figure 1.8 shows an underwater robot developed for removing marine growth and inspection in harsh shallow water environments [Le et al., 2020]. This robot has a body with eight thrusters, four grasping arms, and a number of sensors. It is capable of autonomously navigating to a bridge pile, loosely grasping to a pile, moving around a pile for 3D mapping and marine growth identification, firmly grasping a pile to form a fixed body, and then removing marine growth using waterjet blasting. After the pile surface is cleaned, the robot collects visual information of the surface for inspection.

1.3   Considerations in Infrastructure Robotics Research

A number of specific aspects can be taken into consideration when developing intelligent robotic systems for infrastructure inspection and maintenance.

(1) The specific features of infrastructure environments such as similar structures and geometric shapes. As an example, truss structures are common in steel bridges, electrical transmission towers, telecommunication towers, and offshore oil/gas platforms. Another example is underwater structures such as bridge piles, wharf piles, underwater pipes. Taking these features into consideration and taking advantage of them will help robot design, and at the same time extend the scope of application of a robot.

(2) Robot design: legged robots are more suitable for applications in steel bridges and truss structures compared to wheeled robots. Flying robots (e.g. UAVs) are better for applications with more open spaces and do not require physical interaction with the structure. When a robot is required to physically contact the structure or environment and conduct operations on the surface of a structural member, e.g., removing rust from a steel beam, flying robots need to be able to land on a structure and then firmly grasp the structure member to support the robot and the tools, and resist reaction force from the tools, or wind force.

(3) Co-design: Cooperative design (Co-Design) refers to the creativity of researchers (or robot designers) and potential end-users working together in the design and development process [Sanders and Stappers, 2008]. With Co-Design, researchers, or robot designers view potential end-users as experts in the area of application and involve them in the design process. So, design decisions are made cooperatively.

For intelligent robots to be used in infrastructure maintenance, the application scenario (or work site) is often unique. The end-users' requirements are specific as well. The Co-Design approach is therefore particularly useful. By designing cooperatively with users who have firsthand knowledge of a specific work environment and work tasks, user-friendly results are likely to emerge. Taking a Co-Design approach can aid in reducing anxieties and increasing the levels of trust [Lie et al., 2012].

(4) Knowledge of the application: as the application environments and tasks of robots in infrastructure maintenance are often specific, it is essential to have a good understanding of the application and the requirements. Co-Design is one way; another possible way is site visits to get firsthand knowledge. This has been found to be particularly important because it allows researchers or robot developers to ask the right questions that cover all the aspects of this application. End-users may not know what (or how much details of) information the researchers need. In many cases, after the researchers gain a good understanding of the requirements, the conclusion is that they are not able to design a robot that meets all the requirements until some research questions are answered. Besides identifying research questions, good understanding of the requirements will also help differentiate a proof-of-concept design from a practically deployable prototype robot.

(5) Level of autonomy: The Society of Automotive Engineers (SAE) categorized autonomous vehicles to six autonomous driving levels: level 0 to level 5-from no automation to fully autonomous vehicles. In developing robots for infrastructure inspection and maintenance, the appropriate level of intelligence of a robot is an important factor to consider, for example, capabilities of perception of an unknown environment, awareness of operations, collision avoidance, and motion planning and control. There has been research on autonomous or semi-autonomous robots for infrastructure inspection and maintenance. These robots can operate autonomously for specific tasks and in specific scenarios, for example, the grit-blasting robot, the climbing robot and the underwater robot mentioned earlier.

Due to the complex nature of civil infrastructure, human workers are always needed in robot operation. Careful consideration of the role of human workers will result in human workers' greater willingness to use the robots and in improving human trust in robots. Humans and robots are expected to work together, i.e. human-robot collaboration (HRC) to achieve a shared goal by combining the strength of humans (e.g. perception, decision-making, and uncertainty handling) with the strength of robots (e.g. speed, power, and repeated accuracy). These combined capabilities can relieve humans of arduous tasks such as overhead work or grit-blasting to remove rust and old paint. The level of human involvement, or the roles of a human worker, in robot-facilitated infrastructure maintenance varies significantly, depending on the applications. HRC is another research topic in infrastructure robotics.

1.4 Opportunities and Challenges

The specific applications and their associated requirements make infrastructure robotics research unique and challenging in many ways. There are still a range of research questions and challenges that need to be addressed before robots are widely deployed in inspection and maintenance of civil infrastructure.

(1) Actuation is still considered a bottleneck in developing robots that are lightweight and strong enough for applications in infrastructure environments. The commonly used electrical motors have small torque-to-weight ratios, resulting in heavy and bulky robots. Other actuation methods such as pneumatic, hydraulic, and artificial muscles are useful for some applications, but not for those that require lightweight and powerful robots.

(2) Robot locomotion in an infrastructure environment (e.g. a steel bridge, truss structure) is a challenge. UAVs provide promising solutions for inspection of infrastructure, but they have limited capabilities for interaction with the

environment or structural members. Research on methodologies that enable a UAV to carry a maintenance tool and fly and land to a location in a structure, attach itself to the structure to form a fixed platform, perform maintenance operations, and move to a new location will help address the robot locomotion challenge.

(3) New and reliable sensing technologies are needed for efficient robot navigation in complex structural environments, robot interaction with structural members, and inspection of various types of infrastructure. Currently, robot perception (e.g. sending, localization, 3D map building, and surface condition recognition) of a complex infrastructure environment is arguably the most time-consuming step. In maintenance operations, a robot may need to contact structural members, e.g., using a tool to mechanically remove rust or using a brush to paint a member. Lightweight and low-cost sensors that enable accurate and fast real-time physical robot-environment interaction can be helpful.

(4) Research on HRC for infrastructure maintenance needs to answer the fundamental question of how a robot can physically collaborate with humans. Robot-interpretable human models and control that enable safe, intuitive, and friendly physical HRC are some of the research topics. Human models may include human performance models, human-to-robot and robot-to-human trust models, and human strength models. Other research topics include brain-robot interface, robot learning cognitive conflict recognition, role arbitration human intention recognition dynamics of HRC, and shared control.

(5) When multiple humans and multiple robots are involved at an infrastructure maintenance site, human-robot teaming (HRT) is becoming a topic of research. Dynamic task models, human-skill and performance models, robot capability models, and human-robot task sharing mechanisms need to be developed for optimal collaboration between humans and robots. These models can be comprehensive or simple as long as they can be used in HRT in practical applications. As humans are the centric part in HRT, methodologies need to take social and humans' well-being into consideration. Socio-technical system (STS) principles [Clegg, 2000] can be applied in developing a human-robot team.

(6) Machine learning (ML) algorithms are being implemented in robots. They are useful for infrastructure inspection and maintenance because of the many challenges discussed earlier. With the appropriate ML algorithms, robots can learn from operation and improve their performance over time. Many ML algorithms such as convolutional neural networks, recurrent neural networks, reinforcement learning, and support vector machines, can be used for a robot to detect the environment or objects, segment surfaces, classify conditions, or

predict maintenance operations required. The selection of algorithms depends on the specific task. Attention needs to be paid to ensure that the robot operates safely and effectively in field environments, because the data quality of sensors can be affected by many factors such as weather, lighting conditions, or sensor noise.

1.5 Concluding Remarks

In this chapter an introduction to infrastructure inspection and maintenance, including the current practice and challenges, has been presented. Then, the subject of infrastructure robotics has been introduced, followed by discussions on recent activities and research topics highlighting many research questions and challenges in infrastructure robotics that need to be addressed.

Bibliography

C. Balaguer. A climbing autonomous robot for inspection applications in 3D complex environments. *Robotica*, 18:287–297, 2000.

C. Balaguer, A. Gimenez, and A. Jardon. Climbing robots' mobility for inspection and maintenance of 3D complex environments. *Autonomous Robots*, 18(2):157–169, 2005.

K. Berns, C. Hillenbran, and T. Lucksch. Climbing robots for commercial applications - a survey. In *Proceedings of the Sixth International Conference on Climbing and Walking Robots*, September 2003, pages 17–19, Italy, 2003.

M. Caccia, G. Indiveri, and G. Veruggio. Modeling and identification of open-frame variable configuration unmanned underwater vehicles. *IEEE Journal of Oceanic Engineering*, 25(2):227–240, 2000.

C.W. Clegg. Sociotechnical principles for system design. *Applied Ergonomics*, 31(5):463–477, 2000.

C.H. Hare. Protective coatings for bridge steel, synthesis of HWY, 1987.

Information Services. Lead and you. *HSE Books*, 1998. 7176 1523 5, Caerphilly, UK.

S. Jin, S. Lee, J. Kim, J. Kim, and T.W. Seo. Development of a novel underwater-robotic platform with rotating thrusters for hovering. In *Proceedings of the ICRA 2013*.

B. Joode. *Physical workload in ship maintenance: Using the Observer to solve ergonomics problems*. Noldus Information Technology - Erasmus University of Rotterdam, Rotterdam, Netherlands, 2004.

S. Kim, M. Spenko, S. Trujillo, B. Heyneman, D. Santos, and M.R. Cutkosky. Smooth vertical surface climbing with directional adhesion. *IEEE Transactions on Robotics*, 24(1):1–10, 2008.

Nandha Kizor V, Burhanuddin Shirose, Mainak Adak, Mitesh Kumar, Sudarsana Jayandan J, Arun Srinivasan, and Raashid Sheikh Muhammad. Design of a remotely operated vehicle (ROV) for biofoul cleaning and inspection of variety of underwater structures. In *2021 Ninth RSI International Conference on Robotics and Mechatronics (ICRoM)*, pages 451–457, Tehran, Iran, Islamic Republic of Iran, 2021.

K. Le, A. To, B. Leighton, M. Hassan, and D.K. Liu. The SPIR: An autonomous underwater robot for bridge pile cleaning and condition assessment. In *Proceedings of the 2020 IEEE/RSJ International Conference on Intelligent Robots and Systems (IROS)*, pages 1725–1731, Las Vegas, NV, USA, 2020.

S. Lie, D.K. Liu, and B. Bongers. A cooperative approach to the design of an operator control unit for a semi-autonomous grit-blasting robot. In *Proceedings of the Australasian Conference on Robotics and Automation*, Victoria University, New Zealand, 2012.

D.K. Liu, G. Dissanayake, P.B. Manamperi, and P.A. Brooks. A robotic system for steel bridge maintenance: Research challenges and system design. In *Proceedings of the Australasian Conference on Robotics and Automation*, page 7, Canberra, Australia, 2008.

D.K. Liu, G. Dissanayake, J. Valls Miro, and K.J. Waldron. Infrastructure robotics: Research challenges and opportunities. In *Proceedings of the 31st International Symposium on Automation and Robotics in Construction and Mining (ISARC 2014)*, pages 9–11, Sydney, Australia, 2014.

S.J. Lorenc, B.E. Handlon, and L.E. Bernold. Development of a robotic bridge maintenance system. *Automation in Construction*, 9:251–258, 2000.

F. Lyu, X. Xu, X. Zha, Z. Li, and H. Yuan. A snake eel inspired multi-joint underwater inspection robot for undersea infrastructure intelligent maintenance. In *OCEANS 2022*, pages 1–6. Chennai, India, 2022.

M.C. Nechyba and Y. Xu. SM2 for new space station structure: Autonomous locomotion and teleoperation control. In *Proceedings of the IEEE International Conference on Robotics and Automation*, pages 1765–1770, 1994.

M.C. Nechyba and Y. Xu. Human-robot cooperation in space: SM2 for new space station structure. *IEEE Robotics and Automation Magazine*, 2(4):4–11, 1995.

K. Nguyen and D.K. Liu. Robust control of a brachiating robot. In *Proceedings of the 2017 IEEE/RSJ International Conference on Intelligent Robots and Systems (IROS)*, pages 6555–6560, Vancouver, Canada, 2017.

S.T. Nguyen, A.Q. Pham, C. Motley, and H.M. La. A practical climbing robot for steel bridge inspection. In *Proceedings of the 2020 IEEE International Conference on Robotics and Automation (ICRA)*, pages 9322–9328, Paris, France, 2020.

G. Papadopoulos and H. Kurniawati. Experiments on surface reconstruction for partially submerged marine structures. *Journal of Field Robotics*, 31(2):225–244, 2014.

G. Paul, S. Webb, D.K. Liu, and G. Dissanayake. A robotic system for steel bridge maintenance: Field testing. In *Proceedings of the 2010 Australasian Conference on Robotics and Automation*, page 8, Brisbane, Australia, 2010.

G. Paul, N.M. Kwok, and D.K. Liu. A novel surface segmentation approach for robotic manipulator-based maintenance operation planning. *Automation in Construction*, 29:136–147, 2013.

A.Q. Pham, A.T. La, E. Chang, and H.M. La. Flying-climbing mobile robot for steel bridge inspection. In *Proceedings of the 2021 IEEE International Symposium on Safety, Security, and Rescue Robotics (SSRR)*, pages 230–235, New York, USA, 2021.

H. Preuß, V. Cherewko, J. Wollstadt, A. Wendt, and H. Renkewitz. Blue crawfish goes swimming: Hardware architecture of a crawling skid for underwater maintenance with a blueROV2. In *OCEANS 2022*, pages 1–5. Hampton Roads, VA, USA, 2022.

P. Quin, G. Paul, A. Alempijevic, and D.K. Liu. Exploring in 3D with a climbing robot: Selecting the next best base position on arbitrarily-oriented surfaces. *Proceedings of the 2016 IEEE/RSJ International Conference on Intelligent Robots and Systems (IROS), Daejeon Convention Center*, pages 5770–5775, 2016.

E. Sanders and P.J. Stappers. Co-creation and the new landscapes of design. *CoDesign*, 4:5–18, 2008.

D. Schmidt and K. Berns. Climbing robots for maintenance and inspections of vertical structures - a survey of design aspects and technologies. *Robotics and Autonomous Systems*, 61:1288–1305, 2013.

W. Schmitz. Robotic paint stripping of large aircraft–a reality with the FlashJet® coatings removal process. In *Proceedings of the Aerospace Coating Removal and Coatings Conference*, 2003.

S. Sehestedt, G. Paul, D. Rushton-Smith, and D.K. Liu. Prior knowledge assisted fast 3D map building of structured environments for practical applications. In *Proceedings of the Ninth IEEE International Conference on Automation Science and Engineering (CASE 2013), August 17–21*, Madison, Wisconsin, USA, 2013.

M.J. Spenko, G.C. Haynes, J.A. Saunders, M.R. Cutkosky, and A.A. Rizzi. Biologically inspired climbing with a hexapedal robot. *Journal of Field Robotics*, 25(4–5):223–242, 2008.

P. Ward, P. Manamperi, P. Brooks, P. Mann, W. Kaluarachchi, L. Matkovic, G. Paul, C. Yang, P. Quin, and D. Pagano. Climbing robot for steel bridge inspection: Design challenges. In *Proceedings of the Ninth Austroads Bridge Conference*, Sydney, Australia, 2014.

C. Webster, A. Jusufi, and D.K. Liu. A comparative survey of climbing robots and arboreal animals in scaling complex environments. In *Proceedings of the Fifth IFToMM International Symposium on Robotics & Mechatronics (ISRM2017), 29th November*, Sydney, Australia, 2017.

A. Wendt, H. Preuss, W. Kleinhempel, and H. Renkewitz. Frankenstein goes swimming: Software architecture of a modified BlueROV2 heavy for underwater inspection and maintenance. In *OCEANS 2022*, pages 1–5. Hampton Roads, VA, USA, 2022.

C. Yang, G. Paul, P. Ward, and D.K. Liu. A path planning approach via task-objective pose selection with application to an inchworm-inspired climbing robot. In *Proceedings of the 2016 IEEE International Conference on Advanced Intelligent Mechatronics (AIM)*, pages 401–406, Banff, Alberta, Canada, 2016.

Y. Yoon and D. Rus. Shady3D: A robot that climbs 3D trusses. In *2007 IEEE International Conference on Robotics and Automation*, pages 4071–4076, Roma, Italy, 2007.

2

Design of Infrastructure Robotic Systems

Kenneth Waldron

Robotics Institute, University of Technology Sydney, Sydney, NSW, Australia

2.1 Special Features of Infrastructure

The important feature of infrastructure robotics is that infrastructure is artificial. It has similarities; it has geometric regularities. It is usually constructed of materials entirely different from those found in nature. When designing devices for infrastructure robotics; it is essential that the designer be aware of those features of the environment to take advantage of them in order to simplify the design and improve the performance of the device.

A good example is CROC [Ward et al., 2014]. This device is used for inspection of large steel structures, such as the Sydney Harbour Bridge. It can operate inside steel box girders of varying size. The interior of the girders has a rectangular cross-section of variable height. There are fields of rivet heads. The bottom of the section is often dirty with sediment from old paint and other sources and may be uneven because of rivet heads or plate overlaps. There are diaphragms across the interior of the box at more or less regular intervals. Many of these have manholes of rather small but standard size (Figure 2.1). In other cases, it is necessary to slide through a space under the diaphragm. This environment is problematic for human inspectors and other workers. There is not only a problem extracting a human inspector in the event of an accident, but all of the surfaces are coated with lead-based paint. Obviously, moving about in this environment also presents design challenges for the use of a robotic device.

Infrastructure Robotics: Methodologies, Robotic Systems and Applications, First Edition.
Edited by Dikai Liu, Carlos Balaguer, Gamini Dissanayake, and Mirko Kovac.
© 2024 The Institute of Electrical and Electronics Engineers, Inc. Published 2024 by John Wiley & Sons, Inc.

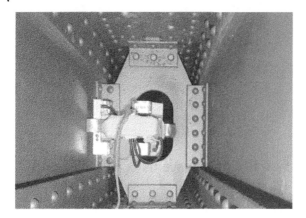

Figure 2.1 The CROC robot in front of a partition plate with a manhole in a confined space.

2.2 The Design Process

Infrastructure robotics deals with applications of robotic technology to development and maintenance of what we term "infrastructure." As such, we are often working on very large systems, and with systems that are not well characterized *a priori* due to structural or maintenance failures, or sometimes simply due to age. This means that the machines used must be adaptable to a wide variety of environments, and capable of handling varied workpieces. They must be able to sense and characterize the objects with which they must interact. All in all, this creates design demands that are much more stringent than for systems that work within relatively well-characterized settings such as in factories.

Machine design is basically an iterative process. If the technology is new it starts with conceptualization: the designer imagines a device concept that may meet a given need. There are formal tools that may be used at this stage of the process such as Quality Function Deployment (QFD), including commonly used tools and procedures such as a brainstorming session, a Pugh Chart, etc. [Ullman, 2009].

Given a design concept and the practical need it must accomplish, a set of performance specifications that the device must meet is next formulated. This should be based not only on the conceived form of the device but on the set of tasks that it is to accomplish. If it is replacing a human worker, as is the case with Submersible Pile Inspection Robot (SPIR) described in Chapter 8 (video), its performance must be measured against what the human can accomplish and should show a definite improvement either in economic performance or by relieving the human of undesirable or unsafe tasks.

The procedure is then to visualize the geometry and mechanics of the device, guess the critical dimensions, then analyze to determine the values of critical parameters, and particularly those performance values that are required by the specification. It is also customary to apply safety factors to the performance values to ensure that they can be met with a high level of reliability. After the analysis is complete, areas of performance that are unsatisfied are identified, and the initially guessed dimensions are adjusted to, hopefully, improve the degree to which the device meets the specifications. The process of analysis and adjustment is repeated as many times as is necessary to satisfactorily meet the values in the specification.

The reason for this iterative process is that our tools for analysis are much better than those for synthesis. Synthesis is a process in which the final dimensions of the device are directly calculated from the specifications. This can be done in some relatively simple situations such as calculating the cross-section of a beam given its length, the load it is to bear and the location of that load. However, real design problems are usually much too complex for this direct approach, so the guess and analysis approach is almost always used.

2.3 Types of Robots and Their Design and Operation

The important distinction in design methodologies for robots is that robots are intended to be multi-functional machines. That does not mean they have to be able to do everything, but they may have to do a variety of tasks within a specified work environment. For example, the robot may be designed to work on electric transmission towers. To move about, it must be able to climb ladders, and/or loco-mote about the varying geometry of the steel trusswork of the tower. In addition, it may need to perform a variety of work tasks ranging from inspecting the structure to grit blasting old paint and applying new paint.

A further complication is that robotic devices are not all the same in terms of their interaction with human operators. Many are what is called teleoperators. These are mechanical devices that are in continuous communication with a human operator who receives visual and other signals from sensors mounted on the robot, and who continuously controls the movements of the device, typically by means of a hand controller and/or pedals. Other robots work autonomously for some periods. When working autonomously, the control of the device is fully self-contained with no input from a human operator. Typically, the device will run autonomously for a set period, with the operator checking on it occasionally and updating whatever it is commanded to do. An example is a driverless car.

Another type of device interacts continuously with a human but not in a remote-control sense like a teleoperator but in a collaborative manner. For example, by sharing a load. An example is the class of devices known as Cobots, typically used to assist humans working in assembly line situations. The Assistance-as-Needed roBOT (ANBOT) described in Chapter 6 is an example of this mode of operation (video). There are examples of devices using all of these modes of operation in the following. Obviously, a very important and early design decision is: Which of these modes of operation best fits the requirements of a given project?

This variety of tasks both makes the formulation of design specifications for a robotic device very much more difficult than for a device with a single function and makes those specifications very much more important to the design process. Formulation of the specifications requires analysis to identify the most extreme service conditions that will be required. It requires the designer to fully understand the operating environment and the tasks that the robot may be required to perform. Given that the robot is an electromechanical device, it is important that the designer be able to visualize its operation in three dimensions. That makes development of animated solid models an important part of the machine design process. The software packages that are used to develop such models, in many ways, make the modern design process possible. Before they were available, it was necessary to build multiple physical prototypes to progressively develop the design concept. It was common to build a succession of prototypes for any complex machine as the design evolved. That was, of course, hugely expensive and so the cost constrained the design process. That, and the lack of control computers, precluded the design of mechanisms with four, five, six, or more degrees of freedom until late in the twentieth century.

A very early design decision is usually the kinematic configuration of the device. This is the arrangement of the joint axes, and the ranges of motion about them, needed to encompass the variety of motions necessary to achieve the device's specified tasks. A good example is the climbing robot (CROC, video) [Ward et al., 2014], described in some detail in Section 2.5. This design went through several kinematic configuration changes before its final form was settled on.

A particular problem here is the possibility of singular positions. These are device positions in which a holonomic serial kinematic chain may lose a degree of freedom, or a parallel mechanism may gain one. For example, if the device assumes a configuration in which three revolute joint axes are parallel, and coplanar, that is an example of a singular position that will require special features in the control software to successfully move through it. A better solution is to configure the joint axes and motion limits so that the device can never traverse a singular position. Another approach, common in design of mobile robots, is to use more than the minimal number of degrees of freedom. This allows for optimization to be applied in software. Of course, adding degrees of freedom may be costly in terms of the weight of the device, since actuators are a significant component of the weight of these devices.

Geometric simulation software is particularly useful at this stage of the design process. The distances between successive joint axes, and their angular relationships must be decided. In other cases, more than six-degrees-of-freedom may be required. An example is the wall-climbing autonomous maintenance robot

(WaumBot), which is the subject of Chapter 13. This robot has a six-degree-of-freedom manipulation system mounted on a three degree of freedom mobility system.

Given a geometric model, along with the loads that the structure might need to resist, or apply, it becomes possible to analyze the system to determine the range of loads to be resisted by individual structural members, the torques and speeds required of the actuators, and the operating conditions that must be covered by other active elements. That, in turn, allows all structural and active elements to be appropriately sized.

An important component of this process is a set of examples of tasks that the device might be asked to perform. Also, in common with any other mechanical design problem, it will be necessary to assign appropriate safety factors applied particularly to required strengths and actuator torques. The combination of position, or position and velocity information along with loads will allow calculation of joint loads and speeds for different candidate geometries. That, in turn, allows the sizing of the actuators and structural members needed. Of course, as was described earlier, it may be necessary to use an iterative process since quantities like the masses of actuators and structural elements must enter into the calculations. Hopefully, each iteration will bring the design closer to performing all the necessary tasks with appropriate safety factors.

2.4 Software System Design

For any robot, the software system is of critical importance. At the present time, this is usually based on a procedural language. One reason for this is that it is required to run on a constant time cycle. That is necessary so that the speeds of motion and actuator torques can be controlled. During each computational cycle, the sensors in the system must be read, and the data thereby acquired must be used to compute appropriate commands for all of the system actuators and other active elements. The computational system must be fast enough to repeat this process hundreds of times each second. Sometimes, there may be a higher-level part of the system that reads more complex sensors, such as a vision system, that runs at slower cycle rates than the actuator cycle rates and may use a different computational architecture such a deep learning structure.

2.5 An Example: Development of the CROC Design Concept

Inspection for the condition of the interior surfaces of the box girders is one of the primary functions of the device. The weight of a robotic device is largely driven by the weights of its actuators. Hence, there is strong motivation to minimize the number of degrees of freedom. Given the likelihood of the use of magnetic adhesion, wheels are unattractive because of the minimal contact areas they provide and their severe limitations when traversing transitions between vertical and horizontal surfaces.

The design of CROC evolved through several different configurations. Early in the design process, the concept pursued was a legged walking robot configuration (Figure 2.2a). This was pursued as far as to build a prototype. However, as was noted earlier, there is strong motivation to minimize the number of actuated degrees of freedom and a four-legged configuration with 12 degrees of freedom (or a 6-legged configuration with 18 degrees of freedom for the limbs, plus two-degrees of freedom in the body) was problematic in terms of negotiating the surface transitions and the small openings noted earlier.

A literature search turned up several examples of climbing robots with an inchworm configuration (Figure 2.2b). In this case, the robot consists of a single,

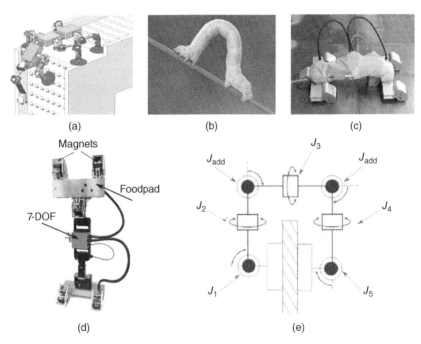

(a) (b) (c)

(d) (e)

Figure 2.2 (a) A legged-climbing robot concept; (b) Inchworm (Source: zhang yongxin/Adobe stock); (c) an inchworm type of climbing robot concept; (d) The final configuration of CROC with Seven revolute joints; (e) kinematic diagram of CROC: each joint axis intersects both its neighbors orthogonally. The "sole" of the top foot is facing forward, showing the camera ports. The lower foot is in support/adhesion position.

serial chain. This minimizes the number of actuated joints as compared to configurations with multiple limbs. It also allows a very low profile suitable for traversing small holes. It was decided to explore the possibilities of using an inchworm configuration with adhesive feet on either end. The concept pursued was a five-degree-of-freedom inchworm. The three middle joint axes (2, 3, 4) are parallel. Joint 1 intersects joint 2 orthogonally, as does joint 5 intersect joint 4. In principle, the axial symmetry of the kinematic chain removed the need for a sixth degree of freedom. However, through simulation, it became apparent that the five degree of freedom device would have serious limitations when transitioning between vertical and horizontal surfaces. It would only be able to accomplish this if it approached the edge orthogonally. For this reason, the decision was made to add actuated degrees of freedom to overcome this problem. In order to preserve the bilateral symmetry of the device two-degrees-of-freedom were added, creating a seven degree of freedom inchworm which is shown on Figure 2.2e. Additional revolute joints were added between joints 2 and 3, and between 3 and 4, of the five degree of freedom configuration. In each case the new joints intersect the previously existing ones orthogonally. The feet on either end are identical. Axes 1 and 5 are normal to the sole planes of the feet. The feet are equipped for magnetic adhesion by means of three magnetic "toes" on each. The cameras are built into the feet. In practice each end of the robot alternates between being the "foot" and the "head" of the device.

As was mentioned, the device must be long enough to reach over at least some fields of rivet heads and must be able to stably cantilever from one foot at full extension in any orientation. The position of the center of gravity at full extension is proportional to the overall length, and, if the structural cross-section remains constant, the weight is also proportional to the length. Thus, the adhesion moment would increase quadratically with the length of the device. Its actually much worse than that because the actuators must increase in weight with length to handle the additional load, and the structure might also need to be strengthened. Obviously, it is imperative to keep the length of the device to the minimum needed to traverse obstacles and locomote effectively.

Since the box girder is made of steel, it is attractive to use magnets to secure the contact between the foot and the structure. However, electromagnets are unattractive because a power interruption would mean the device will fall, and possibly be lost. Permanent magnets remove that problem but cannot be released by simply switching them off. The solution is to use permanent magnets with a mechanical separation mechanism shown in Figure 2.3. Of course, that mechanism carries its own weight penalty [Ward et al., 2013].

CROC does not carry a battery, so it must drag a power cable. That cable is also used to transmit data back to the operator's station. This raises the problem of entanglement of the device in the cable. For a relatively light device like CROC

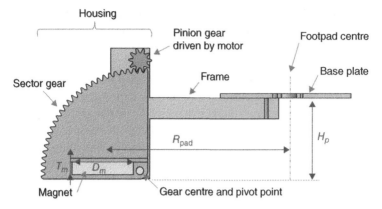

Figure 2.3 The magnet release mechanism contained within each toe of CROC. Source: Ward et al. [2013]/World Scientific Publishing.

cable handling may be left to the operator. In other cases, cable, or hose handling may be a significant problem.

Since the device is intended to perform inspections of the interiors of the box girders, it is natural to think in terms of a telerobot mode of operation with a camera providing the operator with an image of the terrain in front of the device, allowing for continuous operator direction. However, here the geometry of the device and its mode of locomotion enter the picture. The natural mode of locomotion for a device with this geometry is flipping with the underside and top of the chain reversing position with every step. Combining that with the location of the cameras in the feet, this form of locomotion becomes very counter-intuitive for a human operator. Rather, the choice was made to use autonomous operation for most functions. That required the development of sophisticated search algorithms to both explore the space of possible footholds in the environment of the robot and to map the environment [Pagano and Liu, 2017]. Before each step, the algorithm seeks to optimize what might be seen with the robot utilizing each of those footholds in order to select the best movement option [Quin et al., 2017].

The mechanism developed is shown in Figure 2.2. An important capability is that of flexing through a full 360°. This means, for example, that it can adhere to one side of a partition plate and reach through the space under it to place the second foot flat on the other side of the plate in any orientation, as shown in Figure 2.4. Previous inchworm robot designs were unable to do this.

The robot was extensively tested on the Sydney Harbour Bridge before delivering it to the sponsor: the New South Wales Department of Roads and Maritime Services.

The software developed for the autonomous climbing robot (CROC) consists of sensing and exploration, three-dimensional map building, steel surface

Figure 2.4 The geometry involved in CROC's ability to place its feet on either side of a steel plate. Source: Ward et al. [2013].

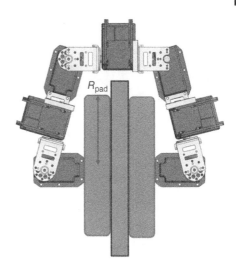

R_{pad}

inspection, situation awareness, motion planning, collision avoidance, and robot motion control (Figure 2.5). For each software module, many algorithms are implemented.

2.6 Some Other Examples

Another example is the WaumBot, described in detail in Chapter 13, as, shown in Figure 2.6. The working environment for this robot is the same as for CROC: the interior of a large box girder of variable height. The motivation here is to provide a device that can do spot repairs by grit blasting and repainting rusted areas. CROC does not have the capability to lift the necessary tooling, so a different solution is necessary. The WaumBot takes advantage of the parallel side walls of the box girder. It has four "legs" with large "feet" that push against the walls to generate enough friction to support the tool arm. The feet are each fitted with four "toes" that allow it to step through rivet fields. Tools needed to perform maintenance functions are carried on the working arm. It walks by pushing one pair of legs against the opposite walls so that the other can be retracted and moved forward. The two bipedal axes are connected by a planar three degree of freedom chain, allowing the foot pairs to be alternately retracted and moved relative to the other foot pair. In this way, the whole robot can move about between the vertical walls of the box girders. A six-degree-of-freedom manipulator is mounted on the mobility platform. This is the portion of the system that uses tooling to perform tasks such as grit blasting, sanding, and painting. Grit blasting requires the device to drag a relatively heavy air hose, as well as power and data cords.

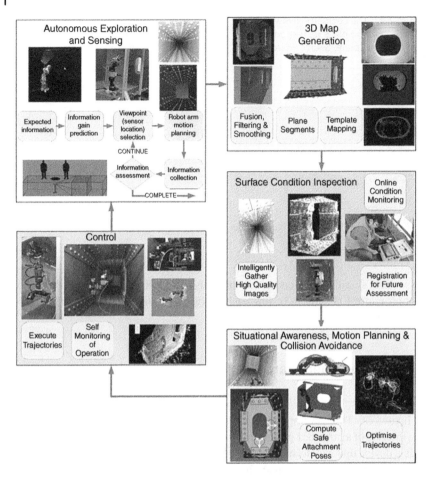

Figure 2.5 Software modules of the bio-inspired climbing robot (CROC).

Yet another example is the SPIR robot described in Chapter 8. Its function is to clean marine growths from bridge and wharf piles and inspect them for structural soundness. For this, it must carry and resist the reaction force from a high-pressure water jet, as well as carrying video cameras. Piles come in different sizes and cross-section geometries, but they all have either constant cross-section, or predictable changes in cross-section. It is necessary for the device to operate under water and to grab and hold the pile to provide a firm base for hydrodynamic cleaning. Thus, the device needs to have a submersible body to swim between piles and enable operation at varied depths, one or more pairs of grabbing arms to grab and secure the device on the pile, and a tool arm to manipulate the water

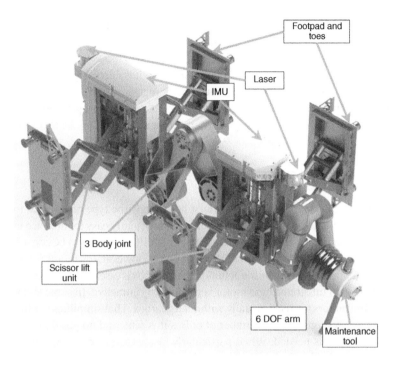

Figure 2.6 This robot is designed to work in the same environment as CROC, but to be able to carry tools for grit blasting and painting. It is described in detail in Chapter 13.

jet and camera. After cleaning a section of pile, the device simply slacks off the grabber arms and moves down, or up, or sideways before re-establishing its hold on the pile. It must also drag a hose to supply the high-pressure water jet, and power and data cables.

Grit blasting in a blast chamber is a different problem. The usual objective is to remove old paint and corrosion from metal objects and structures. The problem is the weight of the tool, whatever force is needed to drag the air hose, and the reaction force from the blasting nozzle. In total, this is of the order of 100 or 200 newtons, depending on the size of the blast nozzle. This load is too large to be resisted by a human operator for an extended period. For this reason, a robot is required to carry the blasting nozzle, so the human operator has only to direct the stream of air and grit. Thus, this is a Cobot problem. The ANBOT is such a device (Chapter 6). It is actually a commercially available UR10 six-degree-of-freedom industrial robot mounted on a fixed but movable base.

Grit blasting of small to medium-sized workpieces is usually done in a blast chamber with a flat floor, so positioning the base is not a problem. The base rests on fixed feet when the machine is in operation. It has rollers that can be lowered to allow it to be easily moved to new locations, but it is then lowered onto the fixed feet before commencing blasting.

In order to properly design and control such a device it is necessary to consider human anatomy and muscular capabilities, including the variation from individual to individual. Thus, the system must be able to adjust for operators with differing physical characteristics: height, reach, strength, etc.

2.7 Actuator Systems

Another important design decision that must be made is the type of actuator to be used in each actuated joint. Electric motors are particularly easy to interface to computer controllers, but then there are a number of different types to be considered. The usual solution is a brushless DC servomotor. However, the motor wants to spin far faster than we need for the actuator, so a gearbox is essential. If a suitable one is available, an integrated gear motor is particularly attractive. In some situations, stepping motors are used, usually without a gearbox. That simplifies control since the motor responds to the number of pulses it is sent, and no position sensor or speed sensor is needed. When particularly fast acceleration is required, a pancake motor might be used.

Their convenience means that electric actuators will be used unless there is some reason to exclude them. Typically, this occurs when very high loads must be controllably accommodated. Infrastructure robots may sometimes be required to perform construction machine-type tasks. The choice then is some form of hydraulic actuation system. The simplest type of system uses a hydraulic fluid reservoir that is maintained at constant pressure by a pump [Stelson, 2011]. The actuators are either hydraulic cylinders or rotary hydraulic actuators. Obviously, hydraulic cylinders function as linear actuators, and rotary actuators are preferable when a continuously rotating output motion is sought.

Unfortunately, conventional valve-controlled hydraulic actuation systems are expensive in terms of energy consumption when applied to robotic systems. Here, a pressurized reservoir is used to supply all the actuators in the system with the flow requirements of each actuator addressed by means of a control valve. Usually, the reservoir pressure is held constant, and the system exhausts to a receptacle at low pressure. The pressure drop between reservoir and exhaust must be large enough to allow the actuators to develop the maximum required force, or torque. The control valve paired with each actuator controls the flow of hydraulic fluid to that actuator. The rate of flow is, of course, proportional to the speed at which the actuator is moving, and there is a pressure drop across the valve proportional to the difference between that required to support the actuator load and the supply pressure drop. A power equal to the product of the pressure drop across the valve, and the flow rate is converted to heat. Typically, that heat is carried off in the hydraulic fluid and must be removed from the system by passing the exhaust fluid through a heat exchanger before returning it to a reservoir. This means, however, that if the actuator is only working against a light load and is moving at substantial speed then a very large power is being converted to heat and wasted. In a robotic system, it is inevitable that some actuators will be running at only a fraction of the load of which they are capable and, likewise, some actuators will be moving fast while others are moving much slower. The net result is that the system consumes much more energy than is actually needed to perform its functions.

At the cost of additional complexity, a more efficient alternative is what has come to be known as hydraulic displacement control. In this mode of operation, each actuator is paired with a variable displacement pump to form a hydrostatic circuit. This system configuration has found its way into some commercial heavy construction machinery. It has been used in several robotic projects, including the Adaptive Suspension Vehicle project [Pugh et al., 1990]. This is a large teleoperated vehicle with a human driver on board. It is intended to carry loads over unimproved terrain.

2.8   Concluding Remarks

In this chapter, we have sought to highlight some of the special features needed when designing robotic systems that must interact with infrastructure for reasons of maintenance, or otherwise. Several detailed examples are laid out in the chapters that follow. Of course, infrastructure takes many forms, presenting problems that require diverse solutions.

Bibliography

David Pagano and Dikai Liu. An approach for real-time motion planning of an inchworm robot in complex steel bridge environments. *Robotica*, 35(6):1280–1309, 2017.

Dennis R. Pugh, Eric A. Ribble, Vincent J. Vohnout, Thomas E. Bihari, Thomas M. Walliser, Mark R. Patterson, and Kenneth J. Waldron. Technical description of the adaptive suspension vehicle. *The International Journal of Robotics Research*, 9(2):24–42, 1990.

Phillip Quin, Gavin Paul, and Dikai Liu. Experimental evaluation of nearest neighbor exploration approach in field environments. *IEEE Transactions on Automation Science and Engineering*, 14(2):869–880, 2017.

Kim A. Stelson. Saving the world's energy with fluid power. In *Proceedings of the Eighth JFPS International Symposium on Fluid Power*, volume 15, 2011.

David Ullman. *EBOOK: The mechanical design process*. McGraw Hill, 2009.

Peter Ward, Dikai Liu, Ken Waldron, and Mahdi Hasan. Optimal design of a magnetic adhesion for climbing robots. In *Nature-Inspired Mobile Robotics*, pages 375–382. World Scientific, 2013.

Peter Ward, Palitha Manamperi, Philip Brooks, Peter Mann, Waruna Kaluarachchi, Laurent Matkovic, Gavin Paul, Chia-Han Yang, Phillip Quin, David Pagano, et al. Climbing robot for steel bridge inspection: Design challenges. In *Austroads Bridge Conference*. ARRB Group, 2014.

3

Perception in Complex and Unstructured Infrastructure Environments

Shoudong Huang, Kai Pan, and Gamini Dissanayake

Robotics Institute, University of Technology Sydney, Sydney, NSW, Australia

3.1 Introduction

An autonomous system requires a map of the environment and its location within that map for efficient operation. Although it is possible to operate a robot purely reactively, e.g., to follow a corridor, almost all practical robot systems used for infrastructure inspection and maintenance are equipped with strategies for building maps of unknown environments and/or estimating robot's location within these maps. For example, the grit blasting robot seen in Figure 3.1a uses the map of the surrounding structure to plan the path of its sand-blasting nozzle such that the paint-removal process is effective. The robot shown in Figure 3.1b uses the estimate of the relative location of its tool with respect to a timber building structure to navigate to a target location and insert a screw. Maps could either be prebuilt or acquired online, whereas location is typically estimated while the robot is in motion. In the grit blasting robot (Figure 3.1a), location of the robot end-effector relative to its fixed robot base is calculated using the measurements of the joint angles. The map of the surroundings is estimated by attaching a camera to the robot end-effector and strategically moving it to capture the 3D structure of the environment relative to the robot base. This map is then stored and used to plan the motion of the blasting nozzle. The timber construction robot (Figure 3.1b) uses a laser range finder to locate itself and navigate in order to insert a screw at a predefined target location. The underwater pylon cleaning robot shown in Figure 8.2 in Chapter 8 uses a stereo camera and an inertial measurement unit to calculate the robot location and map the marine growth on the pylon, in order to clean and then inspect the pylon.

This chapter describes the mathematical background of techniques used for building maps and estimating robot locations. Focus is on how information

Infrastructure Robotics: Methodologies, Robotic Systems and Applications, First Edition.
Edited by Dikai Liu, Carlos Balaguer, Gamini Dissanayake, and Mirko Kovac.

(a)   (b)

(c)   (d)

Figure 3.1 (a) The grit blasting robot. Nozzle attached to the end-effector of the robot emits a stream of grit that removes the paint and rust from the steel bridge structure (BBC News [2013]) (b) Timber construction robot inserting screws into timber floor UTS Robotics Institute [2022]; (c) 2D laser scans and mapping for an unknown environment; (d) Visualization of the data acquired by the camera installed on the robot. Specific camera used captures the image of the scene and the depth of the surfaces present. Source: Meng et al. [2014].

gathered from sensors mounted on a robot to estimate its location and/or generate a description of the surrounding environment. Sensor measurements almost always include noise, therefore, any quantity estimated using this information will include some uncertainties. In practical robot systems, it is important to recognize the presence of errors in maps and location estimates and device action plans such that the desired results are achieved in the presence of such errors. Therefore, this chapter will focus on statistical methods used for estimation under uncertainty as both the estimate of a quantity and the expected error in this estimate are important. This error measure is almost always required for robust decision-making in practical robotic systems.

3.2 Sensor Description

This section will introduce some commonly used sensors in robot perception, including Light Detection and Ranging (LiDAR), sonars, and cameras.

3.2.1 2D LiDAR

LiDAR is a method to determine the distance between an object and the sensor [Mehendale and Neoge, 2020]. 2D LiDAR is a common sensor used in localization, mapping, and SLAM technology. Installed laser range finders calculate the distance in a given direction by measuring the time it takes for a laser beam to arrive at the sensor after being scattered and reflected back by an object in the environment.

Currently, the LiDAR can detect a distance up to 300 meters with an error less than centimeters [Villa et al., 2021]. The angle of view can reach 360° with an accuracy between 0.1 and 1 degree. Therefore, the distance data collected by LiDAR has higher accuracy than other devices, including stereo cameras and GRB-D cameras. Besides, 2D LiDAR also has a relatively small size and low cost, which allows it to be installed on many devices. However, 2D LiDAR can only collect planar data, which restricts it from performing the 3D mapping. Also, particles (e.g. dust and flakes) will affect the accuracy of LiDAR because the light waves cannot penetrate them. Similarly, the LiDAR system is limited in detecting transparent materials (e.g. glass and plastics), so LiDAR is not suggested for working space with particles or transparent materials.

3.2.2 3D LiDAR

With an increasing demand for 3D mappings, like autonomous driving, underground mapping, etc., planar data is insufficient to accomplish such tasks, which requests 3D LiDAR for collecting 3D point clouds. The working principle of 3D LiDAR is similar to 2D LiDAR, but the difference is that 3D LiDAR can emit multiple laser beams simultaneously instead of a single beam. Each laser beam has a predefined angle, which allows it to estimate the elevation of the detected object.

A general 3D LiDAR has a field of view of 40° horizontally and vertically [Raj et al., 2020]. However, due to the sophisticated design, the size, and the cost of 3D LiDAR are higher than other sensors. Besides, multiple laser beams generate massive amounts of data, which significantly increases the computational cost.

In addition, as the 3D LiDAR system also uses the laser beam for detection, it still has limitations for workspace containing some particles and transparent materials.

3.2.3 Sonar

Sound Navigation and Ranging (sonar) is a system to detect an object and measure distance by sound propagation. The working principle of the sonar system is similar to the LiDAR system, which first sends the sound signal and later receives the echo. With the Doppler effect, the distance between the object and the sensor can be calculated by the sound speed and the time sound travels.

Currently, a sonar can detect a range longer than 20,000 yards with speed higher than 2 yards per second [Northardt, 2022]. Based on its working principle and performance, the sonar system has been commonly used in underwater circumstances. Superior to the LiDAR system, the sonar system can detect transparent materials because the sound wave will be reflected after detection. However, it also has limitations since the background noises might interfere with the sound signal, which then affects the accuracy.

3.2.4 Monocular Camera

The camera is another category of sensors other than LiDAR and sonar sensors, which is common in daily life. Instead of collecting point clouds, a monocular camera generates images of the surrounding through lenses. Monocular cameras have various field of view depending on the focal length, but the collected data is in the form of images which does not contain depth information. As a solution, in the mapping process, features of an object from different views can be used for analysis.

Currently, the size of the monocular camera could be very small, and the price can be very low. Cameras have been installed on many devices, like endoscopies and drones, which have been used for the 3D reconstruction of an object. However, as a monocular camera cannot detect the distance between the sensor and the object, mapping using a monocular camera only is still challenging.

3.2.5 Stereo Camera

A stereo camera is a device that installs two monocular cameras with a predefined relative position and orientation. It has a similar working principle to the monocular camera to generate images. However, the two images captured by the stereo camera have a phase difference, which could then estimate the distance by some mathematical calculations.

The stereo camera has been widely used in pose recognition or robot navigation. However, as the depth information obtained by the stereo camera is estimated by mathematical calculation, the accuracy will be affected if the features of an object are not distinct.

3.2.6 GRB-D Camera

GRB-D camera, integrated with a monocular camera and a depth sensor, is a device that not only captures the images like other cameras but also detects and records the depth of the object located at the different pixels in each image. The imaging process is similar to the monocular camera, while the RGB-D camera will also emit an inferred wave and receive it after reflected by objects. By analyzing the time wave traveled and the wave speed, the camera could estimate the distance between the camera and the captured object.

The depth range of the RGB-D camera can now reach 10 meters, and the accuracy can reach 2% of the range or less. With this performance, the GRB-D camera has been used commonly in facial recognition and indoor mapping [Pan et al., 2022]. However, the RGB-D camera also has its limitation of relatively large size and higher cost, which is suggested for indoor mapping.

3.3 Problem Description

localization requires estimating the position and orientation of a point of interest in a robot with respect to a fixed coordinate frame. In the grit blasting robot (Figure 3.1a), this is the location of the camera or the blasting nozzle relative to the fixed robot base. This can be computed at any given time by measuring the angles of the robot joints and knowledge of the robot geometry. Given that the joint angles can be measured very accurately using encoders mounted on joints, uncertainties associated with computed locations is negligible compared to the accuracy requirements of blasting tasks. Therefore, localization in this particular example relies on a set of deterministic kinematic equations. On the other hand, location of the timber construction robot (Figure 3.1b) is computed based on information from a laser range finder that measures the distance to the walls and other objects in the environment in multiple directions with respect to the sensor location.

As mentioned in 3.2, the data collected by sensors has error, and some objects such as glass or mirrors may not provide an adequate return for the LiDAR. An estimation technique that is able to deal with such uncertainties is therefore required in order to make use of these measurements for estimating the robot location.

Map is a geometric description of the environment indicating the locations of features of importance relative to a fixed coordinate frame. In the case of the grit blasting robot, the map consists of the geometry of the surrounding structure that requires cleaning. An example map shown in Figure 3.1d, built using a camera that provides both an image of a scene and the distances to surfaces in the scene. Errors occur due to color and reflectivity of the surfaces in the environment. Also information is never complete due to occlusions and limited field of view of the sensor. Camera is therefore moved to strategic locations and information collected from multiple viewpoints need to be statistically combined to capture the important aspects of the environment. A plan for efficiently cleaning the required surfaces can then be generated.

In the examples presented above, localization relied on a prebuilt map, and mapping relied on the location of the sensor used for observing the environment. In the absence of such information both robot location and the map of the environment need to be estimated from information acquired by sensors. This is known as simultaneous localization and mapping or SLAM. For example, the underwater robot for bridge pylon cleaning (Figure 8.2 in Chapter 8), needs a 3D map of the marine growth on the pylon for cleaning and inspection. As the locations of both the robot and the pylon are unknown, a stereo camera on board the robot is used to capture images of the pylon. These images are then processed to extract points on the pylon that can be identified from different viewpoints. The location of these points, captured while the robot rotates around the pylon are then used together with measurements from an inertial measurement unit on board to construct the map of the pylon while at the same time estimating the locations from where the images were taken.

Localization, mapping, and SLAM require estimating a description of the environment and the location of the robot within this environment as it travels, from multiple observations captured by the sensors on board the robot. This can be mathematically described as follows. Given a 3D environment described by M and a robot whose location described by P moving in it following a motion model with known control input u while collecting observations z using sensors mounted on it.

- Localization is estimating P given observations z, control input u, and the map M;
- Mapping is estimating M given z and P;
- SLAM is estimating both P and M from observations z and control input u.

The next section describes the statistical tools that are typically used for these estimation tasks.

3.4 Theoretical Foundations

Localization, mapping, and SLAM require finding the unknown values of a set of variables (P, M, or both P and M) given a sequence of observations (z) and known control inputs (u). Techniques of statistical estimation are best suited for solving this problem as observations (z) are almost always corrupted by noise and the models that describe the behavior of the robot in response to commands given (u) are never perfect. A useful byproduct of using statistical estimation is that the expected error in the estimates is also become available. As mentioned in the previous section, this information is crucial for robust decision-making. In this section, theoretical foundations of statistical estimation are described together with how these are applied to robot localization and mapping problems.

3.4.1 Extended Kalman Filter

Extended Kalman filter (EKF) is a method for estimating the state x of a dynamic systems given a model that describes how the state evolves with time in response to control commands (u), and an observation model that relates the observations made using sensors (z) to the state x. EKF relies on approximating the nonlinear motion and observation models using linear equations and approximating observation noises using Gaussian distributions. These are reasonable assumptions under many practical conditions and therefore EKF is a suitable choice for solving the robot localization problem and SLAM problem. EKF is best suited for these tasks when the map of the environment can be described using a set of locations and a sensor is available to observe and recognize these locations as it travels. Locations in the environment that can be recognized from multiple viewpoints are known in robotics literature as landmarks.

Consider a discrete-time nonlinear dynamic system (such as a robot motion model) given by

$$\mathbf{x}_{k+1} = \mathbf{f}(\mathbf{x}_k, \mathbf{u}_k, \mathbf{w}_k), \tag{3.1}$$

where $\mathbf{x}_k$ is the system state vector, $\mathbf{f}$ is the system transition function that describes how the state evolves in time in response to control inputs $\mathbf{u}_k$. $\mathbf{w}_k$ is the zero-mean Gaussian process noise $\mathbf{w}_k \sim N(0, Q)$ essentially capturing the differences between the true system and its mathematical model.

Given sensors that can measure some quantity that is a function of the system state (such as the distance from the robot to a landmark), an observation model can be written as

$$\mathbf{z}_{k+1} = \mathbf{h}(\mathbf{x}_{k+1}) + \mathbf{v}_{k+1}, \tag{3.2}$$

where $\mathbf{h}$ is the observation function that relates the system state to the sensor measurements, and $\mathbf{v}_{k+1}$ is the zero-mean Gaussian observation noise $\mathbf{v}_{k+1} \sim N(0, R)$ quantifying the errors in the measurements. The zero-mean assumption essentially indicates that the sensor measurements are correct on the average. In practice a sensor calibration process is used to make sure that this assumption is satisfied.

In EKF, all the estimation results are described by a Gaussian distribution. Let the estimate of the state at time step k, $\mathbf{x}_k$, be

$$\mathbf{x}_k \sim N(\hat{\mathbf{x}}_k, P_k), \tag{3.3}$$

where $\mathbf{x}_k$ is the estimated state, and P_k is the corresponding covariance matrix describing the uncertainty of this estimate. The state estimation problem becomes one of estimating $\mathbf{x}_{k+1}$ at time $k + 1$

$$\mathbf{x}_{k+1} \sim N(\hat{\mathbf{x}}_{k+1}, P_{k+1}) \tag{3.4}$$

where $\hat{\mathbf{x}}_{k+1}, P_{k+1}$ are updated using the information gathered using the sensors. The EKF framework proceeds as follows. To maintain clarity, only the basic equations are presented in the following while more detailed explanation can be found in the references such as Thrun et al. [2005].

Predict the expected state of the system using the system dynamic model and the control input $\mathbf{u}_k$:

$$\overline{\mathbf{x}}_{k+1} = \mathbf{f}(\hat{\mathbf{x}}_k, \mathbf{u}_k, 0) \tag{3.5}$$

$$\overline{P}_{k+1} = \nabla \mathbf{f}_\mathbf{x} P_k \nabla \mathbf{f}_\mathbf{x}^T + \nabla \mathbf{f}_\mathbf{w} Q \nabla \mathbf{f}_\mathbf{w}^T \tag{3.6}$$

where $\nabla \mathbf{f}_\mathbf{x}$ is the Jacobian of function $\mathbf{f}$ with respect to $\mathbf{x}$, $\nabla \mathbf{f}_\mathbf{w}$ is the Jacobian of function $\mathbf{f}$ with respect to $\mathbf{w}$, both are evaluated at $(\hat{\mathbf{x}}_k, \mathbf{u}_k, 0)$.

Update the state estimate using observations $\mathbf{z}_{k+1}$ captured by a sensor and the observation model that relates these observations to the system state:

$$\hat{\mathbf{x}}_{k+1} = \overline{\mathbf{x}}_{k+1} + K(\mathbf{z}_{k+1} - \mathbf{h}(\overline{\mathbf{x}}_{k+1})) \tag{3.7}$$

$$P_{k+1} = \overline{P}_{k+1} - KSK^T \tag{3.8}$$

where the innovation covariance S (here $\mathbf{z}_{k+1} - \mathbf{h}(\overline{\mathbf{x}}_{k+1})$ is called innovation) and the Kalman gain K are given by

$$S = \nabla \mathbf{h} \overline{P}_{k+1} \nabla \mathbf{h}^T + R \tag{3.9}$$

$$K = \overline{P}_{k+1} \nabla \mathbf{h}^T S^{-1}, \tag{3.10}$$

where $\nabla \mathbf{h}$ is the Jacobian of function $\mathbf{h}$ evaluated at $\overline{\mathbf{x}}_{k+1}$.

Recursive application of the above equations at every instant a new observation is gathered yields an updated estimate for the current state and its uncertainty. This recursive nature makes EKF the most computationally efficient algorithm available for state estimation problems.

3.4.2 Nonlinear Least Squares

Consider the problem of finding the solution x to a set of equations $Ax - b = 0$ in which there are more equations than unknowns. Least squares method finds x that minimizes the sum of the squares of the residuals $Ax - b$.

In the robot localization, mapping, and SLAM problems, unknowns are all the robot poses P (for localization), or landmark locations M (for mapping), or both P and M (for SLAM). A set of equations can then be obtained from the process model that relates the states at two successive time instants, and the observations model that relates the observations acquired by a sensor (z) that relates the states robot to each landmark.

Denoting the vector of all the unknowns as $X = [x_1, x_2 \cdots, x_n]^T$, and integrating all the measurements together as a single vector $Z = [z_1, z_2 \cdots, z_m]^T$, and all the available equations can be written together as

$$Z = F(X) + V \tag{3.11}$$

where $V = [v_1, v_2 \cdots, v_n]^T$ is the vector of all the noises involved.

$F(X)$ is the combination of all the functions from the available equations (e.g. from the system dynamic model and the observation functions)

$$F(X) = \begin{bmatrix} f_1(X) \\ f_2(X) \\ \vdots \\ f_m(X) \end{bmatrix} = \begin{bmatrix} f_1(x_1, x_2, \cdots, x_n) \\ f_2(x_1, x_2, \cdots, x_n) \\ \vdots \\ f_m(x_1, x_2, \cdots, x_n) \end{bmatrix}.$$

Since there are more equations than unknowns, $m \geq n$.

The nonlinear least squares (NLLS) method finds unknown parameters X that minimizes

$$\|Z - F(X)\|^2 = [Z - F(X)]^T [Z - F(X)].$$

In general, a closed-form solution to this problem cannot be obtained. Techniques available for solving NLLS problems are, therefore, based on iteration.

Suppose X is close to X_0, by Taylor expansion (linearization),

$$F(X) \approx F(X_0) + J_F(X_0)(X - X_0)$$

where $J_F(X_0)$ is the Jacobian matrix given by

$$J_F(X) = \begin{bmatrix} \frac{\partial f_1}{\partial x_1} & \frac{\partial f_1}{\partial x_2} & \cdots & \frac{\partial f_1}{\partial x_n} \\ \frac{\partial f_2}{\partial x_1} & \frac{\partial f_2}{\partial x_2} & \cdots & \frac{\partial f_2}{\partial x_n} \\ \vdots & \vdots & \vdots & \vdots \\ \frac{\partial f_m}{\partial x_1} & \frac{\partial f_m}{\partial x_2} & \cdots & \frac{\partial f_m}{\partial x_n} \end{bmatrix}$$

evaluated at X_0.

Thus

$$Z - F(X) \approx Z - F(X_0) + J_F(X_0)X_0 - J_F(X_0)X.$$

Let

$$A = J_F(X_0), \quad b = Z - F(X_0) + J_F(X_0)X_0,$$

using the linear least squares solution

$$X^* = (A^TA)^{-1}A^Tb \tag{3.12}$$

for minimizing

$$\|b - AX\|^2 = (b - AX)^T(b - AX), \tag{3.13}$$

we get

$$X_1 = [J_F^T(X_0)J_F(X_0)]^{-1}J_F^T(X_0)[Z - F(X_0) + J_F(X_0)X_0]. \tag{3.14}$$

The above process can be repeated. In general, the iteration step at step k is

$$X_{k+1} = [J_F^T(X_k)J_F(X_k)]^{-1}J_F^T(X_k)[Z - F(X_k) + J_F(X_k)X_k]. \tag{3.15}$$

Iterating until convergence leads to the optimum solution, provided that the initial guess X_0 is sufficiently close to the solution. This is known as the Gauss-Newton iteration.

In practice, different observations have different accuracy; thus, the noise vector V in (3.11) can be generally described as zero-mean Gaussian with covariance matrix P_V, which is the integration of the covariance of all the observation noises and process noises (all assumed to be zero-mean Gaussian). With the observation uncertainty considered, the **weighted nonlinear least squares** problem is to find X that minimizes

$$[Z - F(X)]^T P_V^{-1}[Z - F(X)].$$

Given an initial value X_0, then Gauss-Newton iteration step of weighted nonlinear least squares is:

$$X_{k+1} = [J_F^T(X_k)P_V^{-1}J_F(X_k)]^{-1}P_V^{-1}J_F^T(X_k)[Z - F(X_k) + J_F(X_k)X_k]. \tag{3.16}$$

EKF and NLLS can be applied to many estimation and optimization problems. In the next few subsections, how to apply statistical estimation tools such as EKF and NLLS to the different robot localization, mapping, and SLAM will be discussed.

3.4.3 Environment Representations

Before talking about the robot localization, mapping, and SLAM techniques. The different ways to represent the environment will be briefly discussed. The different localization, mapping, and SLAM algorithms are closely related to the different environment representations.

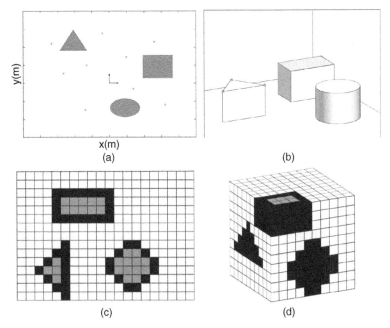

Figure 3.2 Different environment representations. (a) An example of a 2D feature-based map. The environment contains some point features, line segments, and ellipse, (b) An example of a 3D feature-based map. The environment contains some point features, line segments, planes, and pillars. The dots, lines, and planes are some examples of point features, line features, and plane features, respectively, (c) An example of a 2D occupancy grid map. The black grid means occupied, the white grid means free space, and the gray grid means unknown area, (d) An example of a 3D occupancy grid map. The black grid means occupied, the white grid means free space, and the gray grid means unknown area.

Feature-Based Map

One typical way to represent the environment is to use different types of geometric features/landmarks to represent the important parts of the environments. For example, points, lines, and planes are all commonly used features. Figure 3.2a shows an example of 2D feature map, and Figure 3.2b shows an example of 3D feature map.

In the case of using features to represent the environment, the map M can be represented by the parameters (position, orientation, size etc.) of the different features.

Points

Position of the points is the feature parameters. These can be corner points detected by 2D laser scanners or distinct features in the environment extracted by image feature detection techniques obtained from images obtained from cameras.

Lines

Either straight-line segments or curves can be used as features. These usually describe the boundary of the objects of obstacles. These can be extracted by line detection methods from images. Different parameters can be used to describe the line features. For example, a line equation, or two end points of the line segment.

Planes

Planes can also be used as features in 3D environments. For example, surfaces of walls and ceilings can be detected by a depth camera or a 3D laser scanner. The parameters in a plane equation can be used to describe the plane feature.

Occupancy Grid Map

Another way is to use occupancy grids to represent the environment. In this case, the environment is divided into small grids; each grid is either classified as occupied, free, or unknown. Figure 3.2c shows an example of 2D occupancy grid map, and Figure 3.2d shows an example of 3D occupancy grid map.

For occupancy grid mapping, the value of each grid needs to be estimated from the sensor information. The value can be either discrete (e.g. 1 means occupied, 0 means free, and 0.49 means unknown) or continuous (e.g. the probability of being occupied). More details will be provided in Section 3.4.4.

3.4.4 Mapping Techniques

The mapping techniques discussed in this subsection, assuming the sensor locations for taking the observations, are exactly known. In practice, this could be the case when the sensors are mounted on a robot manipulator where the location of the sensor can be obtained accurately from the joint angles and some initial calibrations. The grit blasting robot in Figure 3.1a is such an example.

Feature-Based Mapping

For feature-based mapping. The environment is represented by distinct features such as points, lines, and planes. The mapping problem is to estimate the parameters of these features using the observations detected from the sensors.

Sensor Model for Feature-Based Mapping

The sensor model for feature/landmark-based mapping needs to describe the sensor information using the (known) sensor location and the feature/landmark parameters. If the original sensor information does not contain the feature parameter directly, then some preprocessings are required to extract the information related to the feature parameters.

For example, consider an environment that contains N 2D point features. The position of feature i is denoted as (x_L^i, y_L^i). If the sensor mounted on the robot can obtain the measurement of both the range (distance from the sensor to the feature) and the bearing (angle relative to the robot heading) to landmark i at time step $k + 1$, then the observation model is given by

$$r_{k+1}^i = \sqrt{(x_L^i - x_{k+1}^r)^2 + (y_L^i - y_{k+1}^r)^2} + w_r$$
$$\theta_{k+1}^i = atan\left(\frac{y_L^i - y_{k+1}^r}{x_L^i - x_{k+1}^r}\right) - \phi_{k+1}^r + w_\theta \tag{3.17}$$

where $w_r \sim N(0, \sigma_r^2)$ and $w_\theta \sim N(0, \sigma_\theta^2)$ are zero-mean Gaussian observation noises.

Laser range finders and ultrasonic sensors are the most common sensors used for obtaining range and bearing measurements of landmarks. In case of a sensor that is only able to observe the bearing, for example, a camera, the equation for θ_{k+1}^i becomes the sensor model.

Feature-Based Mapping Algorithm

Both EKF and NLLS can be used to estimate the feature parameters. For the example given above, the parameters to be estimated are the feature positions (x_L^i, y_L^i), $i = 1, \dots, N$.

When EKF is used, the state vector will be $\mathbf{x}_k = (x_L^1, y_L^1, \cdots, x_L^N, y_L^N)^T$. Since features are stationary, and there is no dynamic model; thus, the prediction step is not needed, i.e. $\bar{\mathbf{x}}_{k+1} = \hat{\mathbf{x}}_k, \bar{P}_{k+1} = P_k$. The update step follows the process described in (3.7) and (3.8) where $\mathbf{z}_{k+1}$ is the observation vector containing all the observations $(r_{k+1}^i, \theta_{k+1}^i)$ made at time $k + 1$. Thus EKF will obtain a new updated map after the observations made at each time step.

When NLLS is used to perform the mapping, the parameters to be estimated is the same vector $X = (x_L^1, y_L^1, \cdots, x_L^N, y_L^N)^T$. However, NLLS will use all the observations made to estimate X in one go as a batch process. That means, Z is the vector contains all the observations made from time step 0 to the very end. The solution can be obtained through Gauss-Newton iteration described in (3.16).

When the observations of different features are independent of each other. In the mapping process using EKF and NLLS, each feature can be estimated separately using the information from the observations to this feature only. That is, $X_i = (x_L^i, y_L^i)$ and $X_j = (x_L^j, y_L^j)$ can be estimated separately.

Data Association

For feature-based mapping, one critical preprocessing step is to make the decision on the correspondence between an observation from the sensor and a particular feature. This step is called data association. Data association is crucial to the correct mapping result, since catastrophic failure may result if an observation of one feature is treated as the observation of another feature. Data association is also a critical step for robot localization and SLAM.

Occupancy Grid Mapping
As mentioned in Section 3.4.3, an occupancy grid map assigns a numerical value to each grid representing the occupancy of that small part of the environment.

Sensor Model for Occupancy Grid Mapping
Many sensors now can be used for occupancy grid mapping, including sonar, 2D or 3D Lidar, et al. [Casals, 2012]. These kinds of sensors can detect the range of obstacles by sending and receiving signals. With this information, it is possible to build the occupancy grid map with an appropriate mathematical algorithm.

Figure 3.3a shows the sonar sensor model for building an occupancy grid map.

Evidence Grid Mapping
Evidence grid mapping [Elfes, 1989] is now a popular method for grid-based mapping. In this approach, the status of each grid is described by the **logarithm** of its **odds** value. Here, the odds is the ratio between the probability of being occupied $p(o)$ and the probability of being free $p(\bar{o})$.

$$log_2(odds) = log_2 \left(\frac{p(o)}{p(\bar{o})} \right). \tag{3.18}$$

Suppose there are two sensor readings S_1 and S_2 obtained in a sequence. The odds value of a grid giving both the two sensor readings, $odds(o|S_2 \wedge S_1)$, defined

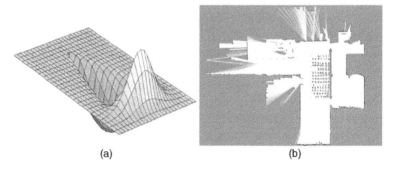

(a)   (b)

Figure 3.3 (a) The red dot in the sonar sensor model is the position of the sensor, and the empty space between the sensor and the bumped place means no obstacles. (b) An example of evidence grid map built from laser data, where the black, white, and gray area means occupied, free, and unknown area, respectively. Source: Adapted from Zhao et al. [2022].

as the ratio between $p(o|S_2 \wedge S_1)$ and $p(\bar{o}|S_2 \wedge S_1)$, can be derived by

$$
\begin{aligned}
odds(o|S_2 \wedge S_1) &= \frac{p(o|S_2 \wedge S_1)}{p(\bar{o}|S_2 \wedge S_1)} \\
&= \frac{p(S_2 \wedge S_1|o)p(o)}{p(S_2 \wedge S_1|\bar{o})p(\bar{o})} \\
&= \frac{p(S_2|o)p(S_1|o)p(o))}{p(S_2|\bar{o})p(S_1|\bar{o})p(\bar{o})} \\
&= \frac{p(S_2|o)p(o|S_1)}{p(S_2|\bar{o})p(\bar{o}|S_1)}
\end{aligned}
\tag{3.19}
$$

where $p(o|S_1)$ is the probability of the grid being occupied given the sensor reading S_1, and $p(\bar{o}|S_1)$ is the probability of the grid being free given the sensor reading S_1. $p(S_2|o)$ is the probability of getting sensor reading S_2 given the grid is occupied, and $p(S_2|\bar{o})$ is the probability of getting sensor reading S_2 given the grid is free.

Equation (3.19) is based on the Bayes' rule and the conditional independence of S_1 and S_2.

From (3.19), we can get the following useful equation in evidence grid mapping:

$$
log_2(odds(o|S_2 \wedge S_1)) = log_2\left(\frac{p(S_2|o)}{p(S_2|\bar{o})}\right) + log_2\left(\frac{p(o|S_1)}{p(\bar{o}|S_1)}\right).
\tag{3.20}
$$

Equation (3.20) shows that the evidence after getting sensor readings S_1 and S_2, is simply the previous evidence after getting sensor readings S_1 only, plus the new evidence from sensor reading S_2, which is the logarithm of the ratio $\frac{p(S_2|o)}{p(S_2|\bar{o})}$.

3.4.5 Localization Techniques

The localization problem aims to find the robot path P given the map M and the observations z, and optionally, the robot motion model together with the control input u.

Robot Motion Model

The motion model of a robot (include manipulator, wheeled robot, legged robot, flying robot, or a mobile manipulator) can be expressed by a differential equation

$$
\dot{\mathbf{x}} = f(\mathbf{x}, \mathbf{u}, \mathbf{w})
\tag{3.21}
$$

where $\mathbf{x}$ is the state of the robot (such as position and orientation), $\mathbf{u}$ is the control signal, $\mathbf{w}$ is the disturbance signal.

Discretizing the continuous-time motion model (3.21) using a sampling time ΔT (decided by the sensor frequency) results in a discrete-time dynamic model in the form of (3.1) where $\mathbf{x}_k$ is the robot location at time step k (Figure 3.4).

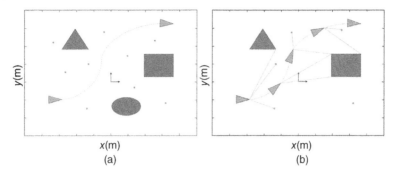

Figure 3.4 (a) The robot localization in a 2D feature-based map. The dark gray triangle illustrates the robot position and orientation. The environment contains some point features, line segments, and ellipse. (b) localization example with 2D feature-based map: The map is defined by a triangle, a rectangle, and few points. The dark gray triangles represent the robot position at the different time step alone with its trajectory, and the gray straight lines indicate the robot observation based on its current location. For example, at the first step, it detected a corner of the light gray triangle and two-point features.

Feature-Based Robot Localization

The main idea of a localization algorithm is to compare the sensor information obtained from the robot with the map, and estimate the location of the robot.

When the map is represented by features, robot localization problem is to estimate the robot location using the sensor information and the provided feature-based map (Figure 3.4).

Observation Model

When the map is given as a 2D point feature map and the observation contains range and bearing, the observation model for localization problem is the same as that for mapping, as given in (3.17). The only difference is that for localization problem, landmark positions (x_L^i, y_L^i) are known while robot locations $(x_{k+1}^r, y_{k+1}^r, \phi_{k+1}^r)$ are unknown.

EKF-Based Robot Localization

The localization problem in a landmark-based map is to find the robot pose at time $k + 1$ given the map, the sequence of robot actions u_i, $(i = 0, \cdots, k)$ and sensor observations z from time 1 to time $k + 1$. In a 2D case, when the robot is operating on a plane, the location of the robot can be represented by its position and orientation

$$\mathbf{x}_{k+1} = (x_{k+1}^r, y_{k+1}^r, \phi_{k+1}^r)^T. \tag{3.22}$$

If the noises associated with the sensor measurements can be approximated using Gaussian distributions and an initial estimate for the robot location at time 0, described using a Gaussian distribution $\mathbf{x}_0 \sim N(\hat{\mathbf{x}}_0, P_0)$ with known $\hat{\mathbf{x}}_0, P_0$ is available, a solution to the robot localization problem can be obtained using EKF. The EKF equations are given in Section 3.4.1. EKF effectively summarizes all the measurements obtained in the past in the estimate of the current robot location and its covariance matrix. When a new observation from the sensor becomes available, the current robot location estimate and its covariance are updated to reflect the new information gathered.

Least Squares for Robot Localization

The localization problem can also be described as estimating the whole robot trajectory $\mathbf{x}_i$ $(i = 0, \cdots, k + 1)$ that best agree with all robot control actions u and all the sensor observations z. This can be formulated as a NLLS problem using the motion and observation models described in the previous section.

The state vector in least squares method is

$$X = (\mathbf{x}_0, \mathbf{x}_1, \cdots, \mathbf{x}_k, \mathbf{x}_{k+1})^T. \tag{3.23}$$

and the vector Z contains all the observations and control inputs.

When formulating the localization as an NLLS problem, the dimensionality of the problem is $3(k + 2)$ for two-dimensional motion and given the sampling rate of modern sensors are in the order for tens of Hertz, this strategy quickly becomes computationally expensive when the robot trajectory is long.

Least squares method provides a way to optimally estimate the robot trajectory using all the available robot motion and observation information up to time $k + 1$. In least squares method, we estimate all the robot poses together instead of only the last pose, as in the case of both the EKF and the particle filter-based robot localization.

Particle Filter for Robot Localization

When the map provided is not represented by features but using an occupancy grid. Particle filter is more suitable for robot localization. In particle filter, the probability distribution of the robot pose is not described by a Gaussian distribution, but by using particles with different weights to represent a more general distribution. Adaptive Monte Carlo Localization (AMCL) [Dellaert et al., 1999] is now a popular and robust method for robot localization which is used in many practical 2D scenarios.

3.4.6 SLAM Techniques

In a SLAM problem, both the robot location P and the map M need to be estimated. There are different SLAM problems depending on the sensors used, the

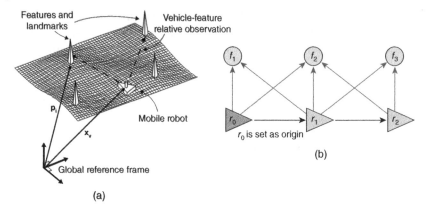

(a)

(b)

Figure 3.5 (a) Illustration of a SLAM problem. A mobile robot moves in an environment containing some features (landmarks). The robot can observe features within the sensor range. The goal is to build a map of the features and localize the robot in a global reference frame. (b) 2D feature-based SLAM using NLLS formulation. The variables are robot poses r_1, r_2, and feature positions f_1, f_2, f_3, the odometry information is shown using black arrows and the observation information is shown using gray arrows.

environment representation, and the SLAM formulations. Figure 3.5a illustrates the SLAM problem [Dissanayake et al., 2001].

Feature-Based SLAM Using EKF

Feature-based SLAM means simultaneous localization and feature map building.

We consider (2D or 3D) point feature-based SLAM problems as an example. Suppose n point features $\{f_i\}_{i=1}^n$ are observed from a sequence of $m + 1$ robot poses $\{r_i\}_{i=0}^m$. We use Z_k^i to denote the observation made from pose r_i to feature f_k. We use O_i ($1 \le i \le m$) to denote the odometry measurement between pose r_{i-1} and pose r_i. Both the observations and odometry are corrupted by zero-mean Gaussian noises with covariance matrices $P_{Z_k^i}$ and P_{O_i}, respectively. X_{f_k} denotes the position of feature f_k. $X_{r_i} = \{R_i, t_i\}$ denotes the rotation matrix and translation vector of robot pose r_i. The coordinate frame is defined by the robot pose r_0. That is, $X_{r_0} = \{I, \mathbf{0}\}$ where I is the 2×2 (2D) or 3×3 (3D) identity matrix.

In order to solve SLAM using EKF, the first step is to formulate the SLAM problem as a state estimation problem of a nonlinear discrete-time dynamic system. The state of the dynamic system at time k include the robot pose at time k and all the feature positions that are already observed from time 0 to time k. Assume all the features are observed from the beginning (time step 0), the state vector of EKF SLAM at time k is

$$\mathbf{X}_k \triangleq \{X_{r_k}, X_{f_1}, \cdots, X_{f_n}\} \tag{3.24}$$

The robot follows the process model and the features are stationary, so the dynamic system model is

$$\begin{aligned}
X_{r_{k+1}} &= f(X_{r_k}, u_k, w_k), \\
X_{f_1}(k+1) &= X_{f_1}(k) \\
&\vdots = \vdots \\
X_{f_n}(k+1) &= X_{f_n}(k)
\end{aligned}$$
(3.25)

This can be written in a compact form as

$$\mathbf{X}_{k+1} = F(\mathbf{X}_k, u_k, w_k)$$
(3.26)

Suppose the sensor can observe the feature and obtain the relative position of the feature respect to the robot pose, then the observation model is given by

$$Z_k^i = H^{Z_k^i}(\mathbf{X}_k) + v_k$$
(3.27)

where

$$H^{Z_k^i}(\mathbf{X}) = R_i^T(X_{f_k} - t_i).$$
(3.28)

and v_k is the observation noise.

Once the process model and the observation model are clearly defined as Eqs. (3.26) and (3.27), EKF can be applied to solve the SLAM problem.

Feature-Based SLAM Using NLLS

Feature-based SLAM can also be formulated as an NLLS problem [Dellaert and Kaess, 2006, Cadena et al., 2016]. A simple 2D point feature-based SLAM with three point features and three robot poses is shown in Figure 3.5b.

Again, consider (2D or 3D) point feature-based SLAM problems. Suppose n point features $\{f_i\}_{i=1}^n$ are observed from a sequence of $m+1$ robot poses $\{r_i\}_{i=0}^m$. We use Z_k^i to denote the observation made from pose r_i to feature f_k. We use O_i ($1 \le i \le m$) to denote the odometry measurement between pose r_{i-1} and pose r_i. Both the observations and odometry are corrupted by zero-mean Gaussian noises with covariance matrices $P_{Z_k^i}$ and P_{O_i}, respectively. X_{f_k} denotes the position of feature f_k. $X_{r_i} = \{R_i, t_i\}$ denotes the rotation matrix and translation vector of robot pose r_i. The coordinate frame is defined by the robot pose r_0. That is, $X_{r_0} = \{I, \mathbf{0}\}$ where I is the 2×2 (2D) or 3×3 (3D) identity matrix.

The NLLS SLAM formulation uses the odometry and observation information to estimate the state containing all the robot poses and all the feature positions

$$\mathbf{X} \triangleq \{X_{r_1}, \cdots, X_{r_m}, X_{f_1}, \cdots, X_{f_n}\}$$
(3.29)

and minimizes the objective function

$$f(\mathbf{X}) = \sum_{i=0}^m \sum_{j=1}^{n_i} \|Z_{k_{ij}}^i - H^{Z_{k_{ij}}^i}(\mathbf{X})\|_{P_{Z_{k_{ij}}^i}^{-1}}^2 + \sum_{i=1}^m \|H^{O_i}(\mathbf{X})\|_{P_{O_i}^{-1}}^2$$
(3.30)

where

$$H^{O_i}(\mathbf{X}) = \begin{bmatrix} O_i^t - R_{i-1}^T(t_i - t_{i-1}) \\ d_{SO}(O_i^R, R_{i-1}^T R_i) \end{bmatrix},$$ (3.31)

where $O_i = \{O_i^t, O_i^R\}$ is the odometry from pose r_{i-1} to pose r_i, O_i^t is the translation part while O_i^R is the rotation part ($1 \leq i \leq m$), $Z_{k_{ij}}^i$ are observations (assume n_i features are observed from robot pose r_i and k_{ij} is the global index of the jth feature observed from pose r_i), and P_{O_i} and $P_{Z_{k_{ij}}^i}$ are the corresponding covariance matrices.

In the above least squares SLAM formulation, $H^{Z_k^i}(\mathbf{X})$ and $H^{O_i}(\mathbf{X})$ are the corresponding functions relating Z_k^i and O_i to the state $\mathbf{X}$. An odometry measurement is a function of two poses $X_{r_{i-1}}$ and X_{r_i} and is given by

$$H^{O_i}(\mathbf{X}) = \begin{bmatrix} O_i^t - R_{i-1}^T(t_i - t_{i-1}) \\ d_{SO}(O_i^R, R_{i-1}^T R_i) \end{bmatrix},$$ (3.32)

where $d_{SO}(\star, \bullet)$ means the distance function on the Lie group $SO(2)$ or $SO(3)$. One example is $\| \log(\star^T \bullet)^\vee \|$ where $\vee$ means the inverse of the skew-symmetric operator.

A single feature observation is a function of one pose rotation and translation X_{r_i} and one feature position X_{f_k} which is given by

$$H^{Z_k^i}(\mathbf{X}) = R_i^T(X_{f_k} - t_i).$$ (3.33)

In particular, since $X_{r_0} = \{R_0, t_0\} = \{I, \mathbf{0}\}$, the odometry function from pose r_0 to pose r_1 is given by

$$H^{O_1}(\mathbf{X}) = \begin{bmatrix} O_1^t - t_1 \\ d_{SO}(O_1^R, R_1) \end{bmatrix}$$ (3.34)

and the observation function from pose r_0 to feature f_k is given by

$$H^{Z_k^0}(\mathbf{X}) = X_{f_k}.$$ (3.35)

Once the SLAM problem is formulated as the NLLS problem as in Eq. (3.30), different algorithms such as Gauss-Newton can be used to solve the problem.

Pose-Graph SLAM

When the environment is not represented by features, there is another way to formulate the SLAM problem, called Pose-graph SLAM [Grisetti et al., 2010].

One example is that the environment is represented by occupancy grid, and the original sensor data are laser scans. In this case, the laser scans obtained from two nearby poses can be used to compute the relative pose between the two robot poses. Once the relative pose information are obtained for many pair of poses, we can formulate an NLLS problem as follows.

We consider 2D or 3D pose-graph SLAM problems. The $m + 1$ robot poses are $\{r_i\}_{i=0}^m$. We use O_{ij} ($1 \leq i, j \leq m$) to denote the relative pose measurement between pose r_i and pose r_j. The relative pose observations are corrupted by zero-mean Gaussian noises with covariance matrices $P_{O_{ij}}$. $X_{r_i} = \{R_i, t_i\}$ denotes the rotation matrix and translation vector of robot pose r_i. The coordinate frame is defined by the robot pose r_0. That is, $X_{r_0} = \{I, \mathbf{0}\}$ where I is the 2×2 (2D) or 3×3 (3D) identity matrix.

The NLLS SLAM formulation of pose-graph SLAM is to use all the relative pose observations to estimate the state containing all the robot poses

$$\mathbf{X} \triangleq \{X_{r_1}, \cdots, X_{r_m}, X_{f_1}, \cdots, X_{f_n}\} \tag{3.36}$$

and minimizes the objective function

$$f(\mathbf{X}) = \sum_{i,j} \|H^{O_{ij}}(\mathbf{X})\|^2_{P_{O_{ij}}^{-1}} \tag{3.37}$$

where

$$H^{O_{ij}}(\mathbf{X}) = \begin{bmatrix} O_{ij}^t - R_i^T(t_j - t_i) \\ d_{SO}(O_{ij}^R, R_i^T R_j) \end{bmatrix}, \tag{3.38}$$

where $O_{ij} = \{O_{ij}^t, O_{ij}^R\}$ is the relative pose observation from pose r_i to pose r_j, O_{ij}^t is the translation part while O_{ij}^R is the rotation part, which is similar to the odometry information in feature-based SLAM.

Gauss-Newton or other algorithms can be used to solve the pose-graph NLLS problem.

SLAM Front-End and SLAM Back-End

To solve a SLAM problem in practice, raw sensor data need to be used. Before formulating the SLAM problem as state estimation problem or least squares optimization problem, some preprocessings are required. The preprocessing steps include feature extraction, scan matching, data association, loop closure detection, etc. These are called SLAM front-end.

Once the estimation problem or NLLS optimization problem is formulated, solving the problem is called SLAM back-end.

3.5 Implementation

Previous sections introduced some theories on localization, mapping, and SLAM, while this section will present some details about the implementations and open-source codes for deeper understanding and practical uses, including Monte Carlo Particle Filter, EKF-based SLAM, optimization-based SLAM, Cartographer, ORB-SLAM, and RGB-D SLAM.

Before selecting an appropriate algorithm to implement, it is necessary to clearly understand the problem, the input and the output of the algorithm. For example, identifying if the problems are related to localization or SLAM and whether the process is online or offline. Besides, the implementing environment, applicable sensors, computation cost, etc., are also the key factors to consider.

3.5.1 Localization

For localization problems, the Monte Carlo Particle Filter algorithm has been widely used. The following link demonstrates the application of Monte Carlo localization and the particle filter in a 2D localization problem, and the input data is collected by 2D LiDAR or sonar. The provided open-source code is compiled in C++, which can be executed after installing OpenCV.
https://github.com/farzingkh/Monte-Carlo-Localization/blob/master/LICENSE

3.5.2 SLAM

One main kind of SLAM problem use LiDAR sensor data, while another main kind of SLAM problems use camera data.

LiDAR-Based SLAM
For LiDAR-based SLAM, the following three links are the applications of the EKF-based, the optimization-based, and the Cartographer method, respectively. The first link uses the EKF-based algorithm to locate the sensor and map Victoria Park, and the data is collected by 2D LiDAR. The provided open-source code is compiled in MATLAB, which can be executed after installing MATLAB. Similarly, the second link also locates the sensor and maps Victoria Park with the data collected by 2D LiDAR, while the method used is the optimization-based algorithm. The provided open-source code is compiled in MATLAB, which can be executed after installing MATLAB. The third link uses the Cartographer algorithm to locate the sensor and map a number of environments, including a factory, a bridge, etc., with the data collected by 2D and 3D LiDAR. The provided open-source code is compiled in C++, which can be implemented in Robot Operating System (ROS).
https://github.com/snehchav
https://drive.google.com/drive/folders/1n0ehQv14lh0m9E_TeHTeDFxcsAY5JRIf
https://github.com/cartographer-project/cartographer_ros

Camera-Based SLAM
The previous three methods locate and map the environment with the sensor of LiDAR, while the orbSLAM and RGB-D SLAM algorithms use the monocular camera or RGB-D camera to solve SLAM problems. The first link uses the

orbSLAM algorithm to map the indoor and outdoor 3D environments with the images collected by a monocular camera. The provided open-source code is compiled in C++, which can be executed in the ROS system. The second link used the RGB-D SLAM algorithm to map the indoor scenario with the data collected by an RGB-D camera. The provided open-source code is compiled in C++, which can be executed in the ROS system.

https://github.com/OpenSLAM-org/openslam_orbslam
https://github.com/OpenSLAM-org/openslam_rgbdslam

3.6 Case Studies

In this section, some practical examples of mapping, localization and SLAM are presented to demonstrate how the basic techniques are used.

3.6.1 Mapping in Confined Space

For the Wall Pushing Maintenance Robot (WAuMBot, see Figure 13.2) described in Chapter 13, the robot needs to build a map of the confined space (Figure 13.1) for robot navigation, condition assessment, and cleaning. For this robot, the sensors used for mapping are two 2D LIDAR sensors which can rotate along a single axis (one LIDAR in the front, one LIDAR in the rear of the robot), thus 3D point clouds can be obtained through rotating the sensor. For this particular application, since the size of the tunnel in which the robot is operating is relatively small, the whole tunnel can be observed using the front and the rear LIDAR. Because the relative pose between the two LIDAR sensors are known, the two 3D point clouds obtained can be fused to get a 3D map of the whole environment. Thus, a map can be obtained from the sensor information at one time step only without fusing the observations from different time steps using EKF or NLLS. In this case, the major processes for mapping are obtaining the 3D point clouds, the detection of the important features of the environment, such as the planes of the tunnel surfaces, the manhole, and the rivets on the surfaces. The details of these processes are described in Chapter 13. If the tunnel is longer than the LIDAR sensor range, then the different local maps obtained in different time steps need to be fused using EKF or NLLS methods presented in Section 2.3.4.1. To obtain a more accurate mapping result, the special property of the environment should be used as the constraints (e.g. the environment is a cuboid that has 6 planes).

3.6.2 Localization in Confined Space

For the localization of the WAuMBot with in a tunnel, the prior knowledge on the shape of the tunnel (a cuboid) is used as the map. When the robot is within a tunnel, two 3D point clouds are obtained from the front and real LIDAR sensors. Comparing these two point clouds with the cuboid shape map will tell the location and orientation of the two LIDAR sensors, thus the location of the robot within the cuboid can be obtained. This localization result is accurate enough for the robot navigation within the tunnel.

When the tunnel is long, the 3D point clouds obtained may not contain an end of the tunnel. In that case, the localization is challenging since the four sides observed can be identical for different robot locations. To solve the localization problem in that scenario, a relatively accurate robot motion estimate is necessary. This could be obtained by the inertial measurement unit (IMU) information and the control signals of the robot. If the accuracy of the robot motion is smaller than the distance between the two adjacent rivets, then accurate robot localization is feasible by using rivets as landmarks. In this case, an accurate map of the tunnel including the locations of all the rivets is needed.

3.6.3 SLAM in Underwater Bridge Environment

The Submersible Pile Inspection Robot (SPIR) described in Chapter 8 is designed for underwater bridge pile inspection and cleaning. In order to understand the size of the marine growth on the pile, a 3D model of the bridge pile needs to be constructed. In this application, the robot is deliberately moved around the pile and the images obtained from the stereo camera on board the robot are used to build the 3D map by applying the publicly available visual SLAM algorithm-ORB-SLAM2 [Mur-Artal and Tardós, 2017].

ORB-SLAM2 is a SLAM system integrating the SLAM front-end and the SLAM back-end. The front-end of ORB-SLAM2 uses images from cameras, like monocular, stereo, and RGB-D camera to detect features at keypoint locations. The back-end of ORB-SLAM uses NLLS optimization to complete the estimation. More details of the SLAM process for SPIR are described in Chapter 8.

3.7 Conclusion and Discussion

This chapter discussed the fundamental techniques used for robot localization, robot mapping and robot SLAM. Statistical tools such as EKF and NLLS are commonly used in solving all three problems. To use practical sensor data in these robot perception problems, data association and other pre-processing are required before the problems can be mathematically formulated.

In the past decades, significant progress has been made on robot perception. Nowadays many localization, mapping, and SLAM algorithms are becoming mature, and open-source codes are available for solving different problems.

There are still some challenges involved, such as the development of efficient SLAM algorithms for handling unknown dynamic or deformable environments, long-term and reliable localization in outdoor environments using visual sensors under different weather and lighting conditions, and semantic mapping in unknown environments with objects never seen before, etc.

Bibliography

Cesar Cadena, Luca Carlone, Henry Carrillo, Yasir Latif, Davide Scaramuzza, José Neira, Ian Reid, and John J. Leonard. Past, present, and future of simultaneous localization and mapping: Toward the robust-perception age. *IEEE Transactions on Robotics*, 32(6):1309–1332, 2016.

Alícia Casals. *Sensor devices and systems for robotics*, volume 52. Springer Science & Business Media, 2012.

Frank Dellaert and Michael Kaess. Square root SAM: Simultaneous localization and mapping via square root information smoothing. *The International Journal of Robotics Research*, 25(12):1181–1203, 2006.

Frank Dellaert, Dieter Fox, Wolfram Burgard, and Sebastian Thrun. Monte Carlo localization for mobile robots. In *Proceedings 1999 IEEE International Conference on Robotics and Automation (Cat. No. 99CH36288C)*, volume 2, pages 1322–1328. IEEE, 1999.

Alberto Elfes. Using occupancy grids for mobile robot perception and navigation. *Computer*, 22(6):46–57, 1989.

M.W.M. Gamini Dissanayake, Paul Newman, Steve Clark, Hugh F. Durrant-Whyte, and Michael Csorba. A solution to the simultaneous localization and map building (SLAM) problem. *IEEE Transactions on Robotics and Automation*, 17(3):229–241, 2001.

Giorgio Grisetti, Rainer Kümmerle, Cyrill Stachniss, and Wolfram Burgard. A tutorial on graph-based SLAM. *IEEE Intelligent Transportation Systems Magazine*, 2(4):31–43, 2010.

Ninad Mehendale and Srushti Neoge. Review on LiDAR technology. *Available at SSRN 3604309*, 2020.

Lili Meng, Clarence W. De Silva, and Jie Zhang. 3D visual SLAM for an assistive robot in indoor environments using RGB-D cameras. In *2014 Ninth International Conference on Computer Science & Education*, pages 32–37. IEEE, 2014.

Raul Mur-Artal and Juan D. Tardós. ORB-SLAM2: An open-source SLAM system for monocular, stereo, and RGB-D cameras. *IEEE Transactions on Robotics*, 33(5):1255–1262, 2017.

Tom Northardt. Observability criteron guidance for passive towed array sonar tracking. *IEEE Transactions on Aerospace and Electronic Systems*, 58(4):3578–3585, 2022.

Fangwei Pan, Jialing Liu, Yueyan Cen, Ye Chen, Ruilie Cai, Zhihe Zhao, Wen Liao, and Jian Wang. Accuracy of RGB-D camera-based and stereophotogrammetric facial scanners: A comparative study. *Journal of Dentistry*, 127:104302, 2022.

Thinal Raj, Fazida Hanim Hashim, Aqilah Baseri Huddin, Mohd Faisal Ibrahim, and Aini Hussain. A survey on LiDAR scanning mechanisms. *Electronics*, 9(5):741, 2020.

Sebastian Thrun, Wolfram Burgard, and Dieter Fox. *Probabilistic robotics*. Cambridge, MA, USA: MIT Press, 2005.

Federica Villa, Fabio Severini, Francesca Madonini, and Franco Zappa. SPADs and SiPMs arrays for long-range high-speed light detection and ranging (LiDAR). *Sensors*, 21(11):3839, 2021.

Liang Zhao, Yingyu Wang, and Shoudong Huang. Occupancy-SLAM: Simultaneously optimizing robot poses and continuous occupancy map. In *Proceedings of Robotics: Science and Systems*, New York City, NY, USA, June 2022. doi: 10.15607/RSS.2022.XVIII.003.

4

Machine Learning and Computer Vision Applications in Civil Infrastructure Inspection and Monitoring

Shuming Liang[1], Andy Guo[1], Bin Liang[1], Zhidong Li[1], Yu Ding[2], Yang Wang[1], and Fang Chen[1]

[1]*Data Science Institute, University of Technology Sydney, Sydney, NSW, Australia*
[2]*School of Computing and Information Technology, University of Wollongong, Wollongong, NSW, Australia*

4.1 Introduction

Machine learning and computer vision techniques have become efficient tools to solve complex engineering problems in civil infrastructure applications ranging from energy and utilities to automotive and transport, such as water infrastructure management [Li et al., 2022], water pipe failure prediction [Liang et al., 2018, 2020, 2021], water quality modeling [Liang et al., 2019], urban traffic analysis [Buch et al., 2011] and traffic sign sensing [Koresh and Deva, 2019]. Thanks to the rapid increase of data availability, as well as increasing computational capacities and simplified programming methods, machine learning and computer vision techniques are being progressively applied in the fields of civil engineering. These methods provide fast and powerful solutions for breaking down complex phenomena into simple mathematical operations.

This chapter will introduce two successful applications of machine learning and computer vision used in water utilities and transport. First, in civil water utilities, the increasing urbanization has driven the great importance of the maintenance of water pipe networks. These networks are composed of a large number of pipes distributed around the whole city area. Any asset failure would lead to a regional or systematic falling, breaks, or even disasters [Li et al., 2014]. Therefore, preventive maintenance for pipes, particularly in aging urban-scale water networks, becomes of vital importance. Due to limited resources, water authorities can hardly afford to comprehensively inspect all assets. Instead, it is necessary to prioritize the pipes that require maintenance. Hence, the ability to identify pipes that are at high risk of failure is a fundamental need of water utilities [Shamir and Howard, 1979, Kumar et al., 2018]. Second, in a railway network, train speed is dictated by the signal conditions observed by the train drivers. Speed can drop significantly when a caution aspect is observed. Scheduled train operations are negatively impacted by speed changes, which in turn is detrimental to the customer experience [Li et al., 2021]. To enhance the intelligence and automation of train driving operation, a computer vision-based object detection methodology is developed that can accurately capture signal aspect states and transitions using video footage collected from front cameras built in the modern train sets.

4.2 GNN-Based Pipe Failure Prediction

4.2.1 Background

In the past decades, both industry and academia have constantly enhanced the ability of machine learning to improve the effectiveness of proactive maintenance in urban water networks. A variety of failure prediction models have been proposed ranging from statistical methods to machine learning algorithms. Traditional statistical methods [Snider and McBean, 2020, Shamir and Howard, 1979, Park et al., 2011, Yan et al., 2013] mainly model the historical failure records and calculate asset failure risk as a function of time. However, many of these methods suffer from specific statistical distributions, which may be limited for real-world applications [Li et al., 2014, Snider and McBean, 2020, Giraldo-González and Rodríguez, 2020]. Recent works use machine learning to predict pipe failure and have achieved promising results through carefully feature engineering [Snider and McBean, 2020, Giraldo-González and Rodríguez, 2020, Kumar et al., 2018, Robles-Velasco et al., 2020].

However, two critical pieces of information about pipe networks (i.e. the structure of pipe connectivity and geographical neighboring effects) are ignored in these models. According to industry practices, if a pipe fails, more failures would be observed on the pipes that are on the same route as this pipe (i.e. connected pipes) due to physical effects such as water hammer [Schmitt et al., 2006]. On the other hand, more failures would also be observed on the nearby pipes even which are not connected to the failed one, since these pipes are exposed to similar environmental factors such as soil properties, ground vibrations, etc. [Barton et al., 2019, Obradović, 2017].

In order to address these challenges, Liang et al. [2021] proposed a graph-based pipe failure prediction framework based on GNNs [Xu et al., 2018] and point process. The whole architecture of the framework is illustrated in Figure 4.1. The framework contains two main parts: data preprocessing and failure prediction.

Figure 4.1   An overview of the pipe failure prediction framework.

Data preprocessing includes geographical graph construction, feature engineering, and temporal failure series extraction. In graph construction, two nodes are treated as being geographically linked if two corresponding pipes are geographically close to each other. Failure prediction contains three modules including GNN module, temporal failure pattern learning, and a predictor. In GNN, an attention mechanism and multi-hop aggregation are used. The failure pattern learning module is used to learn the temporal failure pattern including the base evolutionary effect and historical failures' time-decayed excitement on the current state of a pipe. A multi-layer perceptron (MLP) is used as the final failure predictor.

4.2.2   Problem Formulation

Formally, given a pipe network with N pipes, a graph $\mathcal{G}(\mathcal{V}, \mathbf{A}, \mathbf{X})$ can be constructed by representing pipes as nodes, where $\mathcal{V}$ is the set of nodes, the adjacent matrix $\mathbf{A} \in \mathbb{R}^{N \times N}$ stores the structural information in which rows and columns are indexed by nodes, and $a_{i,j}$ in the i-th row and j-th column of $\mathbf{A}$ indicates the connectivity between node i and j. The value of $a_{i,j}$ is 0 if node i is not linked to j and nonzero if otherwise. The feature matrix $\mathbf{X} \in \mathbb{R}^{N \times f}$ represents the features of nodes where the i-th row of $\mathbf{X}$ is the feature vector $\mathbf{x}_i \in \mathbb{R}^f$ of node i. In addition, we denote $\mathcal{H}_i$ as the set of historical failure events of pipe i. We cast the failure prediction task as a binary classification problem of whether a pipe will fail within a future time window. Hence the task is to learn a model $\mathcal{M} : \left\{ \mathcal{G}(\mathbf{A}, \mathbf{X}), \{\mathcal{H}_i\}_{i=1}^N \right\} \rightarrow \mathbf{y} \in \mathbb{R}^N$ where the i-th entry of $\mathbf{y}$ is the estimated failure risk of the i-th pipe.

4.2.3   Data Preprocessing

Graph Construction
The water network is geographically distributed in a large-scale area. Pipes that are nearby to each other tend to exhibit similar failure trends, regardless of whether the pipes are physically connected or not. Different from the graph construction methods in common networks such as social networks [Backstrom and Leskovec, 2011], recommender systems [Ying et al., 2018], and biochemical interaction networks [Lin et al., 2020, Liang et al., 2021] propose a method of constructing geographical graph structure according to not only the physical connections but

also the distance between pipes. Formally, the geographical graph structure is defined as follows.

Let the centroid of each pipe be the nodes. Geographical graph structure is represented by $\mathbf{A} \in \mathbb{R}^{N \times N}$ that is the adjacency matrix of the graph with N nodes. $a_{i,j}$ in the i-th row and j-th column of $\mathbf{A}$ is assigned as a nonzero value if either node i and node j are connected or $\mathrm{DIST}(i,j) < \rho$, where the function $\mathrm{DIST}(i,j)$ calculates the geographical distance between nodes i and j, ρ is a parameter. Otherwise, $a_{i,j} = 0$.

Feature Engineering

Feature engineering is a commonly used approach to enhance model performance in machine learning. A variety of pipe-wise features are crafted. Pipe basic attributes such as age, length, and diameter are numerical features that can be directly used. Pipe material and external coating material are categorical features that are encoded using the one-hot encoding method. For failures, the number of failures for each pipe is counted in the last 1, 5, and 10 years, respectively. Environmental and physical conditions are important factors leading to pipe failure. A number of soil features are collected, indicating the water capacity, density, composition (including clay, silt, sand, and organic carbon), effective cation exchange, pH, etc. It has been recognized that tree root is the main cause of pipe failure since the roots are more likely to grow around pipes and absorb water and nutrients from pipes [Obradović, 2017, Randrup et al., 2001]. Based on the available tree canopy data, the percentage of a pipe covered by the tree canopy is used as a feature. Two features are designed based on elevation. One feature is the slope of a pipe that reflects the elevation variance of a pipe. The other one is the vertical shape of three adjacent pipes for the middle pipe. In addition, three graph structural properties are used, including node degree, PageRank, and clustering coefficient Hagberg et al. 2008.

Failure Series

For a pipe, the influence of historical failures on its current state shows a strong temporal effect. In the data preprocessing, the failure records of each pipe i are extracted into a time series: $\mathcal{H}_i = \{t_0, \cdots, t_h, \cdots\}$, where t_h is a time interval from the installation date of the pipe i to the h-th historical failure event of pipe i.

4.2.4 GNN Learning

Graph Neural Networks

GNNs have achieved great success in diverse graph applications [Backstrom and Leskovec, 2011, Ying et al., 2018, Lin et al., 2020]. Most GNNs follow a form of neighborhood information aggregation algorithm [Kipf and Welling, 2017,

Li et al., 2020b], where the representation of a node is updated layer-by-layer by aggregating representations of its neighbors and itself. Formally, the representation of a node v of a graph given by the l-th layer of a GNN is:

$$\mathbf{h}_v^{(l)} = \text{AGG}^{(l)} \left(\left\{ \mathbf{h}_u^{(l-1)} \right\}_{u \in \mathcal{N}(v) \cup \{v\}} \right), \tag{4.1}$$

where $\mathcal{N}(v)$ is the set of neighbors of node v. $\mathbf{h}_v^{(0)}$ is initialized with the features of node v. AGG($\cdot$) is an operation over the elements of the set $\{\mathbf{h}_u^{(l-1)}\}$. Various AGG($\cdot$) have been proposed. For example, GraphSAGE [Hamilton et al., 2017] instantiates AGG($\cdot$) function based on MEAN or MAX pooling operation. GAT [Veličković et al., 2018] applies an attention mechanism in AGG($\cdot$) to highlight more relevant neighbors.

GNN Learning for Failure Prediction

In the GNN learning of the failure prediction framework, as shown in Figure 4.1, several GNN techniques are employed, including multi-hop aggregation, attention mechanism, residual connections, and layer-wise aggregation. For the multi-hop aggregation, $\mathcal{N}(v)$ in AGG($\cdot$) in Equation 4.1 is changed to a set of nodes $\{u : \text{SPD}(v, u) \leq \text{k}\}$, where SPD($\cdot, \cdot$) calculates the shortest path distance between two nodes, k is a hyper-parameter. Note that setting $k = 1$ recovers the traditional 1-hop aggregation. Generally, assigning $k = 2$ or 3 is sufficient, which ensures that the aggregation focuses on the local neighboring information and avoids "noisy" information from higher-order neighbors [Xu et al., 2018, Li et al., 2020b].

The neighborhood aggregation in a GNN layer may lead to biased representations of nodes, especially when the adjacent nodes have completely different properties. The GNN module in the failure prediction framework employs an attention mechanism to tackle this issue. It allows the model to learn adaptive importance weight between two adjacent nodes and thereby differentiate neighbors in the aggregation procedure by highlighting the messages of more relevant nodes while suppressing the contributions of less relevant nodes. The attention mechanism follows GAT [Veličković et al., 2018]. Formally, the attention coefficient for the neighboring node u of the node v is:

$$\alpha_{vu}^{(l)} = \frac{\exp \left(\text{LeakyReLU} \left(\mathbf{a}^{(l)} \cdot \text{CONCAT} \left(\mathbf{h}_v^{(l-1)}, \mathbf{h}_u^{(l-1)} \right) \right) \right)}{\sum_{\text{SPD}(v,w)<\text{k}} \exp \left(\text{LeakyReLU} \left(\mathbf{a}^{(l)} \cdot \text{CONCAT} \left(\mathbf{h}_v^{(l-1)}, \mathbf{h}_w^{(l-1)} \right) \right) \right)}, \tag{4.2}$$

where $\mathbf{a}^{(l)} \in \mathbb{R}^{2d}$ is a trainable weight vector for the l-th layer, $\mathbf{h}_v^{(l-1)} \in \mathbb{R}^d$ and $\mathbf{h}_u^{(l-1)} \in \mathbb{R}^d$ are the output states of node v and u from the $(l-1)$-th layer, respectively. CONCAT($\cdot, \cdot$) is the concatenation operation. LeakyReLU is an activation function.

Equipped with the attention mechanism in Eq. 4.2, the output representation of node v given by the l-th GNN layer is formalized as:

$$\mathbf{h}_v^{(l)} = \text{ReLU}\left(\left(\sum \alpha_{vu}^{(l)}\mathbf{h}_u^{l-1}\right) \cdot \mathbf{W}^{(l)}\right),$$
$$\text{where } u : \text{SPD}(v,u) < \text{k or } u = v, \tag{4.3}$$

where $\mathbf{W}^{(l)}$ is the weight matrix for the l-th layer.

Common GNNs based on the neighborhood aggregation algorithm tend to be shallow since deeper GNNs probably suffer from over-smoothing and gradient degeneration [Li et al., 2018, Xu et al., 2018]. Stacking deep layers in the GNN may over-average the information from a too wide range of neighbors, and as a result, the useful information of the central target node may be "washed out." The GNN module adopts the approach of JKGNN [Xu et al., 2018] to tackle the over-smoothing issue. Specifically, a layer is added after the last GNN layer. This layer stores the hidden representations from the previous GNN layers. In other words, the output of each GNN layer has a residual or jump connection to this layer. A layer-wise aggregation is employed on these representations in this layer. Consequently, the final representation $\mathbf{h}_v^{GNN}$ of node v produced by the GNN module is $\text{AGG}\left(\left[\mathbf{h}_v^{(1)}, \cdots, \mathbf{h}_v^{(l)}\right]\right)$, where $\text{AGG}(\cdot)$ can be MEAN, concatenation, etc.

4.2.5 Failure Pattern Learning

Historical pipe failures are critical to failure prediction. The stochastic point process has proven to be an effective tool for dealing with such temporal failure data. In essence, point process is characterized by the conditional intensity function $\lambda(t)$, where $\lambda(t)dt$ is the likelihood for an event occurring within a small window $[t, t + dt]$, given the historical events before t.

To capture the temporal pattern in the failures data, a temporal failure pattern learning module is developed based on the Hawkes process [Hawkes, 1971]. Formally, the node v's representation $\mathbf{h}_v^{TF}$ for temporal failures' effect computed by this module is:

$$h_v^T = \mu + (\omega - \mu)e^{-\delta t},$$
$$h_v^F = \sum_{t_h \in \mathcal{H}_v, t_h < t} \alpha e^{-\beta(t - t_h)}, \tag{4.4}$$
$$\mathbf{h}_v^{TF} = \text{ReLU}\left(\text{CONCAT}\left(h_v^T, h_v^F\right)\right),$$

where $\mu > 0, \omega > 0, \delta > 0, \alpha > 0, \beta > 0$ are trainable parameters that are shared across all pipes, t is the age of pipe v at the time of prediction, and $\mathcal{H}_v$ is the set of historical failures.

h_v^T is the base evolutionary effect that is only a function of pipe age t, where ω is the initial failure risk at $t = 0$, δ is the rate of exponential decay, and μ describes the reversion level. For example, with the age t being increased, h_v^T is exponentially decaying if $\omega > \mu$, while exponentially growing if $\omega < \mu$. h_v^F is the historical failures' time-decayed excitement where α is the size of excitement jump for the exponential decay function. h_v^F is an aggregation of the excitement effects of all historical failures of pipe v on its state at time t.

4.2.6 Failure Predictor

A MLP is used as the predictor of the framework. MLP is a type of neural network used in machine learning for various tasks such as classification, regression, and pattern recognition. The layers in an MLP consist of an input layer, one or more hidden layers, and an output layer. MLP is a feedforward neural network, which means that the information flows in only one direction, from input to output, through a series of interconnected layers of nodes or neurons. Each neuron in an MLP receives input from the previous layer, processes it using a mathematical function called an activation function, and then passes the output to the next layer. The learning process in neurons is known as backpropagation [Rumelhart et al., 1986], where the error in the output is propagated back through the network to adjust the weights of activation functions.

In the failure prediction framework, the predictor MLP takes as input the concatenation of the GNN output and temporal failure representation, i.e., $\text{CONCAT}(\mathbf{h}_v^{GNN}, \mathbf{h}_v^{TF})$. At last, the failure risk of the pipe v is estimated by $\text{MLP}\left(\text{CONCAT}\left(\mathbf{h}_v^{GNN}, \mathbf{h}_v^{TF}\right)\right)$.

It is worth noting that the weights to be trained in the failure prediction framework are shared across all nodes (pipes). This can significantly reduce the computational complexity. On the other hand, it allows the model to be trained with a batched scheme, which is of great importance, especially for real-world large-scale infrastructure networks.

4.2.7 Experimental Study

The failure prediction framework is evaluated on a real-world water pipe network [Liang et al., 2021]. The performance of the framework is compared against three classes of baselines, including (1) statistical models: Weibull model [Duchesne et al., 2013], Hawkes Process model [Yan et al., 2013], Random survival forests (RSF) [Ishwaran et al., 2008]. (2) machine learning models: Random Forests (RF), Support Vector Machine (SVM) [Robles-Velasco et al., 2020], Gradient Boosting Decision Tree (GBDT) [Chen and Guestrin, 2016], and a fully-connected neural network (MLP). (3) graph-based GNNs, including GCN [Kipf and Welling, 2017], GraphSAGE [Hamilton et al., 2017], GAT [Veličković et al., 2018], JKGNN [Xu et al., 2018], DeeperGCN [Li et al., 2020a] and DEA-GNN [Li et al., 2020b].

For brevity, the present failure prediction framework is named Multi-hop Attention-based GNN (MAG) [Liang et al., 2021]. It has three variants, i.e., MAG-Concat, MAG-LSTM, or MAG-MAX according to the method in the aggregation including Concatenation, LSTM [Hamilton et al., 2017], or MAX pooling. The result of MAG-nonGeo which uses the graph structure without geographical links is also reported.

Table 4.1 summarizes the results. MAG-Concat and MAG-LSTM yield the best performance. In practice, MAG-LSTM has more trainable parameters than MAG-Concat and thus requires an additional computational cost. MAG-nonGeo outperforms the best baselines but underperforms MAG-Concat and MAG-LSTM. MAG-nonGeo takes the graph structure that does not consider the geographical neighboring effect. The GNN architecture of MAG is much similar to JKGNN and

Table 4.1 Results of failure prediction on water pipe dataset.

Model	AUC	F1	Precision	Recall
Weibull	0.667 ± 0.032	0.261 ± 0.029	0.167 ± 0.032	0.458 ± 0.033
Hawkes	0.685 ± 0.021	0.278 ± 0.031	0.178 ± 0.041	0.464 ± 0.037
RSF	0.706 ± 0.018	0.269 ± 0.023	0.183 ± 0.029	0.482 ± 0.025
RF	0.765 ± 0.016	0.372 ± 0.011	0.294 ± 0.023	0.586 ± 0.019
SVM	0.787 ± 0.016	0.373 ± 0.014	0.318 ± 0.014	0.591 ± 0.016
MLP	0.769 ± 0.012	0.369 ± 0.015	0.309 ± 0.019	0.581 ± 0.018
GBDT	0.811 ± 0.008	0.411 ± 0.01	0.351 ± 0.011	0.615 ± 0.012
GCN	0.789 ± 0.011	0.368 ± 0.016	0.336 ± 0.013	0.578 ± 0.011
GraphSAGE	0.798 ± 0.01	0.37 ± 0.012	0.336 ± 0.011	0.585 ± 0.012
GAT	0.828 ± 0.014	0.432 ± 0.012	0.379 ± 0.009	0.637 ± 0.01
JKGNN	0.837 ± 0.009	0.439 ± 0.011	0.405 ± 0.012	0.665 ± 0.013
DeeperGCN	0.839 ± 0.015	0.439 ± 0.014	0.391 ± 0.008	0.657 ± 0.011
DEA-GNN	0.806 ± 0.012	0.391 ± 0.018	0.351 ± 0.014	0.607 ± 0.017
MAG-nonGeo	0.842 ± 0.018	0.441 ± 0.013	0.405 ± 0.011	0.669 ± 0.013
MAG-Concat	$\mathbf{0.855 \pm 0.011}$	$\mathbf{0.457 \pm 0.011}$	0.434 ± 0.009	$\mathbf{0.683 \pm 0.012}$
MAG-LSTM	0.851 ± 0.016	0.452 ± 0.012	$\mathbf{0.437 \pm 0.01}$	0.679 ± 0.013
MAG-MAX	0.831 ± 0.014	0.433 ± 0.019	0.371 ± 0.015	0.651 ± 0.018

The evaluation of the models is performed by treating the failure prediction task as a binary classification problem, specifically determining whether a pipe will fail within a future year. The proposed model is referred to as MAG (highlighted in bold). Different types of prediction methods are included as baselines, including traditional statistical methods, machine learning methods, and GNN-based methods.
Source: Liang et al. [2021].

DeeperGCN. MAG-nonGeo is still better than them, indicating that the temporal failure representations learned by MAG have a clear effect on failure prediction. MAG-MAX performs better than GAT while worse than the best results of JKGNN and DeeperGCN. This may be because the MAX pooling in the aggregation of GNN biases the representations and is not suitable for pipe failure prediction.

4.3 Computer Vision-Based Signal Aspect Transition Detection

4.3.1 Background

The objective of this application is to develop a computer vision-based object detection methodology that can accurately capture signal aspect states and transitions. Video footages collected from front cameras built in the modern train sets are used to build the model. A two-stage detection model is developed for this application. Firstly, an object detection model is built for signal detection using a single neural network which performs both classification and prediction of bounding boxes for detected objects. Secondly, as tracks have irregular boundaries compared with signals, the object detection method in the signal detection cannot be used for track detection. Therefore, an image segmentation model is developed based on Bisenet [Yu et al., 2018] in which both low-level details and high-level semantics guide an aggregation layer to enhance mutual connections and fuse both types of feature representation. This model can partition an image into multiple segments (sets of pixels, also known as image objects) and the tracks can be detected from them. At last, the target signals are identified based on the location of the current track.

4.3.2 Signal Detection Model

Object detection is a computer vision technique that involves identifying and localizing objects in an image or video. The algorithm uses deep learning models, such as convolutional neural networks (CNNs), to detect objects and classify them into predefined categories. Object detection typically involves several steps, including image preprocessing, region proposal generation, feature extraction, object classification, and object localization. This algorithm is widely used in applications such as self-driving cars, surveillance systems, and medical imaging to detect and track objects in real-time.

For signal light detection, an object detection algorithm is developed based on YOLOv3 [Redmon et al., 2016, Redmon and Farhadi, 2018], in which a single neural network is used to perform both classification and prediction of bounding boxes

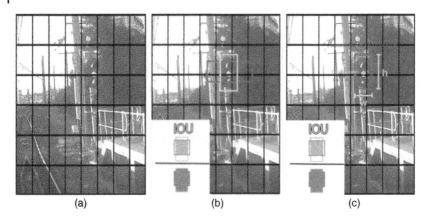

(a)　　　　　　　　　(b)　　　　　　　　　(c)

Figure 4.2　An example of IOU-based object detection. (a) Image grid; (b) IOU; (c) Classification and confidence.

for detected objects. As such, it is heavily optimized for detection performance and can run much faster than running two separate neural networks to detect and classify objects separately. It does this by repurposing traditional image classifiers for the regression task of identifying bounding boxes for objects.

Figure 4.2 shows an example of the process of IOU-based object detection. IOU (Intersection over Union) is an evaluation metric commonly used in object detection and image segmentation. It measures the overlap between the predicted and ground-truth data by dividing the area of overlap by the area of union between the two sets of data. IOU values range between 0 and 1, with 1 indicating a perfect match and 0 indicating no overlap. IOU is used to evaluate the performance of object detection and image segmentation models. The object detection model is based on the idea of segmenting an image into smaller images. The image is split into a square grid of dimensions $S \times S$, as shown in Figure 4.2a. B bounding boxes (BBOX) and C categories of conditional probabilities are, respectively, predicted. Each BBOX has five variables including four values describing the location coordinates and an objectness indicating whether there is a target. In this way, each grid needs to output a vector of $5B + C$ dimension. The tensor of the neural network's final output is $S \times S \times (5B + C)$.

Model Training
During the training of the neural network, for each ground-truth BBOX, its central position, and the cell need to be found. The cell where the center position is located is responsible for the prediction of the object. Therefore, for the output in this cell, its objectness should be increased, and its position coordinates should fit the ground-truth BBOX. As each cell can output multiple alternative BBOXs,

it is necessary to select the predicted BBOX closest to ground truth for tuning. In addition, the category probability vector is optimized according to its actual category. There is no need to optimize the predicted values of the cell that are not responsible for any category.

In this application, more than 1,000 selected typical signal lights for different states (i.e. red, yellow, and green) from video screenshots are manually labeled. The model is trained using a batch size of 64. Data augmentation and dropout are used to prevent overfitting.

The biggest limitation of YOLO-based model is that YOLO predicts a single class per grid cell and will not work well if multiple objects of different kinds are present in the same cell. The second limitation is the grid itself: a fixed grid resolution imposes strong spatial constraints on what the model can do. So when the lighting is too bright or too dark, or when multiple signal lights are densely clustered in the distance, the accuracy of image segmentation algorithms will decrease.

4.3.3 Track Detection Model

As tracks have irregular boundaries compared with signal lights, the object detection method for signal detection cannot be used. Therefore, an image segmentation model is developed based on Bisenet [Yu et al., 2018]. This model can partition an image into multiple segments and the tracks from them can then be identified.

Image segmentation is a computer vision technique that involves dividing an image into multiple segments or regions, each of which corresponds to a specific object or background. This is typically done using deep learning models such as CNNs or fully convolutional networks, which identify and classify the pixels within an image into distinct regions or segments based on their visual features. Image segmentation is widely used in medical imaging and computer vision applications for detecting and tracking objects or identifying different organs or tissues within an image. The result of image segmentation is a set of segments that collectively cover the entire image, or a set of contours extracted from the image. The pixels in a region are similar with respect to some characteristics or computed properties, such as color, intensity, or texture. An image segmentation result for tracks is shown in Figure 4.3.

In this model, a bilateral segmentation network for real-time semantic segmentation is applied. The overall model architecture is shown in Figure 4.4. Specifically, a small-step Spatial Path is designed to save spatial information and generate high-resolution features. At the same time, the Context Path with fast downsampling strategy is adopted to obtain enough receptive fields. On the basis of these two branches, a Feature Fusion Module is developed to fuse the features reasonably.

Figure 4.3   An example of semantic segmentation for tracks.

Spatial Path

In the task of semantic segmentation, it is difficult to achieve both high spatial resolution and large receptive field, especially in the case of real-time semantic segmentation. The existing methods usually use small input images or lightweight backbone models to achieve acceleration. However, compared with the original image, the small image lacks spatial information, while the lightweight model damages the spatial information due to cutting channels [Yu et al., 2018].

Context Path

Context path is used to obtain enough receptive fields which are important in semantic segmentation. The existing methods, such as pyramid pooling [He et al., 2015], atrous spatial pyramid pooling [He et al., 2019], large kernel, etc., are costly in computation. Context path provides a large receptive field with a lightweight model. In Context Path, an Attention Refinement Module (ARM) is used to adjust the characteristics of each stage. As shown in Figure 4.4, ARM uses global average pooling to capture the global context, and calculates attention vector to guide feature learning. This design can refine the output characteristics of each stage in the context path. It can easily integrate global context information without any up-sampling operation. Therefore, its computational overhead can be neglected.

Network Architecture

As shown in Figure 4.4, three convolution layers with step size are used as Spatial Path. A Pre-trained Xception model [Chollet, 2017] is used as the backbone of Context Path. Then, the output representations of these two paths are fused to

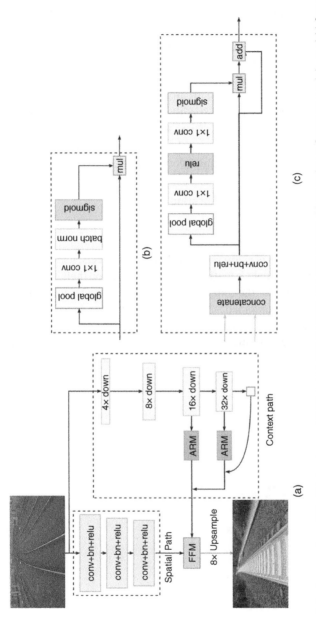

Figure 4.4 An overview of the bilateral segmentation network. (a) Network architecture, (b) attention refinement module, and (c) feature fusion module.

get the final prediction result. Although Spatial Path has a large space size; it only has three convolution layers, so it is not computationally expensive. For Context Path, a lightweight model is used to quickly downsampling. In addition, the two paths are calculated in parallel, which greatly improves computational efficiency.

The Spatial Path encodes rich spatial information, while the Context Path provides a large acceptance field. The features of these two branches are different at the level of feature representation. Most of the spatial information captured by Spatial Path is encoded with rich details. In addition, the output features of Context Path mainly encode the context information. Therefore, a Feature Fusion Module is introduced to fuse these features. Then, Batch Normalization is used to balance the scale of these representations. A weight vector is calculated. This weight vector can re-weight features, which is equivalent to feature selection and combination.

4.3.4 Optimization for Target Locating

In practice, railway tracks criss-cross in a dense network that provides services to different trips. The train front cameras may capture multiple tracks and signal lights at the same time, as Figure 4.5 shows. For train drivers, it is easy for them to recognize which signal light is for the current train based on their rich experiences. However, it is difficult for the computer to identify the target signal light from the complex environment automatically. In this application, a series of methods are proposed to process the detected signal lights and tracks. Consequently, the target signal lights and tracks of current trains can be determined and used for the final signal aspect transition.

Figure 4.5 Sample frames with multiple signals and tracks in the camera simultaneously.

Current Track Location

With the purpose of locating the target signal light, the spatial relationship between signal lights and trains (i.e., how do the train drivers determine which signals they should follow) should be figured out. Based on the expert advice from the train service in this application, the signal light that controls the trains are always on the left of the train. Thus, to locate the target signal light for the current train, the current track should be identified at first. Through the empirical studies, the current track detection can be grouped into the following cases:

Case 1 - Track connected with frame bottom. Figure 4.6a shows the tracks detected typically by the image segmentation model. It can be seen that only the current track is in contact with the bottom of the frame due to the train camera being at the head of the train. Therefore, the current track can be simply located as those who contact the frame bottom.

Case 2 - Tracks that do not connect with the bottom. However, in some special cases as shown in Figure 4.6b, the image segmentation model cannot detect the full track from the bottom due to the brightness problem. For example, when the bottom of the frame is dark, it is hard to detect the tracks. To address these problems, when the current track from the bottom of the frame is failed to be captured, the frame will be scanned from the bottom to the top till the current tracks are captured.

Case 3 - Merging tracks. In the real world, sometimes, two tracks will blend into a one-track to merge routes from different directions into one. It means there will be multi-tracks at the bottom, as shown in Figure 4.6c. In this situation, the actual running tracks cannot be located. Considering that the merging tracks appear suddenly in most cases, if multi-tracks are detected at the bottom of a frame, the location of the detected tracks will be compared with the running tracks in the last frame, and the closest tracks are located as the current tracks.

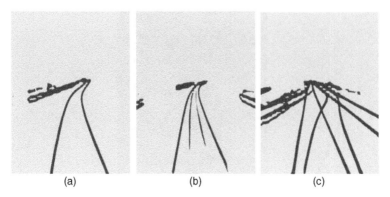

(a)   (b)   (c)

Figure 4.6 Sample of track detection results in different cases. (a) Case 1, (b) Case 2, (c) Case 3.

With these track detection algorithms, the model can locate the current tracks quickly and precisely.

Target Signal Light Location

As mentioned earlier, the target signal light that controls the current train is always on the left side of the train. After locating the current tracks that the train runs on, the target signal light from multiple signal lights can be identified based on this guidance. However, there are various complex situations in the train network. To efficiently and effectively locate the signal lights, two steps containing a series of optimization methods are conducted to deal with different circumstances.

Step 1—Determining the position of the signal lights relative to the current tracks.

To determine the relative position of signal lights, a split line needs to be chosen based on the current track. In this project, the skeletonization algorithm [Saha et al., 2016] is adopted to capture the split line.

The skeletonization algorithm is a method that obtains the skeletons from binary images by thinning regions. Benefits from this algorithm, the split line can be accurately learned, no matter whether the tracks are going straight or turning. Figure 4.7 shows an example of the split line for real-world train tracks. If the signal light is on the left side of the split line, it can be treated as the potential choice of the target signal lights.

Step 2—Identifying target signal light on the left. After determining the detected signal lights on the left side of the current track, the target signal light for the running train can be identified based on their relative positions. To improve the accuracy and speed, the task of target signal light identification is divided into four different cases.

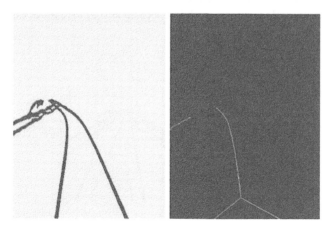

Figure 4.7   Sample of split lines(skeleton) of tracks.

Figure 4.8 The final outputs of the proposed model.

1. Single signal light with nonleft tracks. In this case, the single signal light, will be directly located as the target signal light as there are no more tracks on the left of the current tracks. Thus, the signal light directly controls the current train.

2. Single signal light with left tracks. In this case, the relationship between the signal light and various tracks should be determined first. Similarly, the skeletonization algorithm is used to capture the split line of left tracks. If the signal light is between the current track's split line and its closest left track's split line, it can be identified as the target signal light.

3. Multiple signal lights with nonleft tacks. This situation may be caused by the train camera capturing both the far and near signal lights at the same time, while the train is only controlled by the near signal light. Therefore, the signal light with a bigger size (e.g. larger diagonal distance) is used as the target signal light based on the "near bigger far smaller" law.

4. Multiple signal lights with left tracks. Similar to Case 2, if there are signal lights between the current track and its closest left track, the closest signal light near the current tracks can be determined as the target signal light.

Based on the two steps, the signal lights and tracks can be identified in complex situations. And the target signal light and current tracks can be used for the final signal aspect transition, as shown in Figure 4.8.

4.4 Conclusion and Discussion

This chapter highlights two successful case studies in the fields of water and transportation. Currently, manual inspection remains the primary method for assessing the condition of civil infrastructure. However, leveraging machine learning and computer vision for civil infrastructure inspection and monitoring represents a natural and promising advancement. These technologies have the potential to aid and eventually replace manual visual inspection, driving the next revolution in the inspection and monitoring of civil infrastructure. Despite the abundance of rich spatial, textural, and contextual information in the collected data, extracting actionable insights and identifying relationships from these data poses significant challenges. However, the research community has made notable progress in addressing these challenges through the application of machine learning and computer vision algorithms, including deep learning and GNN. Recent advancements in automated inspections and monitoring have been driven by data-driven solutions, replacing heuristic-based methods with deep learning models trained on large datasets.

The advancements in machine learning and computer vision for civil infrastructure applications, as discussed in this chapter, hold the potential to revolutionize the way we assess and monitor our cities' infrastructure. These technologies offer several benefits, including increased time-efficiency, cost-effectiveness, and the eventual possibility of automation.

One of the significant advantages of using machine learning and computer vision in civil infrastructure applications is the ability to process and analyze large amounts of data quickly. With advanced algorithms, these technologies can rapidly analyze complex datasets, such as images, sensor data, and other forms of information, to extract valuable insights. This can significantly reduce the time required for manual inspections, allowing for quicker decision-making and action.

Moreover, the use of machine learning and computer vision can contribute to cost-effective solutions in civil infrastructure management. Traditional inspection methods may be labor-intensive, costly, and often require significant resources. In contrast, machine learning and computer vision technologies can provide cost-effective alternatives by automating the inspection process, reducing the need for manual labor and associated costs.

Furthermore, the potential for automation in civil infrastructure inspections and monitoring is a promising aspect of these technologies. As machine learning models and computer vision algorithms continue to evolve and improve, the possibility of automating certain tasks, such as routine inspections or continuous monitoring, becomes more feasible. This can result in a more streamlined and efficient process, with reduced human intervention and potential for human error.

Ultimately, the adoption of machine learning and computer vision in civil infrastructure applications has the potential to contribute to safer and more resilient cities. By leveraging the capabilities of these technologies, infrastructure stakeholders can obtain timely and accurate information about the condition of critical assets, enabling proactive maintenance, repair, and replacement decisions. This can result in enhanced safety measures, improved infrastructure performance, and increased resilience against natural disasters and other unforeseen events.

In conclusion, the advances in machine learning and computer vision for civil infrastructure applications offer tremendous potential for time-efficient, cost-effective, and eventually automated processes. The adoption of these technologies has the potential to transform the way we assess, monitor, and manage civil infrastructure, ultimately leading to safer, more resilient, and sustainable cities.

Bibliography

Lars Backstrom and Jure Leskovec. Supervised random walks: Predicting and recommending links in social networks. In *Proceedings of the Fourth ACM International Conference on Web Search and Data Mining*, pages 635–644, 2011.

Neal Andrew Barton, Timothy Stephen Farewell, Stephen Henry Hallett, and Timothy Francis Acland. Improving pipe failure predictions: Factors affecting pipe failure in drinking water networks. *Water Research*, 164:114926, 2019.

Norbert Buch, Sergio A. Velastin, and James Orwell. A review of computer vision techniques for the analysis of urban traffic. *IEEE Transactions on Intelligent Transportation Systems*, 12(3):920–939, 2011.

Tianqi Chen and Carlos Guestrin. XGBoost: A scalable tree boosting system. In *Proceedings of the 22nd ACM SIGKDD International Conference on Knowledge Discovery and Data Mining*, pages 785–794, 2016.

François Chollet. Xception: Deep learning with depthwise separable convolutions. In *Proceedings of the IEEE Conference on Computer Vision and Pattern Recognition*, pages 1251–1258, 2017.

Sophie Duchesne, Guillaume Beardsell, Jean-Pierre Villeneuve, Babacar Toumbou, and Kassandra Bouchard. A survival analysis model for sewer pipe structural deterioration. *Computer-Aided Civil and Infrastructure Engineering*, 28(2):146–160, 2013.

Mónica Marcela Giraldo-González and Juan Pablo Rodríguez. Comparison of statistical and machine learning models for pipe failure modeling in water distribution networks. *Water*, 12(4):1153, 2020.

Aric Hagberg, Pieter Swart, and Daniel S. Chult. Exploring network structure, dynamics, and function using NetworkX. Technical report, Los Alamos National Lab.(LANL), Los Alamos, NM (United States), 2008.

Will Hamilton, Zhitao Ying, and Jure Leskovec. Inductive representation learning on large graphs. In *Advances in neural information processing systems*, pages 1024–1034, 2017.

Alan G. Hawkes. Spectra of some self-exciting and mutually exciting point processes. *Biometrika*, 58(1):83–90, 1971.

Kaiming He, Xiangyu Zhang, Shaoqing Ren, and Jian Sun. Spatial pyramid pooling in deep convolutional networks for visual recognition. *IEEE Transactions on Pattern Analysis and Machine Intelligence*, 37(9):1904–1916, 2015.

Hao He, Dongfang Yang, Shicheng Wang, Shuyang Wang, and Yongfei Li. Road extraction by using atrous spatial pyramid pooling integrated encoder-decoder network and structural similarity loss. *Remote Sensing*, 11(9):1015, 2019.

Hemant Ishwaran, Udaya B. Kogalur, Eugene H. Blackstone, and Michael S. Lauer. Random survival forests. *Annals of Applied Statistics*, 2(3):841–860, 2008.

Thomas N. Kipf and Max Welling. Semi-supervised classification with graph convolutional networks. In *International Conference on Learning Representations (ICLR)*, 2017.

M.H.J.D. Koresh and J. Deva. Computer vision based traffic sign sensing for smart transport. *Journal of Innovative Image Processing (JIIP)*, 1(01):11–19, 2019.

Avishek Kumar, Syed Ali Asad Rizvi, Benjamin Brooks, R. Ali Vanderveld, Kevin H. Wilson, Chad Kenney, Sam Edelstein, Adria Finch, Andrew Maxwell, Joe Zuckerbraun, et al. Using machine learning to assess the risk of and prevent water main breaks. In *Proceedings of the 24th ACM SIGKDD International Conference on Knowledge Discovery & Data Mining*, pages 472–480, 2018.

Zhidong Li, Bang Zhang, Yang Wang, Fang Chen, Ronnie Taib, Vicky Whiffin, and Yi Wang. Water pipe condition assessment: A hierarchical beta process approach for sparse incident data. *Machine Learning*, 95(1):11–26, 2014.

Qimai Li, Zhichao Han, and Xiao-Ming Wu. Deeper insights into graph convolutional networks for semi-supervised learning. In Sheila A. McIlraith and Kilian Q. Weinberger, editors, *Proceedings of the 32nd AAAI Conference on Artificial Intelligence, (AAAI-18)*, New Orleans, Louisiana, USA, February 2–7, 2018, pages 3538–3545. AAAI Press, 2018.

Guohao Li, Chenxin Xiong, Ali Thabet, and Bernard Ghanem. DeeperGCN: All you need to train deeper GCNs. *arXiv preprint arXiv:2006.07739*, 2020a.

Pan Li, Yanbang Wang, Hongwei Wang, and Jure Leskovec. Distance encoding: Design provably more powerful neural networks for graph representation learning. *arXiv preprint arXiv:2009.00142*, 2020b.

Boyu Li, Ting Guo, Ruimin Li, Yang Wang, Yuming Ou, and Fang Chen. Delay propagation in large railway networks with data-driven Bayesian modeling. *Transportation Research Record*, 2675(11):472–485, 2021.

Zhidong Li, Bin Liang, and Yang Wang. Machine learning for efficient water infrastructure management. In *Humanity Driven AI*, pages 37–61. Springer, 2022.

Bin Liang, Zhidong Li, Yang Wang, and Fang Chen. Long-term RNN: Predicting hazard function for proactive maintenance of water mains. pages 1687–1690, October 2018. doi: 10.1145/3269206.3269321.

Bin Liang, Zhidong Li, Ronnie Taib, George Mathews, Yang Wang, Shiyang Lu, Fang Chen, Tin Hua, Andrew Peters, Dammika Vitanage, et al. Predicting water quality for the Woronora delivery network with sparse samples. In *2019 IEEE International Conference on Data Mining (ICDM)*, pages 1210–1215. IEEE, 2019.

Bin Liang, Sunny Verma, Jie Xu, Shuming Liang, Zhidong Li, Yang Wang, and Fang Chen. A data driven approach for leak detection with smart sensors. In *2020 16th International Conference on Control, Automation, Robotics and Vision (ICARCV)*, pages 1311–1316. IEEE, 2020.

Shuming Liang, Zhidong Li, Bin Liang, Yu Ding, Yang Wang, and Fang Chen. Failure prediction for large-scale water pipe networks using GNN and temporal failure series. In *Proceedings of the 30th ACM International Conference on Information & Knowledge Management*, pages 3955–3964, 2021.

Xuan Lin, Zhe Quan, Zhi-Jie Wang, Tengfei Ma, and Xiangxiang Zeng. KGNN: Knowledge graph neural network for drug-drug interaction prediction. In *Proceedings of the 29th International Joint Conference on Artificial Intelligence, IJCAI-20 (International Joint Conferences on Artificial Intelligence Organization)*, pages 2739–2745, 2020.

Dino Obradović. The impact of tree root systems on wastewater pipes. In *Zbornik radova, Zajednički temelji*, pages 65–71, 2017.

Suwan Park, Hwandon Jun, Newland Agbenowosi, Bong Jae Kim, and Kiyoung Lim. The proportional hazards modeling of water main failure data incorporating the time-dependent effects of covariates. *Water Resources Management*, 25(1):1–19, 2011.

Thomas B. Randrup, E. Gregory McPherson, and Laurence R. Costello. Tree root intrusion in sewer systems: Review of extent and costs. *Journal of Infrastructure Systems*, 7(1):26–31, 2001.

Joseph Redmon and Ali Farhadi. YOLOv3: An incremental improvement. *arXiv*, 2018.

Joseph Redmon, Santosh Divvala, Ross Girshick, and Ali Farhadi. You only look once: Unified, real-time object detection. In *Proceedings of the IEEE Conference on Computer Vision and Pattern Recognition*, pages 779–788, 2016.

Alicia Robles-Velasco, Pablo Cortés, Jesús Mu nuzuri, and Luis Onieva. Prediction of pipe failures in water supply networks using logistic regression and support vector classification. *Reliability Engineering & System Safety*, 196:106754, 2020.

David E. Rumelhart, Geoffrey E. Hinton, and Ronald J. Williams. Learning representations by back-propagating errors. *Nature*, 323(6088):533–536, 1986.

Punam K. Saha, Gunilla Borgefors, and Gabriella Sanniti di Baja. A survey on skeletonization algorithms and their applications. *Pattern Recognition Letters*, 76:3–12, 2016.

C. Schmitt, G. Pluvinage, E. Hadj-Taieb, and R. Akid. Water pipeline failure due to water hammer effects. *Fatigue & Fracture of Engineering Materials & Structures*, 29(12):1075–1082, 2006.

Uri Shamir and Charles D.D. Howard. An analytic approach to scheduling pipe replacement. *Journal-American Water Works Association*, 71(5):248–258, 1979.

Brett Snider and Edward A. McBean. Improving urban water security through pipe-break prediction models: Machine learning or survival analysis. *Journal of Environmental Engineering*, 146(3):04019129, 2020.

Petar Veličković, Guillem Cucurull, Arantxa Casanova, Adriana Romero, Pietro Liò, and Yoshua Bengio. Graph attention networks. In *International Conference on Learning Representations*, 2018.

Keyulu Xu, Chengtao Li, Yonglong Tian, Tomohiro Sonobe, Ken-ichi Kawarabayashi, and Stefanie Jegelka. Representation learning on graphs with jumping knowledge networks. In *International Conference on Machine Learning*, pages 5453–5462. PMLR, 2018.

Junchi Yan, Yu Wang, Ke Zhou, Jin Huang, Chunhua Tian, Hongyuan Zha, and Weishan Dong. Towards effective prioritizing water pipe replacement and rehabilitation. In *23rd International Joint Conference on Artificial Intelligence*. Citeseer, 2013.

Rex Ying, Ruining He, Kaifeng Chen, Pong Eksombatchai, William L. Hamilton, and Jure Leskovec. Graph convolutional neural networks for web-scale recommender systems. In *Proceedings of the 24th ACM SIGKDD International Conference on Knowledge Discovery & Data Mining*, pages 974–983, 2018.

Changqian Yu, Jingbo Wang, Chao Peng, Changxin Gao, Gang Yu, and Nong Sang. BiSeNet: Bilateral segmentation network for real-time semantic segmentation. In *Proceedings of the European Conference on Computer Vision (ECCV)*, pages 325–341, 2018.

5

Coverage Planning and Motion Planning of Intelligent Robots for Civil Infrastructure Maintenance

Mahdi Hassan* and Dikai Liu*

Robotics institute, University of Technology Sydney, Australia

5.1 Introduction to Coverage and Motion Planning

Infrastructure maintenance often requires surface preparation using abrasive blasting, high-pressure cleaning, surface coating, or spray painting. Complete coverage is an integral part of such applications, i.e. all surface areas of interest need to be operated on to achieve complete coverage. Figure 5.1 shows an example, in which each Autonomous Industrial Robot (AIR) has a nozzle attached to its end-effector, and the stream of grit coming out of the nozzle is directed to follow the paths on the surfaces of the object. If all the paths are covered by the AIRs, then complete coverage is obtained. In this chapter, an AIR is defined as an industrial robot, with or without a mobile platform, that has the intelligence needed to operate autonomously in a complex and unstructured environment. This intelligence includes self-awareness, environmental awareness, and collision avoidance.

Utilizing multiple AIRs can provide greater capacity and flexibility. As an example, consider the two mobile AIRs shown in Figure 5.1 deployed for a complete coverage task, such as grit-blasting. For such applications, utilizing multiple AIRs can help minimize the overall completion time of the task and potentially maximize coverage of the target object, particularly if the robots have different capabilities.

As an example application, the steel bridge maintenance task can be considered [Liu et al., 2008]. As part of this task, grit-blasting is performed to remove old paint, rust, and other debris. Then, spray painting is carried out to protect the surfaces and preserve the integrity of the structure. Both grit-blasting and spray painting are examples of complete coverage tasks. With the enormous number of steel bridges in the world and the cost associated with maintaining these bridges,

*Email: Mahdi.Hassan@alumni.uts.edu.au, dikai.liu@uts.edu.au

Infrastructure Robotics: Methodologies, Robotic Systems and Applications, First Edition.
Edited by Dikai Liu, Carlos Balaguer, Gamini Dissanayake, and Mirko Kovac.

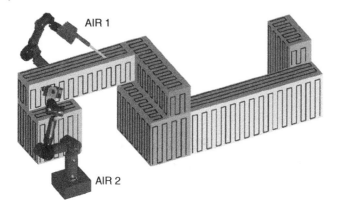

Figure 5.1 Two mobile AIRs performing grit-blasting on a complex structure. Source: Adapted from Hassan et al. [2016].

it is self-evident how grit-blasting or spray-painting AIRs can be beneficial in reducing maintenance cost, increasing productivity, and reducing human exposure to unpleasant and risky environments. In Australia alone, there are over 30,000 bridges [Liu et al., 2008].

A challenge in the deployment of AIRs is to develop methodologies that enable the AIRs to achieve optimal complete coverage. Optimality may be with respect to a single objective (such as minimizing completion time) or multiple objectives (such as minimizing completion time while simultaneously minimizing energy consumption and robot joint torques). Constraints, such as operating without any collisions, also need to be considered. If multiple AIRs are deployed, then the AIRs need to decide on their base placements, partition and allocate the surface areas among themselves, and perform collision-free motion planning. The AIRs may also be required to perform real-time coverage path planning (CPP) while taking into account changes that occur in the environment.

5.2 Coverage Planning Algorithms for a Single Robot

This section presents an offline and a real-time coverage algorithm for single-robot applications where there are no moving obstacles in the environment.

5.2.1 An Offline Coverage Planning Algorithm

An offline algorithm, Squircular-CPP [Hassan et al., 2020a], is devised for CPP, with the aim of generating smooth coverage path and robot motion, and reducing aggressive acceleration and deceleration.

Problem Definition

Let $\mathcal{X} \subset \mathbb{R}^2$ represent a target surface area for coverage. The coverage problem is made flexible by allowing the robot to cover a slightly larger area $\Omega \subset \mathbb{R}^2$, $\mathcal{X} \subseteq \Omega$. However, this flexibility comes with an expectation of generating a path within Ω that is smooth and not excessively long.

Given $\mathcal{X}$, the problem is to find an area Ω, $\mathcal{X} \subseteq \Omega$, that approximates $\mathcal{X}$ and enables a smooth path within Ω.

Convex Decomposition

At first, convex decomposition of the surface area may be needed. That is, the surface area, which may be nonconvex or contain holes and surface obstacles (e.g. object on the surface), will be decomposed into a number of convex subareas (e.g. through the method in [Ren et al., 2008]).

Fitting a Minimum Bounding Rectangle

Let an area $A \subseteq \mathcal{X}$ be represented as a set of points $X = \{x_1, x_2, \ldots, x_\eta\}$ (e.g. from point cloud or centers of a uniform grid). The next step is then to fit a Minimum Bounding Rectangle (MBR) to the points in X.

Let S-space (square) be defined as $S = \{(x, y) \in \mathbb{R}^2 | |x| \leq 1, |y| \leq 1\}$ where x and y are coordinates in S. The MBR is scaled along both axes to occupy the entire S-space. Let the scaled MBR and the scaled points within this MBR be termed as MBR† and $X^\dagger$, respectively.

Fitting a Squircle

The scaling of the MBR helps with finding a bounding squircle within the MBR†. A squircle is an intermediate shape between the square and the circle, as shown in Figure 5.2. A squircle (based on Fernandez-Guasti squircle or FG-squircle in short) is defined as [Fong, 2015]:

$$x^2 + y^2 - \frac{s^2}{k^2}x^2y^2 = k^2 \tag{5.1}$$

where $s \in [0, 1]$ is the squareness parameter, and k, analogous to the radius of a circle, is the length from the centroid to the point on the boundary that intersects

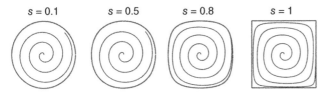

Figure 5.2 Squircles with spirals within them created using varying values of s and constant value of k. Source: Adapted from Hassan et al. [2020a].

the x or y axis (considering axis-aligned squircle, centered at origin). As shown in Figure 5.2, s defines how close a squircle is to a square or a circle.

Given a set of points $X^\dagger$ within the MBR†, Squircular-CPP can analytically fit an FG-squircle with minimum s value that is axis-aligned with respect to MBR†'s axes and that is upper bounded in size by the MBR†.

Rearranging Eq. (5.1) to make s the subject gives:

$$s = \sqrt{\frac{(x^2 + y^2 - k^2)k^2}{x^2 y^2}}. \tag{5.2}$$

For a given value of k, the squareness value s of the squircle can be calculated by substituting the x and y coordinates of a point b on the boundary of the squircle. In Squircular-CPP, b is the farthest point, $x^f \in X^\dagger$, from the centroid.

The point $x^f = x^\dagger_{l^*} \in X^\dagger$ where l^* is simply

$$l^* = \underset{l \in \{1,2,\dots,\eta\}}{\operatorname{argmax}} \|x^\dagger_l - c\| \tag{5.3}$$

where c is the centroid of the S-space which is at the origin.

By definition, the MBR† would have a side length of 2 since it occupies the entire S-space. The bounding squircle (henceforth referred to as BS†), that fits the points $X^\dagger$ in MBR† shares boundary with the MBR† and would always have a k value of 1 (half of MBR† side length). Thus, given $k = 1$, the s value of BS† can be calculated from Eq. (5.2) by substituting the x and y coordinates of x^f. Once BS† is defined, a spiral path is generated within this BS† and later scaled and transformed to the original size and pose of the MBR.

Deforming a Spiral Path into a Squircle

Similar to the S-space, let the O-space (circle) be defined as $O = \{(u, v) \in \mathbb{R}^2 | u^2 + v^2 \leq 1\}$ where u and v are coordinates in O, as shown in Figure 5.3(Left).

A squircular mapping technique, named Fernandez-Guasti squircular mapping (FG-squircular mapping) [Fong, 2015], is used to map a nonuniform arithmetic spiral created within the O-space to a deformed spiral within the S-space, as shown

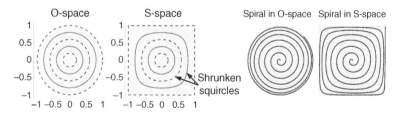

Figure 5.3 (Left) FG-squircular mapping; (Right) Mapping of a spiral path. Source: Adapted from Hassan et al. [2020a].

in Figure 5.3(Right). Using the FG-squircular mapping, given u and v coordinates of a point, $\boldsymbol{p}$, in O-space, the corresponding x and y coordinates of p in S-space can be calculated [Fong, 2015]:

$$
\begin{aligned}
\boldsymbol{p}_x &= \text{O2S}(u, v) = \frac{\text{sgn}(u, v)}{v\sqrt{2}}\sqrt{u^2 + v^2 - \sqrt{(u^2 + v^2)(u^2 + v^2 - 4u^2v^2)}}, \\
\boldsymbol{p}_y &= \text{O2S}(u, v) = \frac{\text{sgn}(u, v)}{u\sqrt{2}}\sqrt{u^2 + v^2 - \sqrt{(u^2 + v^2)(u^2 + v^2 - 4u^2v^2)}}.
\end{aligned}
\tag{5.4}
$$

Conversely, to map from S-space to O-space:

$$
\begin{aligned}
\boldsymbol{p}_u &= \text{S2O}(x, y) = x\sqrt{x^2 + y^2 - x^2y^2}/\sqrt{x^2 + y^2}, \\
\boldsymbol{p}_v &= \text{S2O}(x, y) = y\sqrt{x^2 + y^2 - x^2y^2}/\sqrt{x^2 + y^2}.
\end{aligned}
\tag{5.5}
$$

The goal is to generate a spiral within a region of O-space such that, when mapped into the S-space it occupies BS†. This goal is achieved using the shrunken FG-squircle [Fong, 2015]:

$$
x^2 + y^2 - x^2y^2 = s^2
\tag{5.6}
$$

by making $k = s$ in Eq. (5.1). Using this equation, both the size and the squareness of the squircle are controlled using the value of s, as shown in Figure 5.3(Left). The larger the s value, the bigger the squircle and the closer it is to a square.

The s values of these shrunken FG-squircles (in S-space) are in fact the radii of the circles (in O-space) [Fong, 2015]. Thus, a circle in the O-space with a radius of r would map to a shrunken FG-squircle (in S-space) with $s = r = k$. Thus, the spiral path is first generated within the circle of radius $r = s$ in the O-space where s equals to the s value of the BS†. The spiral path is then mapped to the shrunken FG-squircle in the S-space. Since $k = s$ for shrunken FG-squircle, then the spiral needs to be scaled up by a factor of s to occupy BS†. Example spiral paths are shown in Figure 5.4.

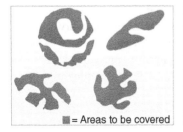
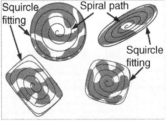

Figure 5.4 (Left) Areas to be covered; (Right) Coverage using Squircular-CPP. Source: Adapted from Hassan et al. [2020a]/IEEE.

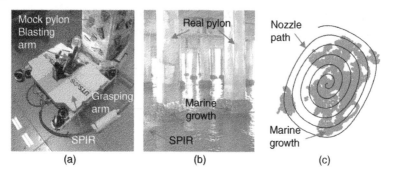

(a)   (b)   (c)

Figure 5.5 (a) Testing using (SPIR); (b) SPIR in real-world environment; (c) An example path using Squircular-CPP. Source: Hassan et al. [2020a]/IEEE.

A Case Study

Suppose that an intervention autonomous underwater vehicle (I-AUV), such as submersible pile inspection robot (SPIR) [Woolfrey et al., 2020], is to remove marine growth from surfaces of underwater structures using high-pressure water jet blasting (Figure 5.5). Slightly covering outside of the target area is acceptable for this application as a trade-off to generating smoother path. An example coverage path for removing one layer of marine growth is shown. The work in Mei et al. [2004] has proved, through analytical and experimental studies, that "spirals become the most energy-efficient because the robot can continuously move without stopping and turning" as compared to square spiral paths and boustrophedon paths.

5.2.2 A Real-Time Coverage Planning Algorithm

A real-time CPP algorithm, named Predator-Prey Coverage Path Planning (PPCPP) [Hassan and Liu, 2020], is presented in this section.

Problem Description

The CPP problem addressed here involves unexpected changes. For example, the coverage area may change in size/shape or unknown moving/stationary obstacles may suddenly become present in the environment. Unexpected obstacles make the coverage problem challenging since the robot needs to make quick decisions to avoid obstacles and efficiently re-plan the path while still aiming to achieve complete coverage with minimal cost.

Predator-Prey Inspiration for CPP

The approach is inspired by the concepts of *foraging* and *risk of predation* in predator-prey relation. When foraging in an area "prey must select their optimal

Figure 5.6 A grazing path of a prey. Source: Adapted from Hassan and Liu [2020].

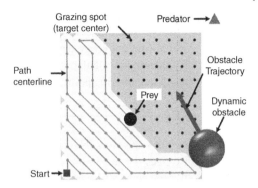

level of vigilance in response to their perceptions of a predator's whereabouts" [Brown et al., 1999]. This means that while foraging, animals continue to assess the risk of predation. Figure 5.6 shows a simple example where a prey assumes a square-shaped area to be a satisfying area for grazing due to the abundance of food and its faraway distance from a perceived ambushing predator. Since animals have to accomplish more than just avoiding predation (e.g. search for food); they need to continuously assess their current risk level of predation. This means that while foraging, the prey will aim at visiting all regions of the target area in search of food (achieving complete coverage), but it will do so by first consuming the food in regions farthest away from the predator and gradually taking greater risks to move closer and closer to the predator.

The PPCPP Approach

The approach is named PPCPP. The *prey* is designed to be the coverage spot with a size equivalent to the coverage size of the robot's end-effector tool. The prey's location at step k is denoted as $\boldsymbol{o}_k$. The *predator*, denoted as $\boldsymbol{\Psi}$, is designed to be a virtual point outside the coverage area of the robot. The predator is considered to be stationary (e.g. an ambushing predator). The predator enforces a spatial order for the prey, and as a result, a shorter path can be obtained in real time. The predator may be placed closest to a region of the coverage area where it is preferred for the coverage path to end, as shown in Figure 5.6.

At first, by scanning the environment, obstacles are detected, and the surface area, which is represented using discrete points, is updated. Then from prey's neighboring points, $\boldsymbol{N}(\boldsymbol{o}_k)$, the set of obstacle-*free* and *uncovered* target points, $\boldsymbol{N}^{uf}(\boldsymbol{o}_k)$ are determined.

From $\boldsymbol{N}^{uf}(\boldsymbol{o}_k)$, the index j_k^* of the neighboring point that results in maximum reward at step k is calculated and the corresponding best neighbor, $\boldsymbol{o}_{j_k^*}$, is selected for the prey to move to. R_j is the total reward for moving to a neighboring point $\boldsymbol{o}_j \in \boldsymbol{N}^{uf}(\boldsymbol{o}_k)$, and is calculated as per Eqs. 5.7 to 5.10.

Predation Avoidance Reward

At each step, the prey maximizes its reward by moving toward a neighboring point that is uncovered (not yet covered) and that has the farthest distance from the predator. The function for calculating the predation avoidance reward for the prey moving to the jth neighboring point, o_j is formulated as follows:

$$R^d\left(o_j\right) = \frac{D(o_j) - D_{min}(o_k)}{D_{max}(o_k) - D_{min}(o_k)} \tag{5.7}$$

where $D(o_j) = \|o_j - \Psi\|$ gives the distance from o_j to the predator Ψ, $D_{max}(o_k) = \max_j \|o_j - \Psi\|$ gives the maximum distance from one of the neighboring points of the current prey point to the predator, and similarly, $D_{min}(o_k) = \min_j \|o_j - \Psi\|$ gives the minimum distance.

Smoothness Reward

Having a path that has more straight lines (or fewer turns) can be beneficial, e.g. saving energy or time due to less frequent turns.

The second reward function is formulated as follows:

$$R^s\left(o_j\right) = \frac{\angle o_{k-1}o_k o_j}{180°} \tag{5.8}$$

where $R^s\left(o_j\right) \in (0,1]$ is the reward associated with the jth neighboring point, o_j, of the current prey point, o_k, due to the angle $\angle o_{k-1}o_k o_j \in (0°, 180°]$ which is the angle between the vectors $\left(o_{k-1} - o_k\right)$ and $\left(o_k - o_j\right)$, and o_{k-1} is the target point covered by the prey at the previous step $(k-1)$.

Boundary Reward

This reward is related to covering the boundary target points. At each step, the prey will be given extra reward for covering a boundary target point.

The third reward function is formulated as follows:

$$R^b\left(o_j\right) = \frac{n^{N_{max}} - n^N(o_j)}{n^{N_{max}}} \tag{5.9}$$

where $R^b\left(o_j\right) \in [0,1]$ is the reward associated with the jth neighboring point, o_j, of the current prey point, o_k, and $n^N(o_j)$ calculates the number of uncovered neighbors of the target point o_j. $n^{N_{max}}$ is the maximum possible number of neighbors for a target point, and can be determined based on the decomposition of the surface. For this reward function, the smaller the number of uncovered neighbors for the target point o_j, the higher the reward.

Total Reward (Sum of All Rewards)

The total reward for o_j is the sum of all the rewards, i.e.:

$$R\left(o_j\right) = R^d\left(o_j\right) + \omega^s\left(R^s\left(o_j\right)\right) + \omega^b\left(R^b\left(o_j\right)\right) \tag{5.10}$$

where ω^s and ω^b are the weighting factors for the smoothness and the boundary reward functions, respectively. The weighting factors are optimized only once for a given environment, prior to the deployment of the robot. Various optimization algorithms may be employed to optimize the weighting factors.

Thus, at step $k(k = 1, 2, \ldots, n^k)$, the index of the neighboring point that gives the maximum reward can be found ($R(\boldsymbol{o}_j)$ is based on Eq. (5.10)):

$$j_k^* = \arg\max_j \left(R\left(\boldsymbol{o}_j \in \boldsymbol{N}^{u^l}(\boldsymbol{o}_k) \right) \right). \tag{5.11}$$

It needs to be noted that a dead end (or local minimum) may occur, for example, when the prey reaches a target where all neighboring targets have already been covered. To continue its coverage task, the prey must revisit a certain number of targets until it reaches an uncovered target. To do so, a path planner like A* or Dijkstra's algorithm can be utilized. The uncovered target to reach is an intermediate goal target and is a target closest to both the already-covered targets and the prey. Once this target is reached, the coverage task resumes.

Case Study 1 – Stationary Obstacles

In this case study, eight scenarios are considered. In each scenario, a number of obstacles with different shapes and sizes are arbitrarily placed in the environment, as shown in Figure 5.7. The real-time path for each scenario using PPCPP is shown in the figure.

Case Study 2 – Moving Obstacles

The effect of each of the three obstacles shown in Figure 5.8 continuously moving back-and-forth along its trajectory is studied. When obstacle 1 is moving, the path

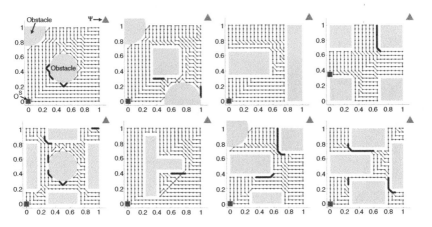

Figure 5.7 Eight different scenarios and a path created for each scenario in real-time using PPCPP. Source: Adapted from Hassan and Liu [2020]. In each scenario, a number of obstacles with different shapes and sizes are arbitrarily placed in the environment.

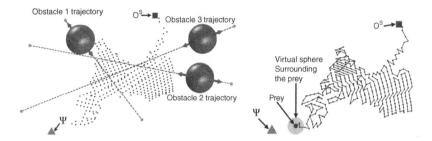

Figure 5.8 (Left) the trajectory of each obstacle; (Right) an example path when obstacle 1 is moving. Source: Adapted from Hassan and Liu [2020].

length is 23.86 m (the path is shown in Figure 5.8). Similarly, when obstacles 2 and 3 are moving, the path lengths are 22.52 m, and 22.75 m, respectively.

5.3 Coverage Planning Algorithms for Multiple Robots

5.3.1 Base Placement Optimization

An algorithm [Hassan et al., 2016] for optimizing the base placements of AIRs relative to the objects under consideration and the environment is presented in this section.

Problem Description
Optimizing base placements, i.e., finding optimal base location and/or orientation for each AIR, is crucial so as to enable the AIRs to collectively cover all surfaces

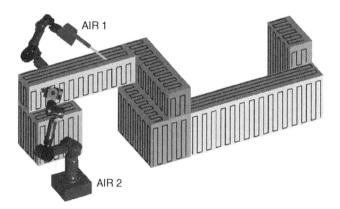

Figure 5.9 Optimizing base placement of two AIRs relative to complex structure. Source: Hassan et al. [2016]/IEEE.

of objects. As shown in Figure 5.9, this problem is further complicated when the object is large and has a complex geometric shape, and hence multiple base placements for each AIR are required to achieve complete coverage.

The OMBP Approach

The approach to Optimization of Multiple Base Placements (OMBP) starts by discretizing the search space. Simple preliminary filtering of the base placements can prevent poorly performing base placements from becoming candidates, hence reducing the size of the search space. The rest are anticipated to have high coverage (referred to as favored base placements, FBPs) as shown in Figure 5.10. Then, multi-objective optimization is performed to select a subset of FBPs for each AIR and to determine the visiting sequence of these FBPs.

Design Variables

Let $B_i^{FBP} = \{\beta_{i1}, \beta_{i2}, \ldots, \beta_{i(n_i^F)}\} \subseteq B_i$ be the FBPs associated with the ith AIR, for $i = 1, 2, \ldots, n$ where β_{i1} is a base placement (base location and/or orientation) for the ith AIR. The design variables are $Z_{ik} \in \{0, 1, \ldots, n_i^F\}$ with constraints $Z_{ij} \neq Z_{ik} \iff Z_{ik} > 0, i = 1, 2, \ldots, n, j = 1, 2, \ldots, n_i^F, k = 1, 2, \ldots, n_i^F, k \neq j$. As an example, the design variable $Z_{ik} = 3$ means that the kth base placement of the ith AIR is the third FBP, i.e. the ith AIR is to visit $\beta_{i(Z_{ik})} = \beta_{i3}$ for its kth base placement. Hence, the ith AIR can visit up to a maximum of n_i^F FBPs where n_i^F is the total number of FBPs. If a design variable is given a value of zero, i.e. if $Z_{ik} = 0$, then one less FBP will be visited by the ith AIR and the AIR will move from the $(k-1)$th base placement $\beta_{i(Z_{ik-1})}$ to the $(k+1)$th base placement $\beta_{i(Z_{ik+1})}$. Let Z be a set containing all the design variables that have a value greater than 0, i.e. $Z = \{Z_{ik}|Z_{ik} > 0, \forall i, k : i = 1, \ldots, n, k = 1, \ldots, n_i^F\}$.

Objective Functions

Objective 1–Maximal Coverage To achieve complete surface coverage, the base placements of all AIRs should be selected such that all targets (discrete points

Figure 5.10 All base placements and FBPs of one of the AIRs. Source: Hassan et al. [2015]/IEEE.

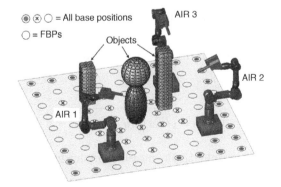

on a surface) representing the surfaces can be reachable by feasible poses of the AIRs. This objective is therefore to minimize missed-coverage:

$$\min_{Z} F_1(Z) = 1 - \sum_{i=1}^{n} \frac{\sum_{k=1}^{n_i^v} N^f(Z_{ik})}{n_i^T} \tag{5.12}$$

where Z is a set containing all the design variables that have a value greater than zero, n is the number of AIRs, n_i^v is the total number of base placements that the ith AIR needs to visit, n_i^T is the total number of targets associated with the ith AIR which represents all surfaces, and $N^{f(Z_{ik})}$ calculates the number of targets that can be reached with feasible AIR poses at a base placement.

Objective 2–Minimal Makespan The second objective is to minimize the makespan (i.e. the overall completion time of the task). Thus, this objective is:

$$\min_{Z} F_2(Z) = \max \{T_1(Z), T_2(Z), \ldots, T_n(Z)\} \tag{5.13}$$

where $T_i(Z)$ is the completion time of the ith AIR, which can be calculated as

$$T_i(Z) = \left(\sum_{k=1}^{n_i^v} N^f(Z_{ik}) \cdot \frac{d_i}{v_i} \right) + n_i^v \cdot t_i^s \tag{5.14}$$

where d_i is the distance between two adjacent targets along ith AIR's path, v_i is the chosen end-effector speed of the ith AIR suitable for the application, and t_i^s is the set-up time for ith AIR to move to the next base placement.

Objective 3–Maximal Manipulability Measure Manipulability measure [Yoshikawa, 1985] can be used to obtain a measure for a manipulator pose corresponding to a target in the environment. Therefore, this objective is to maximize the sum of manipulability measures for all AIR poses corresponding to all targets representing the surface. That is,

$$\max_{Z} F_3(Z) = \sum_{i=1}^{n} \sum_{k=1}^{n_i^v} \sum_{j=1}^{N^f(Z_{ik})} W(q_{ikj}^f) \tag{5.15}$$

where

$$W(q_{ikj}^f) = \sqrt{\det \left(J(q_{ikj}^f) J^T(q_{ikj}^f) \right)} \tag{5.16}$$

is the manipulability measure (a value from 0 to 1), and $J(q_{ikj}^f)$ is the Jacobian of the pose q_{ikj}^f.

Objective 4–Minimal Torque To improve the operating condition of an AIR, it is preferable to minimize the torque experienced by the joints of the AIR.

Let the torque ratio of joint m of a feasible AIR pose $\boldsymbol{q}^f_{ikj}$ be the amount of torque the joint has experienced divided by its torque capacity. The maximum torque ratio $\mathcal{T}^{Rmax}(\boldsymbol{q}^f_{ikj})$ for the pose $\boldsymbol{q}^f_{ikj}$ is the largest torque ratio from all the joints of the ith AIR, i.e.

$$\mathcal{T}^{Rmax}(\boldsymbol{q}^f_{ikj}) = \max_m \left| \frac{\mathcal{T}_{im}(\boldsymbol{q}^f_{ikj})}{\tau^{cap}_{im}} \right| \tag{5.17}$$

where $\mathcal{T}_{im}(\boldsymbol{q}^f_{ikj})$ is the torque experienced by joint m of the ith AIR at pose $\boldsymbol{q}^f_{ikj}$, and τ^{cap}_{im} is the torque capacity of the mth joint.

This objective is to minimize the sum of maximum torque ratios of the AIR that experiences the most amount of torque. That is,

$$\min_Z F_4(Z) = \max_i \left(\sum_{k=1}^{n^v_i} \sum_{j=1}^{N^f(Z_{ik})} \mathcal{T}^{Rmax}(\boldsymbol{q}^f_{ikj}) \right). \tag{5.18}$$

Constraint Functions

The distance between AIRs should be above a threshold δ (e.g. based on workspace size of AIRs). Thus, the AIRs' base placement should be chosen such that:

$$\left\| \boldsymbol{\beta}^{AIR}_i(t) - \boldsymbol{\beta}^{AIR}_l(t) \right\| > \delta \tag{5.19}$$

$\forall i, l : i = 1, \dots, n, l = 1, \dots, n, i \neq l$, and $\forall t : t = 0, \dots, t^c$ where $\boldsymbol{\beta}^{AIR}_i(t)$ and $\boldsymbol{\beta}^{AIR}_l(t)$ are the base placements of the ith and lth AIRs at time t, respectively, n is the total number of AIRs, and t^c is the overall completion time of the task.

During the selection of FBPs, the base placements that are in close proximity to the objects are discarded. Thus, the distance between any AIR and the objects (obstacles) is already considered.

A Case Study

A mock environment, as shown in Figure 5.11, is created using real data obtained from a part of a steel bridge maintenance site. Simulations were then performed on the data obtained from the mock environment (Figure 5.11). Two identical simulated AIRs modeled upon real AIRs are used to perform the task of grit-blasting for removing rust and old paint from steel surfaces. A solution from the Pareto front is selected which is shown in Figure 5.11. Based on the chosen solution, over 96% of the reachable targets can be covered.

5.3.2 Area Partitioning and Allocation

An algorithm [Hassan et al., 2014] for partitioning and allocating the objects' surface areas is presented in this section.

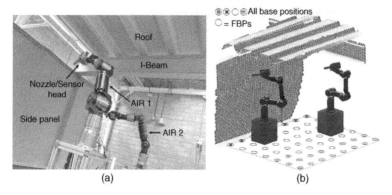

Figure 5.11 (a) AIRs applied for steel bridge maintenance; (b) placement of the AIRs based on a solution. Source: Hassan et al. [2015]/IEEE.

Problem Description

After optimizing for base placements of AIRs, it may happen that some surface areas are reachable by multiple AIRs at their current base placements. These areas are referred to as *overlapped areas* and need to be partitioned and allocated appropriately among the AIRs such that relevant objectives are optimized. Figure 5.12 shows an example.

The APA Approach

The Area Partitioning and Allocation (APA) approach makes use of Voronoi partitioning [Okabe et al., 2009] to partition the overlapped areas into n Voronoi cells,

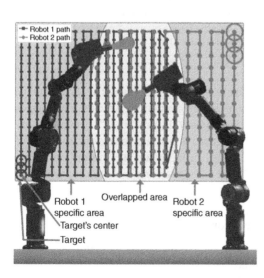

Figure 5.12 Overlapped and specific areas as well as the final paths associated with two AIRs. Source: Hassan et al. [2014]\IEEE.

Figure 5.13 (Left) overlapped areas; (Middle and Right) two examples of Voronoi partitioning. Source: Hassan et al. [2014]/IEEE.

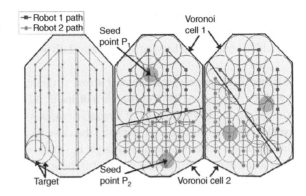

where n is the number of AIRs. This is shown in Figure 5.13. Each cell is allocated to an AIR, and the size of the cells is dependent on the location of the Voronoi graph's seed points $\{\boldsymbol{p}_1^s, \boldsymbol{p}_2^s, \dots, \boldsymbol{p}_n^s\}$. Thus, the seed points are considered to be the design variables of the multi-objective optimization problem. The goal is to obtain locations for the seed points that result in optimal solutions with respect to a set of objectives.

Design Variables

From Figure 5.13, it can be seen that the targets in each Voronoi cell are closest to the seed point of the cell; therefore, $O_i^{al} = \{\boldsymbol{o} \in O_i \mid \|\boldsymbol{o} - \boldsymbol{p}_i^s\| \le \|\boldsymbol{o} - \boldsymbol{p}_i^s\|, \forall i \in \{1, 2, \dots, n\}\}$ where O_i^{al} is the set containing the allocated targets to the ith AIR and O_i is the set containing the targets representing the overlapped areas of the ith AIR. Thus, the design variables are:

$$Z = \left\{\boldsymbol{p}_1^s, \boldsymbol{p}_2^s, \dots, \boldsymbol{p}_n^s\right\}. \tag{5.20}$$

Objective Functions

Objective 1–Makespan The first objective is to minimize the variance of the completion times:

$$\min_{Z} F_1(Z) = \frac{1}{n} \sum_{i=1}^{n} \left(T_i(Z) - \bar{t}\right)^2 \tag{5.21}$$

where $\bar{t}$ is the average of the completion times of the n AIRs, and $T_i(Z)$ returns the completion time of the ith AIR. The time $T_i(Z)$ is expressed as

$$T_i(Z) = \frac{L_i^o(Z) + l_i^s}{v_i} \tag{5.22}$$

where $L_i^o(Z)$ and l_i^s are the lengths of the paths generated on the overlapped and specific areas of the ith AIR, respectively, and v_i is the ith AIR's end-effector speed relative to the surface.

Minimal difference in completion times of AIRs will result in optimal makespan due to the fixed coverage area. Minimizing the difference in completion times has the added benefit of achieving fair workload division between the AIRs.

Objective 2–Closeness of the Allocated Areas If the areas allocated to the ith AIR are closer to other AIRs than the ith AIR itself, then the motion of the ith AIR can be affected by other AIRs or irregular motions of the AIR can be caused to avoid a potential collision. Minimizing the closeness can be done by minimizing the sum of distances between each target o_{ij} (in the set of targets that represented the allocated areas O_i^{al}) and the centroid of the specific areas c_i^s associated with the ith AIR. Thus, the second objective is:

$$\min_Z F_2(Z) = \sum_{i=1}^{n} \sum_{j=1}^{N_i^o(Z)} \left\| c_i^s - o_{ij} \right\|. \tag{5.23}$$

If there are no specific areas for an AIR, then minimizing the distance between the areas to be allocated and AIR's base location can be considered.

Objective 3–Minimal Torque Let $\mathscr{T}^{Rmax}(q_{ij}^f)$ be the function that calculates the maximum torque ratio experienced by a joint $k(k = 1, \ldots, n^J)$ of the ith AIR, where q_{ij}^f is a feasible AIR pose that can reach the target $o_{ij} \in O_i^{al}$ with an acceptable end-effector position and orientation, and j is the AIR's pose index as well as the target's index. $\mathscr{T}^{Rmax}(q_{ij}^f)$ can be expressed as

$$\mathscr{T}^{Rmax}(q_{ij}^f) = \max_k \left| \frac{\mathscr{T}_{ik}(q_{ij}^f)}{\tau_{ik}^c} \right| \tag{5.24}$$

where $\mathscr{T}_{ik}(q_{ij}^f)$ gives the torque experienced by joint k of the ith AIR at pose q_{ij}^f, and τ_{ik}^c is the torque capacity of the kth joint.

The third objective is therefore to minimize the sum of all the maximum torque ratios corresponding to all the AIR poses generated for the targets allocated to each AIR, i.e.

$$\min_Z F_3(Z) = \sum_{i=1}^{n} \sum_{j=1}^{N_i^o(Z)} \mathscr{T}^{Rmax}(q_{ij}^f). \tag{5.25}$$

Objective 4–Maximal Manipulability Measure Manipulability measure of an AIR pose q_{ij}^f can be calculated as

$$W(q_{ij}^f) = \sqrt{\det \left(J(q_{ij}^f) J^\top(q_{ij}^f) \right)} \tag{5.26}$$

where $J(q_{ij}^f)$ is the Jacobian matrix associated with the pose q_{ij}^f.

The fourth objective is to maximize the sum of the manipulability measures, i.e.

$$\max_{Z} F_4(Z) = \sum_{i=1}^{n} \sum_{j=1}^{N_i^o(Z)} W(\boldsymbol{q}_{ij}^f). \tag{5.27}$$

A Case Study

This case study demonstrates the effectiveness of the APA approach for conditions where AIRs' capabilities are different. Assume that the overlapped and specific areas of each AIR are as shown in Figure 5.14 and are calculated based on the base locations and workspace size of the AIRs. A solution from the Pareto front that results in the minimal overall completion time (makespan) of the task is chosen. The time t (in seconds) taken for each robot to cover its path, the length L of the path (in meters), and the distance d between neighboring targets for each robot are shown on the figure.

5.3.3 Adaptive Coverage Path Planning

This section presents an extension to the PPCPP approach presented in Section 5.2.2. The extension enables the approach to be applied to multiple robots [Hassan et al., 2020b]. The base placement optimization approach as well as the APA approach could be applied to offline planning; however, the multi-robot decentralized approach (Dec-PPCPP) presented here is for real-time adaptive coverage by multiple robots.

Problem Description

The multi-robot decentralized coverage problem is defined as follows: Given the set of targets O (discrete point representing the surface), how to devise a computationally tractable real-time planner that enables each robot to traverse a path with

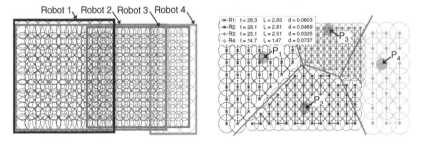

Figure 5.14 Four AIRs with different capabilities to cover a flat surface: (Left) overlapped and specific areas of each AIR; (Right) partitioning result and final path of each AIR. Source: Hassan et al. [2014]/IEEE.

the aim of achieving: (1) a minimal cost path, (2) swift adaptation to a nearby detected obstacle $x_j \in X_i$ for collision avoidance without drastic impact on the overall performance, and (3) coverage of all targets O in collaboration with other robots. The cost of the path could consider the path execution time, path length, smoothness, etc.

The DEC-PPCPP Approach
The approach is named Decentralized PPCPP (Dec-PPCPP). In brief, during the real-time deployment, each robot iteratively determines the next best neighboring target that will be visited based on the local environment and the information received from other robots. This one-step planning based on local information makes the approach efficient for real-time coverage and swift for adaptation with respect to obstacles. Reward functions act as heuristics for evaluating the performance of uncovered and obstacle-free neighboring targets, and the neighbor with maximum reward is selected for visiting next.

The ith robot will be calculating its next best target (a point on the surface) while it is moving from its current *position* p_i to its *destination* target o_i^d. For the ith robot to calculate its next best target, it will first communicate with other robot $\mathcal{R}_j, j :$ $\{1, 2, \dots, n\} \backslash i$ to receive updates. If a moving obstacle is within an α_i-distance to the ith robot, then the robot's priority is to move away from the obstacle; otherwise, the robot continues the coverage task.

Reward Functions
Reward 1 – Stationary Predator Avoidance Each robot considers itself a prey that needs to avoid two types of predators: (1) a stationary virtual predator, and (2) dynamic predators (other robots).

The function for calculating the stationary predator avoidance reward for moving to the jth neighbor, o_j is:

$$R^{sp}\left(o_j\right) = \frac{D(o_j) - D_{min}(o_i)}{D_{max}(o_i) - D_{min}(o_i)} \tag{5.28}$$

where $D(o_j) = \|o_j - \Psi_i^s\|$ is the distance from o_j to the predator Ψ_i^s, $D_{max}(o_i) = \max_j \|o_j - \Psi_i^s\|$ is the maximum distance to the predator from one of the neighbors of the prey o_i, and similarly, $D_{min}(o_i) = \min_j \|o_j - \Psi_i^s\|$ is the minimum distance. Note that $R^{sp}(o_j) \in [0, 1]$. Thus, the prey will aim to move to the farthest neighbor to achieve the maximum reward $R^{sp}(o_j) = 1$.

Reward 2 – Dynamic Predator Avoidance Unlike the stationary predator, the dynamic predators (other robots) do *not* affect the prey's motion for the entire coverage task. Instead, they only cause a local temporary effect on the prey's motion when they are close to the prey.

Let $\Psi_{i,k}^d$ denote the kth dynamic predator ($k \in \{1, 2, \ldots, n\} \setminus i$) for the prey representing the ith robot. The function for calculating the dynamic predator avoidance reward for the prey (ith robot) moving to the jth neighbor, $\mathbf{o}_j$:

$$R^{dp}\left(\mathbf{o}_j, \Psi_{i,k}^d\right) = S(\mathbf{o}_i) \frac{D(\mathbf{o}_j) - D_{min}(\mathbf{o}_i)}{D_{max}(\mathbf{o}_i) - D_{min}(\mathbf{o}_i)} \tag{5.29}$$

where $D(\mathbf{o}_j)$, $D_{max}(\mathbf{o}_i)$, and $D_{min}(\mathbf{o}_i)$ are calculated in the same way as in Section 5.3.3.3 except that the predator is now the location of the kth robot, i.e. $\Psi_{i,k}^d = \mathbf{p}_k$. For brevity, the notation $\Psi_{i,k}^d$ is dropped from above functions.

The function $S(\mathbf{o}_j)$ is the inverted Sigmoid function:

$$S(\mathbf{o}_i, \Psi_{i,k}^d) = \frac{1}{1 + \exp^{\kappa\left(a - \frac{b}{2}\right)}} \tag{5.30}$$

where κ determines the slope of the sigmoid function, $a = \|\mathbf{o}_i - \Psi_{i,k}^d\|$ is the distance from the prey's current destination target $\mathbf{o}_i = \mathbf{o}_i^d$ to the predator $\Psi_{i,k}^d$, and b is the effective range. When the kth predator $\Psi_{i,k}^d$ is far away (i.e., when $a > b$), then $S(\mathbf{o}_i, \Psi_{i,k}^d) \approx 0$ meaning that when dynamic predators are far away then this reward becomes negligible and the prey emphasizes more on maintaining the global behavior of keeping a spatial order (or an overall direction of motion) enforced by the stationary predator. The larger the effective range b, the farther away the robots will be from each other when covering the target area. Note that $R^{dp}\left(\mathbf{o}_j, \Psi_{i,k}^d\right) \in (0, 1)$.

Reward 3–Smoothness Another relevant reward for robotic coverage applications is the smoothness of the path. This reward function is formulated as follows:

$$R^s\left(\mathbf{o}_j\right) = \frac{\Theta(\mathbf{o}_j) - \Theta_{min}(\mathbf{o}_i)}{\Theta_{max}(\mathbf{o}_i) - \Theta_{min}(\mathbf{o}_i)} \tag{5.31}$$

where $R^s\left(\mathbf{o}_j\right) \in (0, 1]$ is the reward associated with the jth neighbor, $\mathbf{o}_j$, of the current prey target, $\mathbf{o}_i$, due to the angle $\Theta(\mathbf{o}_j) = \angle \mathbf{o}^p \mathbf{o}_i \mathbf{o}_j \in (0°, 180°]$ which is the angle between the vectors $\left(\mathbf{o}^p - \mathbf{o}_i\right)$ and $\left(\mathbf{o}_i - \mathbf{o}_j\right)$, and $\mathbf{o}^p$ is the previous target covered by the prey. $\Theta_{max}(\mathbf{o}_i) = \max_j \angle \mathbf{o}^p \mathbf{o}_i \mathbf{o}_j$ is the maximum possible angle considering all neighbors, and similarly $\Theta_{min}(\mathbf{o}_i)$ is the minimum angle.

Reward 4–Boundary Boundary targets are those that are closest to the boundary of the surface and the *covered* regions (i.e. lie on the boundary of the *uncovered* region).

This reward function is formulated as follows:

$$R^b\left(\mathbf{o}_j\right) = \frac{B_{max}(\mathbf{o}_i) - B(\mathbf{o}_j)}{B_{max}(\mathbf{o}_i) - B_{min}(\mathbf{o}_i)} \tag{5.32}$$

where $R^b\left(\boldsymbol{o}_j\right) \in [0,1]$ is the reward associated with the jth neighbor $\boldsymbol{o}_j$ of the current prey target $\boldsymbol{o}_i$, $B(\boldsymbol{o}_j)$ returns the number of *uncovered* neighbors of $\boldsymbol{o}_j$, $B_{max}(\boldsymbol{o}_i) = \max_j B(\boldsymbol{o}_j)$ for $j = \{1, \ldots, |\boldsymbol{N}_j^{u,f}|\}$ where $\boldsymbol{N}_j^{u,f}$ is the set of uncovered and obstacle-free neighbors, and similarly $B_{min}(\boldsymbol{o}_i) = \min_j B(\boldsymbol{o}_j)$.

Total Reward (Sum of All Rewards) Total reward is the sum of all the rewards previously stated, i.e.:

$$R\left(\boldsymbol{o}_j\right) = R^{sp}\left(\boldsymbol{o}_j\right) + \omega^p\left(R^{dp}\left(\boldsymbol{o}_j, \Psi_{i,1}^d\right) + R^{dp}\left(\boldsymbol{o}_j, \Psi_{i,2}^d\right) + \cdots + R^{dp}\left(\boldsymbol{o}_j, \Psi_{i,n}^d\right)\right)$$
$$+ \omega^s\left(R^s\left(\boldsymbol{o}_j\right)\right) + \omega^b\left(R^b\left(\boldsymbol{o}_j\right)\right) \tag{5.33}$$

where ω^p, ω^s and ω^b are the weighting factors associated with the dynamic predator avoidance, the smoothness and the boundary reward functions, respectively. These weighting factors need to be optimized using one of many optimization algorithms that can be implemented. Thus, for $j = \{1, \ldots, |\boldsymbol{N}_j^{u,f}|\}$, the index j^* of the uncovered neighbor with maximum reward is:

$$j^* = \arg\max_j\left(R\left(\boldsymbol{o}_j \in \boldsymbol{N}_j^{u,f}\right)\right). \tag{5.34}$$

Hence, the prey will move to the target $\boldsymbol{o}_{j^*} \in \boldsymbol{N}_j^{u,f}$ next, and the process is repeated.

A Case Study
Suppose four robots are operating in a 3D environment with moving obstacles, as illustrated in Figure 5.15. To make the scenario harder, robot 3 is made to start at t = 20, robot 2 is made to stop at t = 50, and obstacles continuously go back-and-forth on the trajectories shown. For comparison's sake, three cases are considered: (1) with no moving obstacles, (2) with moving obstacles having half the speed of the robots, and (3) with moving obstacles having the same speed as the robots. The results are paths with lengths of 99.8, 110.8 and 112.1

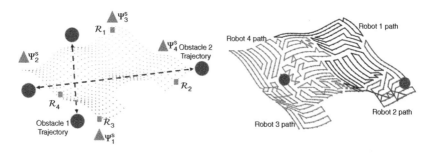

Figure 5.15 (Left) the scenario; (Right) the resulting paths. Source: Adapted from Hassan et al. [2020b].

units, respectively. Thus, the performance of Dec-PPCPP, amid moving obstacles continuously running through the surface, is *not* substantially worse than when there are no obstacles. Figure 5.15(Right) is related to case 2.

5.4 Conclusion

This chapter presented methodologies for achieving optimal coverage by a single AIR or multiple AIRs for both offline and real-time coverage.

More specifically, the problems discussed in this chapter are as follows: (1) smooth CPP (2) partitioning and allocation of surface areas among AIRs; (3) determining optimal base placements for each AIR; and (4) adaptability of CPP for dynamic environments and unexpected changes.

A summary of the methodologies is as follows:

- The Squircular-CPP algorithm (in Section 5.2.1) is developed to achieve smooth coverage paths for offline coverage. The novelty of the algorithm lies within the squircular shape fitting which is not only simple, fast and analytical, but also enables a smooth spiral path to be appropriately deformed within the fitted shape. The best fit squircle, which is an intermediate shape between the square and the circle, is determined analytically. A simple arithmetic spiral path is then generated within a circle and then deformed to fit within the squircle using Fernandez- Guasti squircular mapping.

- The PPCPP approach in Section 5.2.2 and the Dec-PPCPP appraoch in Section 5.3.3 were designed for adaptive coverage of dynamic environments using single robot and multiple robots, respectively. Due to the stationary virtual predator, each robot maintains a direction of overall motion causing the overall path to be reasonably organized (less back-and-forth between regions) despite moving obstacles being present. Each robot perceives other robots as dynamic predators, which results in the robots repelling each other causing better task allocation and collision avoidance. A number of reward functions were designed as heuristics to guide the robots throughout the coverage task.

- In Section 5.3.1, a mathematical model is developed for the problem of base placement such that the AIRs collectively cover the entire object while optimizing team's objectives and accounting for relevant constraints. This method enables the AIRs to find: (i) the minimal number of base placements for each AIR to operate from, (ii) the locations of the base placements for each AIR, and (iii) the visiting sequence of the base placements associated with each AIR. The model accounts for AIRs with different capabilities and objects with complex geometric shapes.

- In Section 5.3.2 A mathematical model is developed for simultaneously partitioning and allocating surface areas that can be reached by multiple AIRs at their current base placements. The approach utilizes Voronoi partitioning to partition objects' surfaces. Multi-objective optimization is conducted to allocate the partitioned areas to AIRs by optimizing the team's objectives.

The methodologies discussed earlier exhibit flexibility and are not restricted to a specific collision avoidance algorithm. A variety of collision avoidance algorithms, including potential field and virtual force field, can be employed. The selection of an appropriate collision avoidance algorithm should be based on the specific requirements of the application at hand.

Furthermore, the objective functions presented in this chapter are generic to suit most applications. However, additional objective functions, such as minimal energy, can be integrated into the methodologies for a particular application. In doing so, it is essential to perform a thorough investigation of potential conflicts between the new objective functions and the original ones.

For further reading and in-depth analysis and results related to the methodologies presented in this chapter, readers are advised to refer to the publications in the reference list that include authors' name.

Bibliography

J.S. Brown, J.W. Laundré, and M. Gurung. The ecology of fear: Optimal foraging, game theory, and trophic interactions. *Journal of Mammalogy*, 80:385–399, 1999.

C. Fong. Analytical methods for squaring the disk. *arXiv preprint arXiv:1509.06344*, 2015.

M. Hassan and D. Liu. A prey-predator behavior based approach to adaptive coverage path planning. *IEEE Transactions on Robotics*, 36:284–301, Feb 2020.

M. Hassan, D. Liu, S. Huang, and G. Dissanayake. Task oriented area partitioning and allocation for optimal operation of multiple industrial robots in unstructured environments. In *International Conference on Control Automation Robotics Vision*, pages 1184–1189, 2014.

M. Hassan, D. Liu, G. Paul, and S. Huang. An approach to base placement for effective collaboration of multiple autonomous industrial robots. In *IEEE International Conference on Robotics and Automation (ICRA)*, pages 3286–3291, May 2015.

M. Hassan, D. Liu, and G. Paul. Modeling and stochastic optimization of complete coverage under uncertainties in multi-robot base placements. In *IEEE/RSJ International Conference on Intelligent Robots and Systems (IROS)*, pages 2978–2984, 2016.

M. Hassan, D. Liu, and X. Chen. Squircular-CPP: A smooth coverage path planning algorithm based on squircular fitting and spiral path. In *International Conference on Advanced Intelligent Mechatronics (AIM)*, pages 1075–1081, 2020a.

M. Hassan, D. Mustafic, and D. Liu. A decentralized predator–prey-based approach to adaptive coverage path planning amid moving obstacles. *IEEE/RSJ International Conference on Intelligent Robots and Systems (IROS)*, pages 11732–11739, 2020b.

D.K. Liu, G. Dissanayake, P.B. Manamperi, P.A. Brooks, G. Fang, G. Paul, S. Webb, N. Kirchner, P. Chotiprayanakul, N.M. Kwok, et al. Quake. A robotic system for steel bridge maintenance: Research challenges and system design. In *Australasian Conference on Robotics and Automation*, pages 3–5, Dec 2008.

Y. Mei, Y.H. Lu, Y.C. Hu, and C.S.G. Lee. Energy-efficient motion planning for mobile robots. *IEEE International Conference on Robotics and Automation*, 5:4344–4349, Apr 2004.

A. Okabe, B. Boots, K. Sugihara, and S.N. Chiu. *Spatial tessellations: Concepts and applications of Voronoi diagrams*. John Wiley & Sons, page 501, 2009.

Z. Ren, J. Yuan, and W. Liu. Minimum near-convex shape decomposition. *IEEE Transactions on Pattern Analysis and Machine Intelligence*, 35:2546–2552, Oct 2013.

J. Woolfrey, D. Liu, and M. Carmichael. Kinematic control of an Autonomous Underwater Vehicle-Manipulator System (AUVMS) using autoregressive prediction of vehicle motion and Model Predictive Control. In *International Conference on Robotics and Automation (ICRA)*, 4591–4596, Feb 2016.

T. Yoshikawa. Manipulability of robotic mechanisms. *The International Journal of Robotics Research*, 80:3–9, 1985.

6

Methodologies in Physical Human–Robot Collaboration for Infrastructure Maintenance

Marc G. Carmichael*, Antony Tran, Stefano Aldini, and Dikai Liu

Robotics Institute, University of Technology Sydney, Broadway, Ultimo, NSW 2007, Australia

6.1 Introduction

There have been many robotic systems developed that aim to improve the safety and efficiency of managing and maintaining civil infrastructure. The advantages of robots lend themselves to tasks common in infrastructure maintenance, such as inspection and maintenance operations that are tedious, laborious, or dangerous. Common solutions utilize robots that are remotely controlled by human operators or, in some cases, employing autonomous robots that can operate with minimal human intervention. An alternate approach, which has not been widely adopted for infrastructure inspection and maintenance, is physical human–robot collaboration (pHRC). The pHRC paradigm sees humans and robots working physically together, combining the complementary strengths of the human and the robot to perform a task.

In this chapter, we discuss methodologies and opportunities for pHRC in infrastructure maintenance tasks. To the author's knowledge, a summary of pHRC applied specifically to the area of infrastructure has not been performed. Section 6.2 begins by explaining pHRC in comparison to teleoperation and fully autonomous paradigms. In Section 6.3, we summarize the common control strategies that are used for pHRC. Section 6.4 then focuses on adaptive assistance strategies that adapt robot performance based on a human worker's needs. Section 6.5 presents a framework for safety in pHRC. Section 6.6 discusses methods for measuring human performance during pHRC and how this may be utilized to improve the collaboration. Section 6.7 details examples of pHRC systems successfully applied to infrastructure and related tasks. Finally, in Section 6.8, we

**Email: Marc.Carmichael@uts.edu.au

Infrastructure Robotics: Methodologies, Robotic Systems and Applications, First Edition.
Edited by Dikai Liu, Carlos Balaguer, Gamini Dissanayake, and Mirko Kovac.

discuss the state of pHRC and opportunities for its utilization in infrastructure maintenance.

6.2 Autonomy, Tele-Operation, and pHRC

There is a wide variety of robotic solutions and techniques that have been successfully applied to infrastructure management and maintenance. The level of human involvement in the operation of these robotic solutions is also widely varied, as depicted in Figure 6.1. At one end of the spectrum, fully autonomous systems operate with little to no human intervention. At the other end, robots that are tele-operated or operate via pHRC are highly dependent on the skills of a human operator. These different approaches each present unique advantages and disadvantages that need to be understood when applying robotics to infrastructure applications.

6.2.1 Autonomous Robots

Autonomous robots operate without the need for human intervention. Autonomy requires the robot to have the ability to make decisions about which actions it should take, which may require advanced capabilities including environmental sensing and perception, planning, navigation, and locomotion. Traditional industrial robots, like those in manufacturing plants, are able to operate automatically due to the highly structured environment and repetitive preplanned tasks. A challenge for autonomous robots in infrastructure maintenance is, somewhat amusingly, the unstructured nature of the environment. Uncontrollable environmental factors that are inherent in environments such as bridges, façades, tunnels, water systems, and the like present significant challenges. As a result, there are limited examples of fully autonomous robots used in civil infrastructure environments.

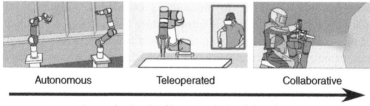

Autonomous Teleoperated Collaborative

Increasing levels of human-robot collaboration

Figure 6.1 Examples of robotic systems with different levels of human–robot collaboration. Left: Fully autonomous robots capable of operation with minimal human interaction. Center: Teleoperated robots that operate remotely from humans but are controlled, either fully or partially, by a human operator. Right: Robots that operate via physical human–robot collaboration with the human operator in close proximity to the robot and the task being performed.

There is great motivation to pursue autonomous robots despite the challenges. Autonomous operation means that human workers can be completely removed from the environment. This is a substantial benefit for infrastructure maintenance that have inherent physical risk or Occupational Health and Safety (OH&S) concerns, for example due to hazardous materials like asbestos. Another advantage is that multiple autonomous robots may be deployed in parallel, providing a scalable solution to increasing the productivity of tasks including inspection and maintenance.

An example is the autonomous abrasive blasting robot (Figure 6.2a) presented by Paul et al. [2011] to remove hazardous substances during bridge maintenance. This robotic system first performs a scan of the initially unknown environment, generates a 3D map of the surroundings, plans a robot trajectory for blasting, then executes this trajectory. Apart from being initially set up and monitored by a team of workers, the system would perform the blasting operation without human intervention.

Another example of autonomous robotic bridge maintenance is CROC [Quin et al., 2016] (Figure 6.2b). This seven-degree-of-freedom climbing robot is capable of autonomously exploring and navigating complex 3D environments, such as narrow arch tunnels. Along its path, CROC collects high-definition images that can then be viewed remotely to assess the surface conditions of bridges.

Aerial drones are already widely used for monitoring and inspecting civil infrastructures. Although they are commonly teleoperated remotely by human operators, Kabbabe Poleo et al. [2021] present the benefits of using a fully autonomous

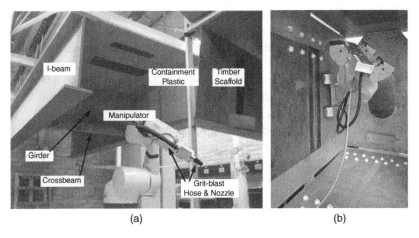

(a)   (b)

Figure 6.2 Examples of autonomous robots for bridge maintenance: (a) An autonomous abrasive blasting robot for bridge maintenance [Paul et al., 2011]/with permission from ELSEVIER. (b) Climbing RObot Caterpillar (CROC), an autonomous climbing robot for bridge condition assessment. Source: Quin et al., [2016]/IEEE.

solution for drone inspections of offshore wind farms. Using an autonomous drone fleet would increase the business scalability.

6.2.2 Teleoperated Robots

Teleoperated robots are controlled remotely by the human operator, who has either full or partial control over the robot's actions. The physical separation between robot and operator might be small, for example both within the same room, or large with extreme cases being lunar or Mars rovers. Conceptually, teleoperated robots can be separated into their *local* (master) and *remote* (slave) subsystems. The subsystem that is local to the operator both receives inputs about how the robot should perform, and provides feedback about the robot and its environment. Input modalities vary from simple keyboard, mouse and joystick inputs, to non-contact hand tracking or wearable exoskeletons. Common modalities for operator feedback include visual displays (e.g. monitors and virtual reality), force feedback (e.g. vibrotactile), and audio.

The level of autonomy in teleoperated robots varies greatly. *Supervisory control* is a form of tele-operation where a robotic system is capable of autonomous operation, only requiring occasional high-level input from a human operator. Supervisory control is suited to applications with large communication delays. Less autonomous is *shared control* where the operator and robot share portions of the task control. This may be preferred in safety critical or high risk applications where human judgment is necessary. *Direct* tele-operation refers to the robot having no autonomy and directly following the commands of the operator. Manually piloting an aerial drone for bridge inspection is an example of direct tele-operation.

The robot-operator separation creates several technical challenges. Generally, with increased separation comes greater communications latency, which can lead to system instabilities or poor quality of service. It is also critical the operator be provided with system and environmental information of appropriate fidelity to facilitate successful tele-operation. Since the operator is part of the control loop, human factors are also considerations.

Despite the challenges, tele-operation has benefits for infrastructure inspection and maintenance. Since teleoperated systems can leverage human cognitive capabilities, the need for robotic intelligence can be significantly reduced or eliminated. Additionally, the physical separation means dangerous tasks can be performed without placing the operator at risk. Finally, a single human operator can control a fleet of robots by only intervening when human input is required.

An example is the teleoperated robot used to inspect water and waste pipes [Gunatilake et al., 2021] shown in Figure 6.3. The robot would enter and obtain detailed 3D maps of the pipes' internal structure, showing all pipe defects to the remote human operator controlling the robot from outside the pipe.

Figure 6.3 Teleoperated pipe inspection robot. Source: Gunatilake et al., [2021]/IEEE.

Another example is the SPIR (Submersible Pylon Inspection Robot) [Le et al., 2020]. This submersible robot is remotely controlled to monitor and maintain underwater pylons using high-pressure water jets to remove marine growth, a task typically performed manually by human divers.

6.2.3 Physical Human–Robot Collaboration

pHRC sees the human and the robot share the same physical space as they work together toward a common goal. An example of pHRC is depicted in Figure 6.4. Since the robot workspace is shared with human operators, then safety is of primary concern. Generally, the robot's mechanical power is limited, and the robot joints present some level of compliance (by control or hardware) to reduce the risk of human injury. The requirements for safety in pHRC are detailed further in Section 6.5.

In infrastructure maintenance, pHRC has seen relatively little use compared to autonomous and teleoperated robots. This is not surprising as a common

Figure 6.4 Depiction of Physical Human–Robot Collaboration. Unlike autonomous or teleoperated robots, the human operator works in close proximity to the task and the robot which is controlled via direct physical interaction.

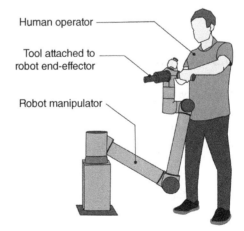

Human operator

Tool attached to robot end-effector

Robot manipulator

motivation for robotics in infrastructure inspection and maintenance is to totally remove human workers from hazardous environments. In pHRC, the operator is co-located with the robot where the task is being performed, negating this benefit. Despite this, there are other unique benefits that pHRC provides. Much like teleoperated robots, pHRC can leverage the cognitive abilities of the human operator. However, since the operator is local to the task, pHRC avoids the inherent challenges of tele-operation where it can be difficult to render environmental information such that the human's perception and decision-making abilities can be effectively leveraged. Additionally, teleoperated systems have both local and remote subsystems with a communication architecture connecting them, whereas with pHRC there is commonly just one single system, which may reduce system complexity. The key benefit of pHRC is the ability to combine the complementary skills of both the human and the robot. The robot does not become fatigued and is able to support sustained loads while maintaining its speed, accuracy, and repeatability. Combined with the cognitive capabilities of the human operator, this human-robot dyad can be applied to many infrastructure maintenance tasks.

An example of pHRC applied to infrastructure maintenance is the ANBOT [Carmichael et al., 2019] which used a collaborative robot manipulator to support the physical loads of an abrasive blasting task. The human operator would control the motions of the blasting nozzle attached to the end-effector directly through physical interaction. The robot made no decisions about how the task was performed, rather a skilled worker would perform the task as normal but with reduced physical burden. More details on this example of pHRC is provided in Section 6.7.

Another example is the WallMoBot, a robot designed to lift, transport, and install glass panels in buildings, in physical collaboration with a human operator [Yousefizadeh et al., 2019]. For large movements, the operator uses a joystick, but for fine movements, requiring dexterity and precision, the robot used impedance control to achieve a compliant behavior for the human to manually move the glass panel.

6.3 Control Methods

Robotic systems that operate in direct physical contact with humans have unique requirements that need to be considered when developing control solutions. Primary considerations include safety, operator intuitiveness, and assistance paradigms. In this section, we present methodologies for controlling robots in pHRC.

6.3.1 Motion Control

With traditional industrial robots, the priority of a control system is to execute motions as accurately and repeatedly as possible. A basic example of a robot

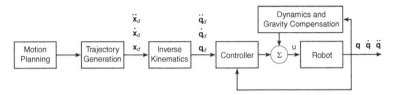

Figure 6.5 Basic motion control of a robotic system.

motion control system is shown in Figure 6.5. A motion plan, typically created offline by a trained human technician, is converted into a trajectory that defines how the robot should move its end-effector over time. The terms $\mathbf{x}_d$, $\dot{\mathbf{x}}_d$, and $\ddot{\mathbf{x}}_d$ represent the desired position, velocity, and acceleration of the end-effector. Using inverse kinematics, the desired end-effector motion is transformed into a desired joint trajectory that will facilitate this motion ($\mathbf{q}_d$, $\dot{\mathbf{q}}_d$, and $\ddot{\mathbf{q}}_d$). These aforementioned steps commonly take into consideration factors such as collisions with the environment and robot self-collisions, kinematic singularities, speed and acceleration limits, or other constraints on the system and task. This approach requires a detailed description of the robotic system and the environment to be known, making it well suited for repetitive and well-defined tasks in structured environments like a factory. With a trajectory generated, a control system commands the robot to follow the desired motion, compensating for the robots own dynamics, gravity, and other factors.

In pHRC applications, the control system is instead designed to facilitate interaction, often unpredictable, with a human operator. For example, the operator may control the motions of a tool attached to the robot end-effector, with measured forces between the robot and human being used to interpret the intended motion. Because the intentions of the operator are not known beforehand, preplanned motion strategies are not suitable. However, motion-based control is often used within more complex control schemes, such as admittance control or hybrid force-motion control, as explained in the following subsections.

Instead of executing preplanned trajectories, the robot can be controlled by a human operator who issues commands from a separate input device, such as a joystick or robot teach pendant, a process often called *jogging*. Unlike pHRC, jogging places the operator at a distance from the task and can require additional cognitive load regarding the mapping from joystick input to robot motion. It can be desirable to instead control the robot with direct physical interaction, placing the operator local to the task and providing an intuitive input modality [Bicchi et al., 2008].

6.3.2 Force Control

Force control aims to modulate the forces resulting between the robot and its environment, which includes humans. The control of interaction forces is essential in achieving robust behavior in poorly structured environments, and for safe operation in the presence of humans [Luigi Villani and Joris De Schutter, 2008]. Force control can be broadly classified as either *direct* or *indirect*.

Direct force control uses measurements from force sensors as feedback to create a closed force-feedback loop. An example is hybrid control [Raibert and Craig, 1981] where the force and position in orthogonal directions are controlled simultaneously. This scheme requires a detailed model of the environment which in most practical situations is not available, especially during pHRC where human motions are unpredictable.

Indirect force control uses the motion of the robot to indirectly control interaction forces. The relationship between movements and forces is usually modeled as a linear mass-spring-damper, and the control implemented using either an impedance or admittance controller [Hogan, 1985]. Impedance control uses deviations in position measurements to calculate the reaction forces the desired dynamic system would produce. These forces are then actuated by the control scheme. Admittance control operates the opposite way, reacting to measured external forces by generating appropriate motions.

The choice and performance of force control schemes rely largely on the construction of the robot and its actuation capabilities. Impedance control requires a robot that has the ability to render accurate forces based on measured movements, implying actuators that have low friction, low inertia, or contain the hardware necessary to implement closed-loop force control at each joint. Alternatively, admittance control requires accurate positioning capabilities based on measured external forces, implying the use of a robot with high power actuators and rigid construction [Kooij et al., 2006].

Infrastructure maintenance tasks where loads might be excessive, the environments can be challenging, or the workspace can be large, a robot with substantial physical capabilities is needed. This lends itself toward robots that are closer to traditional industrial robots; hence, admittance-based controllers may be preferable given that inbuilt joint torque sensing is not available. However, recent advances in collaborative robots (or cobots) mean that manipulators with joint torque sensing and moderate payload capacities are available from most major robotics manufacturers. Such robots may utilize impedance-based control strategies.

A typical admittance control scheme is depicted in Figure 6.6. The interaction force between the human and robot is measured ($\mathbf{f}_H$) and used with a predefined admittance model commonly represented in *mass-damper-spring* form shown in Equation 6.1. The behavior of the system is tuned by altering the parameters of the virtual admittance model. The output of the admittance model is a motion that the robot is then controlled to execute using techniques commonly used for robot trajectory tracking.

$$m\ddot{x} + c\dot{x} + kx = f \qquad (6.1)$$

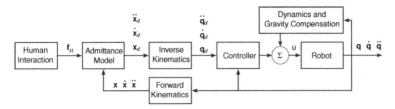

Figure 6.6 An admittance control scheme.

These force control schemes require the robot to measure interaction forces. If interaction only occurs at the robot's end-effector, then a multi-axis force-torque sensor (typically 6-axis) can be mounted at the robot tool flange. A downside to this approach is interaction forces can only be measured distally from the sensor. Robots with several points of interaction along the device (e.g. a full-body exoskeleton) require force measurements at each interaction point. Alternatively, robots with high-precision torque sensing integrated directly into the joints can allow physical interactions with the robot to be reconstructed [Haddadin et al., 2017]. With a force control scheme implemented, the robot can be controlled in a compliant fashion, responding in real-time to the forces applied by a human operator.

6.4 Adaptive Assistance Paradigms

When designing a robotic system that provides human workers with physical assistance, an important consideration is how this assistance is to be administered to the operator. Robotic physical assistance often comes in the form of shielding the operator from large loads intrinsic to the task being performed. At first glance, controlling the robot to fully support the loads of the task is an appealing idea. This can, in theory, remove all of the physical burden from the tasks, and require the human operator to provide only cognitive input that dictates how the task is performed. However, it has been shown that in some applications, this fixed *total assistance* approach may not be as effective in achieving certain desired outcomes compared to utilizing approaches that dynamically adapt the assistance provided. In this section, we specifically focus on paradigms which adapt the robotic assistance provided and how these may benefit applications in infrastructure maintenance.

There are many modalities in which assistance can be provided and can be utilized during infrastructure maintenance operations. Robotic guidance can help the operator guide a payload into a target location. Bicchi et al. [2008] describe a system utilizing virtual walls as a funnel-shaped guide to assist the operator in moving a payload without collision. Erden and Marić [2011] present a collaborative robot to help suppress abrupt motions during welding using a damping impedance controller. The results showed a significant improvement in the performance of welders when they are assisted by the robot. As well as physical strength, collaborative robots can be used to enhance the precision of human operators.

Force-feedback schemes control the force that the operator experiences as a modulated version of the robot's interaction forces with the environment. It is common for the forces felt by the operator to be a scaled-down version of the forces required by the task. This provides the operator with haptic feedback while less effort is required to perform the task. The strength and endurance of the robot is leveraged to assist the operator. This paradigm is often used in applications that require large physical effort, for example, materials handling. Figure 6.7 depicts an admittance controller presented in Carmichael and Liu [2013a]. The level of assistance is governed by parameter $A \in [0,1]$, with $A = 0$ meaning zero assistance is provided and the full task load is reflected back to the operator via the robot, and $A = 1$ means the robot provides all the force required to perform the task. Values of A in between mean the operator and robot share the task load. This level of assistance can be set fixed, or changed dynamically during robot operation. The next sections detail methodologies for adapting the level of assistance provided to the human operator.

6.4.1 Manually Adapted Assistance

The simplest approach to adaptive assistance is to simply let the operator determine their own needs and choose the appropriate level of assistance. As well as being simple to implement, it can provide the operator with an increased sense of control and transparency over the system's behavior. Gopinath et al. [2017] proposed a framework for an assistive robot that adapts the amount that the user controls the robot motion using an arbitration function. Verbal cues from the operator were translated to adjustments in the assistance level. It was shown that this approach led to higher levels of user satisfaction.

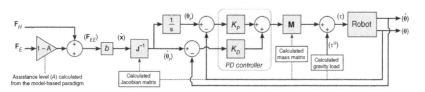

Figure 6.7 Example of an admittance control scheme used in Carmichael and Liu, [2013a]/IEEE. The variable *A* dictates the level of task force that is felt by the human operator, which may be adapted in real-time during task execution.

6.4.2 Assistance-As-Needed Paradigms

The Assistance-As-Needed (AAN) paradigm aims to provide an operator that is utilizing a robot for physical benefit, a level of assistance that is adapted to meet their needs. What exactly are the *needs* of the operator depends on the use case, and can depend largely on both the task being performed and the operator who is being assisted.

The AAN paradigm has its roots in robotic rehabilitation, which involves a robot providing assistance to the patient as they perform exercises using their impaired limb to promote recovery. Depending on the severity of the patient's impairment, exercises are often unable to be performed without external assistance. In stroke therapy, it has been shown that the quality of recovery is largely dependent on utilizing an appropriate amount of assistance. Too much assistance and the patient remains passive as the robot maneuvers their limb, which does not facilitate motor neuron recovery [Hogan et al., 2006]. Too little assistance and the exercises cannot be performed, leading to patient frustration and demotivation. The AAN paradigm seeks to adapt the assistance so that active engagement is achieved and recovery is maximized. It has been applied to a variety of therapies including rehabilitation of the upper limb [Krebs et al., 2003; Wolbrecht et al., 2007, 2008] and lower limb [Emken et al., 2005].

The primary challenge when utilizing AAN paradigms is to determine the actual needs of the user. In rehabilitation, this can be done using performance measures associated with the exercises, for example movement smoothness or time-to-completion. In infrastructure and other applications, determining the worker's need for assistance is not as straight forward, and hence utilization of AAN is less common. However, AAN paradigms for industrial applications are starting to gain the attention of researchers with several approaches having been developed to achieve desired outcomes during pHRC.

6.4.3 Performance-Based Assistance

An approach to implementing AAN is to adapt the assistance based on the observed performance of the operator. Using information such as the motion of the robot or the force interactions, operator performance is critiqued based on some predefined criterion relevant to the task. Based on how well the operator is deemed to have performed, the assistance is adjusted. If the operator performs the task well, then they are provided less assistance, and vice versa. The intention is that over time, the assistance will converge to an appropriate level.

Performance-based methods may not be well suited for many industrial tasks. Firstly, having a method of measuring and quantifying task performance is complicated, with metrics such as time-to-completion often not capturing the full story. There may also be ethical concerns with measurements of employee performance in a workplace setting, leading to worker anxiety for example. Additionally, the main benefit of pHRC (as opposed to autonomous robots) is being able to leverage the capabilities of a skilled worker. If the robotic system itself is able to distinguish good work from bad work, this implies that the robot itself is skilled, which negates this benefit. In reality, measuring operator skill or performance in complex tasks is not yet solved. Some tasks, however, may be suitable for performance-based AAN. In a robotic-aided welding scenario, Erden and Billard [2014] found that impedance measurements of welder's hands could be used as a measure of skill. The authors proposed that this technique could also be applied to other tasks such as painting. Although this study was small, it does suggest that AAN strategies in welding and similar applications could be utilized.

6.4.4 Physiology-Based Assistance

Adapting assistance based on the physiology of the operator is another approach to AAN. In an application described by Lynch et al. [2002] robotic trolleys are used to guide heavy loads maneuvered by workers with the objective of reducing the forces between the load and worker in the directions they require greater assistance due to higher risk of physiological strain. Work done by Carmichael and Liu [2013b] developed a framework based on a musculoskeletal model to estimate the physical strength capacity of the human upper limb. Comparing this modeled strength against the known physical needs of the task, the robot assistance could be adapted to fill this gap.

An interesting approach to AAN is ergonomic-based assistance. This approach utilizes the posture of the operator to adapt the robotic assistance provided based on ergonomic needs. The approach presented by Kim et al. [2018] uses a static joint torque model of human workers to calculate their center-of-pressure and ground reaction force. From this, overloaded joints are evaluated in real time, and an online optimization technique used to adjust robot trajectories to promote more ergonomic body poses of the operator during the collaboration. Other factors including worker stability, human and robot work spaces, and task constraints are considered. The approach was successfully demonstrated in co-assembly and transportation scenarios.

A common challenge for any AAN methods based on human physiology, bio-mechanics, or ergonomics, is modeling of the human operator. The human body and its physical capabilities are extremely challenging to model accurately, with large differences between people, and even the same person over time due to numerous factors, for example, fatigue. This presents a challenge for researchers to understand and develop solutions.

6.5 Safety Framework for pHRC

Safety in pHRC is always critical, whether it be for infrastructure maintenance or any other application. The human proximity to moving mechanical parts represents a challenge, and this often leads to control methods and strategies specific to pHRC applications. The International Standard ISO 10218 is the principal resource used to design, integrate, and certify safe robots and robotic systems. It consists of two parts, one intended mostly for robot manufacturers [International Organization for Standardization, 2011b] and one for people and companies integrating robots into bigger systems or environments [International Organization for Standardization, 2011a]. This document has represented the main guideline for manufacturers and integrators but is intended for generic industrial robots. With collaborative robots becoming more common, a Technical Specification was released in 2016, which addresses safety challenges specific to robots for HRC [International Organization for Standardization, 2016]. This Technical Specification has the purpose of extending the specifications outlined in ISO 10218 for collaborative robots.

A risk assessment and a plan to manage the identified risk are critical to ensure safety for the operator, the robot, and the environment. The risk associated with a hazard is a function of the probability of occurrence and its resulting severity. Reducing one or the other will mitigate the risk. In pHRC, many safety hazards are introduced by the proximity of the human to the robot. Those hazards cannot always be completely eliminated, but the associated risk can be managed and reduced to an acceptable level. Some hazards can only be identified once the workspace, its environment, and the application are clearly defined. Therefore, risk management is often the responsibility of the system integrator. The solution resulting from a risk management process is always a compromise between system performance, residual risk and cost, in terms of time and money.

To reduce the occurrence of hazardous situations, signs, barriers (where possible), and visual or haptic feedback are often used in pHRC. Due to their unexpected behavior, the human operator is often the cause of many hazards. In some applications, the hazard occurrence might be difficult to reduce. In those cases, the goal of a safety framework is to reduce the severity resulting from that hazard happening. In pHRC, the human is in physical contact and exchanging forces with the robot. The robot hitting the human is a serious hazard, and it may not be possible to prevent the occurrence of this hazard; thus, the severity of the hazard must be reduced. There are several methods to reduce the severity of a collision. A robot for pHRC should not have any sharp edges or protrusions and have an ergonomic design. A common approach is to limit the mechanical power transferred during the impact, by making the robot compliant. In recent years, several manufacturers released hardware specifically designed to physically interact with humans.

Another approach to achieve an acceptable risk is to address safety at a control level. Lasota et al. [2017] provide a comprehensive review of modern safety methods to address safety in pHRC. Many industrial applications present dynamic and complex environments, with many hazards to humans that share that environment with a robot. Therefore, ensuring safety often leads to many implementation challenges. This results in the integrators having to compromise, due to the cost associated with safety-related technology. Many software solutions might also not be feasible due to limits in terms of computational power.

Figure 6.8 presents common means to reduce the risk in pHRC applications, organized in modules. In order to achieve an acceptable level of risk, those safety modules can be combined. To fully appreciate the extent of physical and psychological harm that a hazard presents, models of the human are often used to characterize the human–robot interaction.

Figure 6.8 Safety framework for risk reduction in applications involving pHRC.

6.6 Performance-Based Role Change

In Human–Robot Collaboration, the human and robot work together to achieve a shared goal. How much autonomy the robot has in performing the task dictates the role that each party plays. For example, a robot controlled via direct tele-operation leaves all decision-making to the human operator. Contrast this with a fully autonomous robot that makes all the decisions on its own and is able to on-the-fly adapt to changes in the environment. True human–robot collaboration lies somewhere between these two extremes with both the human and robot working together, at times taking and ceding control of the interaction based on their understanding of the task, the environment, and the participants of the interaction. In this section, we consider varying levels of robot autonomy in human–robot collaboration and present a method of facilitating this role change based on the human performance.

Role arbitration is crucial in ensuring that each member in a collaboration is assigned tasks that match their abilities and limitations and that their actions are compatible with those of the other team members. One factor that influences when a change in role is actioned is how much trust one collaborator has in the other. When considering human–human interactions, if a human has high trust in their co-worker, they are less likely to intervene, and if they do, the intervention would usually take the form of a warning or direction. In contrast, if a human has low trust in their co-worker, they would be more likely to intervene and take a leading role.

To achieve human–robot interaction that is comparable to human–human interaction, a similar modeling of trust in the other party is required. For humans to build an internal trust model and act on it is natural, for example, taking control over the interaction when the robot makes a mistake. What is missing is a mechanism for the robot to know when to take over control of an interaction. This raises the question of how a robot's trust in its human co-worker can be modeled. This is where performance comes into the picture. If the observed performance of the human is low, the robot's trust in the human should decrease, and likewise, as the human's performance increases the robot's trust in the human should also increase.

In previous work [Tran et al., 2018], the robot's perception of its human co-worker's performance in a task was modeled using a Fluid Stochastic Petri Net (FSPN), which intuitively models accumulated rewards and penalties. At the conceptual level, a FSPN can be envisioned as a container with fluid within it, as depicted in Figure 6.9. In the performance model, the fluid level within the container represents the robot's perception of the human's performance in a task. The container has a number of pipes and valves which control the flow of fluid in and out of the container modeling the robot's perception of the human's

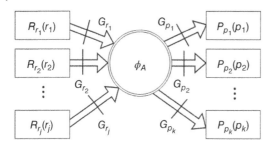

Figure 6.9 Example of an FSPN model for a generic task *A* described in Tran et al., [2018]/IEEE. *R* and *P* represent task-related rewards and penalties that increase or decrease the robot's perception of the human co-worker's performance respectively. *G* represents conditional requirements for rewards and penalties.

performance increasing and decreasing, respectively. The flow rate of the fluid correlates to the changes in robot's perception of performance.

Consider the previous lane assist safety mechanism; lane assist activates when the car determines that the human is unable to stay inside the boundaries of the lane. Modeling this using the FSPN model, the driver's deviation from the center of the lane can be used to model one of the penalties *P*, the further the driver is from the center, the more the robot will penalize the performance. Because the FSPN model is a cumulative model, if the driver continues to deviate from the center of the lane, the car's perception of the driver's performance will continue to decrease. However, a condition *G* might be added to the penalty where if the driver is signaling in a specific direction the penalty does not apply. Modeling the reward *R* for the lane assist can also be based on the position of the car in the lane, or it can be a constant rate, rewarding the driver for driving in the middle of the lane.

Using this model, the car is able to model its perception of the driver's performance, and thus, some measure of the car's trust in the driver can be obtained. In simpler interactions, the performance can directly be used as a measure of how and when robot should intervene. In this scenario, when the performance of the human drops below a threshold, lane assist can be activated, and the strength of the resistance in the steering wheel can be proportional to its loss of trust. As the car returns to the center (through the human's actions or via lane assist) the fluid in the FSPN will increase again turning off the lane assist.

In more complex interactions, more than one aspect needs to be taken into account, allowing for more nuanced decision-making on the robot's part. For example, continuing the previous scenario, the car also took into account other factors such as how long the driver has been on the road or their focus on the road by tracking the driver's eyes. It can also defer to the driver based on its trust in its own sensors, for example, if the lighting conditions are preventing it from seeing the lines on the road.

By taking this information into account, the safety of the driver and car, or the human and robot, can be improved. Modeling the performance of the human and the robot's trust in the human is just one more module that can be used in a safety framework.

6.7 Case Study

This section presents an example of a physical human–robot collaborative system being successfully applied to infrastructure maintenance and other related tasks. The ANBOT (Assistance-As-Needed Robot), as shown in Figure 6.10, is a portable collaborative robot designed to provide physical assistance to its human co-worker when performing grit-blasting tasks [Carmichael et al., 2019]. Grit-blasting is a physically intensive activity in a hazardous environment. Usually, a human would have to endure strong reaction forces from abrasive material such as garnet being shot out of a hose for long periods of time. When the abrasive material hits a surface, it ricochets, resulting in a low visibility environment where a human worker would have to wear thick protective clothing to prevent injuries from the projectiles.

The ANBOT was proposed and developed in collaboration with end users to improve the efficiency and safety of the grit-blasting task. By introducing a robot into the task, the physically demanding aspects of the grit-blasting task would be performed by the robot while the human provides the task direction.

The core of the ANBOT is a UR10 collaborative robot from Universal Robots. A custom end-effector with mounting for the blasting nozzle was fabricated. Handles allowed the human operator to control the robot's motion using measurements from a force-torque sensor that was embedded between the end-effector and the handles. Figures 6.10a and 6.10b show an operator controlling the robot intuitively through direct physical interaction, which is enabled using an admittance control scheme shown in Figure 6.11.

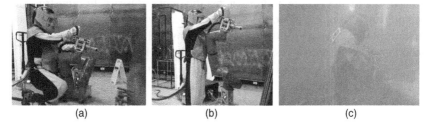

| (a) | (b) | (c) |

Figure 6.10 The Assistance-As-Needed Robot allows skilled workers to perform abrasive blasting tasks while shielding them from large and fatiguing reaction forces.

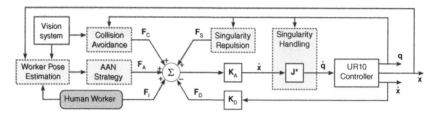

Figure 6.11 Control system of the ANBOT that utilized an admittance control scheme.

The ANBOT itself is an intelligent robot with sensors to detect the state of the environment, the human and the task. This allows the ANBOT to accurately adapt to the human's intentions and interrupt the task under certain conditions such as someone approaching the blasting zone or preventing collisions with the environment or itself. Even in low-visibility environments, the ANBOT is capable of detecting the position of the human and their pose as well as its own configuration. With this information, it is able to configure its joints in such a way that based on the human's intentions, would prevent its joints from entering singularity, self-collisions, or other conditions that interrupt the interaction.

The ANBOT was tested in the field under real blasting conditions, as shown in Figure 6.10c, receiving positive feedback from the end-users. The goal of the ANBOT was not to replace the human performing the task, but rather to augment them. At this point, although robotics has advanced significantly in recent years, it is not at the point of being able to remove humans from the loop for all applications. This also applies to many aspects of infrastructure maintenance where robots can be used to perform the hazardous or physically intensive aspects of the task and present information to a human co-worker to determine the best action.

6.8 Discussion

As shown, the opportunities and benefits of pHRC have been successfully demonstrated in a number of applications across different industries. In infrastructure maintenance tasks, however, pHRC has not been widely utilized. There is an opportunity to leverage pHRC to reduce the physical burden inherent in many laborious tasks that are necessary to maintain the world's civil infrastructure.

The use of pHRC can go a long way in improving the efficiency and safety in infrastructure maintenance. The capabilities and intelligence of robots have significantly increased, allowing for more nuanced interactions to complete more complex operations. In this chapter, different paradigms have been presented along with examples of how these paradigms have been incorporated in infrastructure maintenance and related applications.

The use of phRC in infrastructure maintenance is still young and there is still a largely unexplored potential to uncover. As robotics research continues to make leaps and bounds, the incorporation of robots into the field as both tools and co-workers will become more common.

Acknowledgements

The authors would like to extend their deepest gratitude to the industry partners, funding bodies, academics, and engineers who were critical to the success of the case studies discussed in this Chapter. In particular we would like to thank Burwell Technologies Pty Ltd, SABRE Autonomous Systems, and the Australian Research Council.

Bibliography

Antonio Bicchi, Michael A. Peshkin, and J. Edward Colgate. Safety for Physical Human-Robot Interaction. In Bruno Siciliano and Oussama Khatib, editors, *Springer Handbook of Robotics*, pages 1335–1348. Springer-Verlag, Berlin, Heidelberg, 2008. ISBN 978-3-540-30301-5. doi: 10.1007/978-3-540-30301-558.

Marc G. Carmichael and Dikai Liu. Admittance control scheme for implementing model-based assistance-as-needed on a robot. In *Proceedings of the Annual International Conference of the IEEE Engineering in Medicine and Biology Society, EMBS*, pages 870–873, 2013a. ISSN 1557170X. doi: 10.1109/EMBC.2013.6609639.

Marc G. Carmichael and Dikai Liu. Estimating physical assistance need using a musculoskeletal model. *IEEE Transactions on Biomedical Engineering*, 60(7):1912–1919, 2013b. ISSN 00189294. doi: 10.1109/TBME.2013.2244889.

Marc G. Carmichael, Stefano Aldini, Richardo Khonasty, Antony Tran, Christian Reeks, Dikai Liu, Kenneth J. Waldron, and Gamini Dissanayake. The ANBOT: An intelligent robotic co-worker for industrial abrasive blasting. In *IEEE International Conference on Intelligent Robots and Systems*, pages 8026–8033, Nov 2019. ISSN 21530866. doi: 10.1109/IROS40897.2019.8967993.

Jeremy L. Emken, James E. Bobrow, and David J. Reinkensmeyer. Robotic movement training as an optimization problem: Designing a controller that assists only as needed. In *Proceedings of the 2005 IEEE Ninth International Conference on Rehabilitation Robotics*, pages 307–312, 2005. doi: 10.1109/ICORR.2005.1501108.

Mustafa Suphi Erden and Aude Billard. End-point impedance measurements at human hand during interactive manual welding with robot. In *Proceedings - IEEE International Conference on Robotics and Automation*, pages 126–133, Sep 2014. ISSN 10504729. doi: 10.1109/ICRA.2014.6906599.

Mustafa Suphi Erden and Bobby Marić. Assisting manual welding with robot. *Robotics and Computer-Integrated Manufacturing*, 27(4):818–828, Aug 2011. ISSN 0736-5845. doi: 10.1016/J.RCIM.2011.01.003.

Deepak Gopinath, Siddarth Jain, and Brenna D. Argall. Human-in-the-loop optimization of shared autonomy in assistive robotics. *IEEE Robotics and Automation Letters*, 2(1):247–254, Jan 2017. ISSN 23773766. doi: 10.1109/LRA.2016.2593928.

Amal Gunatilake, Lasitha Piyathilaka, Antony Tran, Vinoth Kumar Vishwanathan, Karthick Thiyagarajan, and Sarath Kodagoda. Stereo vision combined with laser profiling for mapping of pipeline internal defects. *IEEE Sensors Journal*, 21(10):11926–11934, May 2021. ISSN 15581748. doi: 10.1109/JSEN.2020.3040396.

Sami Haddadin, Alessandro De Luca, and Alin Albu-Schäffer. Robot collisions: A survey on detection, isolation, and identification. *IEEE Transactions on Robotics*, 33(6):1292–1312, Dec 2017. ISSN 15523098. doi: 10.1109/TRO.2017.2723903.

Neville Hogan. Impedance control: An approach to manipulation: Part III– Applications. *Journal of Dynamic Systems, Measurement, and Control*, 107(1):17–24, Mar 1985. ISSN 0022-0434. doi: 10.1115/1.3140701. URL https:// asmedigitalcollection.asme.org/dynamicsystems/article/107/1/17/400610/ Impedance-Control-An-Approach-to-Manipulation-Part.

Neville Hogan, Hermano I. Krebs, Brandon Rohrer, Jerome J. Palazzolo, Laura Dipietro, Susan E. Fasoli, Joel Stein, Richard Hughes, Walter R. Frontera, Daniel Lynch, et al. Motions or muscles? Some behavioral factors underlying robotic assistance of motor recovery. *Journal of Rehabilitation Research and Development*, 43(5):605–618, 2006. ISSN 07487711. doi: 10.1682/JRRD.2005.06.0103.

International Organization for Standardization. ISO 10218-2:2011 - Robots and robotic devices - Safety requirements for industrial robots - Part 2: Robot systems and integration, 2011a. URL https://www.iso.org/standard/41571.html.

International Organization for Standardization. ISO 10218-1:2011 - Robots and robotic devices - Safety requirements for industrial robots - Part 1: Robots, 2011b. URL https://www.iso.org/standard/51330.html.

International Organization for Standardization. ISO/TS 15066:2016 - Robots and robotic devices - Collaborative robots, 2016.

Khristopher Kabbabe Poleo, William J. Crowther, and Mike Barnes. Estimating the impact of drone-based inspection on the Levelised Cost of electricity for offshore wind farms. *Results in Engineering*, 9:100201, Mar 2021. ISSN 2590-1230. doi: 10.1016/J.RINENG.2021.100201.

Wansoo Kim, Jinoh Lee, Luka Peternel, Nikos Tsagarakis, and Arash Ajoudani. Anticipatory robot assistance for the prevention of human static joint overloading in human-robot collaboration. *IEEE Robotics and Automation Letters*, 3(1):68–75, Jan 2018. ISSN 23773766. doi: 10.1109/LRA.2017.2729666.

H. van der Kooij, J.F. Veneman, and R. Ekkelenkamp. Compliant Actuation of Exoskeletons. In Aleksandar Lazinica, editor, *Mobile Robots: Towards New Applications*, pages 129–148. IntechOpen, Mammendorf, Germany, Dec 2006. doi: 10.5772/4688. URL https://research.utwente.nl/en/publications/compliant-actuation-of-exoskeletons-chapter-7.

H.I. Krebs, J.J. Palazzolo, L. Dipietro, M. Ferraro, J. Krol, K. Rannekleiv, B.T. Volpe, and N. Hogan. Rehabilitation robotics: Performance-based progressive robot-assisted therapy. *Autonomous Robots*, 15(1):7–20, 2003. ISSN 09295593. doi: 10.1023/A:1024494031121.

Przemyslaw A. Lasota, Terrence Fong, and Julie A. Shah. A survey of methods for safe Human-Robot interaction. *Foundations and Trends in Robotics*, 2017. ISSN 1935-8253. doi: 10.1561/2300000052.

Khoa Le, Andrew To, Brenton Leighton, Mahdi Hassan, and Dikai Liu. The SPIR: An autonomous underwater robot for bridge pile cleaning and condition assessment. In *IEEE International Conference on Intelligent Robots and Systems*, pages 1725–1731, Oct 2020. ISSN 21530866. doi: 10.1109/IROS45743.2020.9341687.

Luigi Villani and Joris De Schutter. Force Control. In Bruno Siciliano and Oussama Khatib, editors, *Springer Handbook of Robotics*, pages 161–185. Springer-Verlag, Berlin, Heidelberg, 2008. ISBN 978-3-540-30301-5. doi: 10.1007/978-3-540-30301-58.

Kevin M. Lynch, Caizhen Liu, Allan Sørensen, Songho Kim, Michael Peshkin, J. Edward Colgate, Tanya Tickel, David Hannon, and Kerry Shiels. Motion guides for assisted manipulation. *The International Journal of Robotics Research*, 21(1):27–43, 2002. doi: 10.1177/027836402320556467.

Gavin Paul, Stephen Webb, Dikai Liu, and Gamini Dissanayake. Autonomous robot manipulator-based exploration and mapping system for bridge maintenance. *Robotics and Autonomous Systems*, 59:543–554, 2011. doi: 10.1016/j.robot.2011.04.001. URL www.elsevier.com/locate/robot.

Phillip Quin, Gavin Paul, Alen Alempijevic, and Dikai Liu. Exploring in 3D with a climbing robot: Selecting the next best base position on arbitrarily-oriented surfaces. In *IEEE International Conference on Intelligent Robots and Systems*, 2016-November, pages 5770–5775, Nov 2016. ISSN 21530866. doi: 10.1109/IROS.2016.7759849.

M.H. Raibert and J.J. Craig. Hybrid position/force control of manipulators. *Journal of Dynamic Systems, Measurement, and Control*, 103(2):126–133, Jun 1981. ISSN 0022-0434. doi: 10.1115/1.3139652. URL https://asmedigitalcollection.asme.org/dynamicsystems/article/103/2/126/400298/Hybrid-Position-Force-Control-of-Manipulators.

Antony Tran, Dikai Liu, Ravindra Ranasinghe, and Marc Carmichael. A method for quantifying a robot's confidence in its human co-worker in human-robot cooperative grit-blasting. In *ISR 2018; 50th International Symposium on Robotics*, pages 1–8, 2018.

Eric T. Wolbrecht, Vicky Chan, Vu Le, Steven C. Cramer, David J. Reinkensmeyer, and James E. Bobrow. Real-time computer modeling of weakness following stroke optimizes robotic assistance for movement therapy. In *Proceedings of the Third International IEEE EMBS Conference on Neural Engineering*, pages 152–158, 2007. doi: 10.1109/CNE.2007.369635.

Eric T. Wolbrecht, Vicky Chan, David J. Reinkensmeyer, and James E. Bobrow. Optimizing compliant, model-based robotic assistance to promote neurorehabilitation. *IEEE Transactions on Neural Systems and Rehabilitation Engineering*, 16(3):286–297, 2008. ISSN 15344320. doi: 10.1109/TNSRE.2008. 918389.

Shirin Yousefizadeh, Juan De Dios, Flores Mendez, and Thomas Bak. Trajectory adaptation for an impedance controlled cooperative robot according to an operator's force. *Automation in Construction*, 103:213–220, 2019. doi: 10.1016/ j.autcon.2019.01.006.

Part II

Robotic System Design and Applications

Part II presents 11 case studies of robotic systems developed for maintaining various civil infrastructures. Chapter 7 introduces a robot for climbing steel bridge structures. Two motion modes, i.e. mobile mode for climbing flat surfaces and worming mode for joint transitions, enable this robot to traverse complex geometries on steel structures. Chapter 8 presents an autonomous underwater robot developed for removing marine growth that covers the surface of underwater structures such as bridge piles, wharf piles, and underwater pipelines. For underground tunnel inspection and development, a robotic system for tunnel structure inspection is presented in Chapter 9, and a robotic drilling system for developing small-size subterranean tunnel networks is presented in Chapter 10. For underground pipe condition assessment, Chapter 11 presents sensing technologies and a robot for assessment of metallic water pipelines, and Chapter 12 describes the development of an in-pipe robotic sensing system for inspecting and assessing the condition of concrete wastewater pipe walls.

Chapter 13 presents an autonomous legged climbing robot for cleaning, vacuuming, and painting steel surfaces in confined spaces. Chapter 14 discusses the application of multiple UAVs for inspecting electrical power lines, including methods for multi-UAV planning and mapping of vegetation near the electrical lines. Chapter 15 discusses the application of robotic systems for inspection of oil refineries. Chapter 16 presents a summary overview of drone-based photovoltaic module inspection and a case study, including the use of a convolutional neural network (CNN) for detecting defective solar cells. The last chapter of the book, Chapter 17, presents the broader context of aerial repair and aerial additive manufacturing.

Infrastructure Robotics: Methodologies, Robotic Systems and Applications, First Edition.
Edited by Dikai Liu, Carlos Balaguer, Gamini Dissanayake, and Mirko Kovac.
© 2024 The Institute of Electrical and Electronics Engineers, Inc. Published 2024 by John Wiley & Sons, Inc.

7

Steel Bridge Climbing Robot Design and Development
*Hung M. La**

Advanced Robotics and Automation Lab, Department of Computer Science and Engineering, University of Nevada, Reno, 1664 N. Virginia Street, MS 0171, Reno, Nevada, USA, NV 89557

7.1 Introduction

As per the Federal Highway Administration (FHWA) [USF, 2019], approximately one-third of the 607,380 bridges in the United States are made of steel. The National Bridge Inventory (NBI) [USF, 2019] indicates that 25% of these steel bridges are either deficient or functionally obsolete, posing an increasing threat to transportation safety. Recent bridge collapses, such as the I-5 Skagit River Bridge collapse in 2013 [Skagit, 2013], have underscored the need for more frequent inspections.

The rising number of bridge collapses has significant implications for traveler safety. Current inspection practices for steel bridges predominantly rely on manual and visual evaluations, which are both time-consuming and hazardous. However, due to a shortage of qualified personnel, the demand for adequate inspection and maintenance of these bridges remains unmet [USF, 2019; McCrea et al., 2002]. Therefore, there is a pressing societal need for alternative solutions that can address the growing demand for safe, cost-effective, and accurate inspections. One such solution is automated inspection using robots.

Typically, the maintenance process for civil infrastructures like bridges involves inspectors physically reviewing the condition of each steel member by tapping a hammer on the steel to collect impact echo waves for fatigue crack inspection [Purna Chandra Rao, 2017], or visually examining all surfaces of the steel members to detect shallow surface cracks.

The described procedure is labor-intensive, time-consuming, and often poses risks to inspectors, especially when dealing with inaccessible areas of bridges

*Email: hla@unr.edu

Infrastructure Robotics: Methodologies, Robotic Systems and Applications, First Edition.
Edited by Dikai Liu, Carlos Balaguer, Gamini Dissanayake, and Mirko Kovac.

[Golden Gate Bridge, 2018]. For instance, the Golden Gate Bridge, a prominent landmark in the San Francisco Bay area, requires a team of 12 rope-certified bridge engineers to manually inspect its high steel structures by climbing and hanging on them [Golden Gate Bridge, 2018]. However, conventional inspection methods may not be effective in reaching all areas of the structure or may be entirely inaccessible.

As a result, there has been a growing effort to automate the inspection process, particularly through the development of climbing robots in recent years. These robots aim to mitigate the risks associated with manual inspections and improve the efficiency of the process. By utilizing climbing robots, the inspection of hard-to-reach areas can be performed more effectively, reducing the need for human inspectors to put themselves in hazardous situations.

There are several designs based on conventional mobile robots. Some make use of tank-like tracks to enhance the friction of the robot on the steel surfaces they adhere to Shen et al. [2005] and Nguyen and La [2019a]. Others function as magnetic wheeled robots [Pham et al., 2020; Wang and Kawamura, 2014; La et al., 2019] or roller chain-liked robot [Nguyen and La, 2019b]. Notable developments of climbing robots for steel bridge/structure inspection can be seen in Ward et al. [2015], Pham and La [2016], Wang and Kawamura [2016], Pham et al. [2016, 2022], Takada et al. [2017], Nguyen et al. [2022], Motley et al. [2022], Nguyen and La [2021], and Otsuki et al. [2022]. The adhesion force for tank-like types of inspection robots is typically created by magnets attached on the robot's roller chains [Nguyen and La, 2019b], on the robot's wheels/sprockets, or on the robot body (untouched magnets). These magnets are kept in close proximity to the steel surfaces that these robots adhere to. Each of the measures has particular merits in different working conditions. Touched magnets on a roller chain or wheels allow the robot to transfer seamlessly between surfaces at 90° angles as well as other sharp angles, which might appear on a bridge [Eich and Vögele, 2011; Nguyen and La, 2019a; Leibbrandt et al., 2012; Leon-Rodriguez et al., 2012; and San-Millan, 2015]. Tank-like robots with untouched permanent magnets help to allow the robots to pass small struggles like nuts and prevent loss of adhesion force when crossing rusty areas of steel [Versatrax, n.d.]. Some improved wheel robots with soft frames make them more adaptive on a wider range of surfaces [BridgeBot, 2017].

Imitation of mobility of climbing creatures is another approach. A spider-like robot with electromagnets on its feet was reported [Bandyopadhyay et al., 2018], and a legged robot was developed [Mazumdar and Asada, 2009]. An inch-worm-like robot [Ward et al., 2015] was an efficient design and creation, for Sydney bridge, Australia. This robot excels in its ability to transfer smoothly 360° to other surfaces. It is equipped with a camera and sensors for structure 3D mapping.

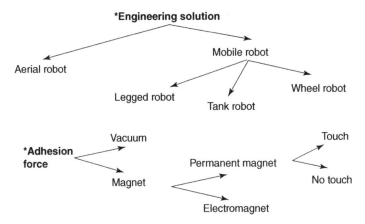

Figure 7.1 Overall approach for steel bridge inspection robot.

In summary, the current automated bridge inspection methods face challenges as depicted in Figure 7.1. Robots that rely on flying, electromagnets, or vacuum adhesion often encounter energy issues and cannot operate for extended periods in the field. Tethering these robots with power wires is not feasible due to the complex structures commonly found on steel bridges. Climbing mobile robots work well on simple and continuous surfaces but struggle with the varied geometries of steel bridge structures, such as cylinders (Figure 7.2(a, c)) and I-beams (Figure 7.2(b, d)). Conventional-wheeled robots also face challenges in navigating these surfaces. Legged robots have mobility advantages but may encounter difficulties on normal surfaces, and worm or spider-like designs can be complex compared to wheeled robots. Additionally, each transportation method has its own challenges in navigating nuts (Figure 7.2(b, d)) while maintaining adhesion. Although flying robots offer a promising approach without surface adhesion limitations, they currently lack the ability to perform in-depth structural analyses required for steel bridge maintenance.

This chapter presents the recent development of a practical climbing robotic system by the ARA Lab [Nguyen et al., 2020; Bui et al., 2020b], which combines the methods and advantages of previous steel inspection robots. The robot is capable of adapting to various types of bridge surfaces, including flat, curving, and rough surfaces. The hybrid approach implemented in the robot allows it to switch between a mobile mode and a transforming mode (Figure 7.5), enabling it to efficiently inspect complex steel bridge surfaces. The robot utilizes adhesion force generated by permanent magnets in two modes-untouched magnets for mobile mode and touched magnets for transforming mode. The flexible magnet array enables the robot to overcome obstacles such as nuts and bolts. The robot's working principle

Figure 7.2 (a) Cylinder steel bridge, (b) Complex I bar steel bridge with nuts, (c) A bridge with curving and plat surfaces, (d) A complicated joint.

has been demonstrated through climbing on indoor and outdoor steel structures as well as steel bridges.

In addition, this chapter presents a control framework for the autonomous navigation of the ARA robot on steel bridge structures. The control framework includes a switching control mechanism that allows the robot to transition between two different transformations: mobile and worming. The switching control takes into account the availability of planar surfaces, their area, and height to determine the appropriate transformation. An area estimation algorithm is proposed using 3D point cloud data from the robot's stereo camera to determine if the available area is sufficient for the robot's foot transition. Based on the height estimation from the switching control, the robot chooses its transformation mode. In worming transformation, the robot performs an inch-worm jump. For navigation in mobile transformation mode, a path planning control is applied. To enable visual inspection, an encoder–decoder-based convolutional neural network [Hafsa et al., 2020] is integrated. Additionally, magnetic array-based distance control is used for autonomous magnetic adherence to the steel surface. The rest of the chapter is organized as follows: Section 7.3 provides details on the overall design and implementation of the robot. Section 7.4 elaborates on the proposed control and navigation methodology. Section 7.5 presents experimental results, and Section 7.6 concludes the work.

Videos of the robot demonstration can be seen on the ARA lab website: https://ara.cse.unr.edu/?page_id=11 or youtube links: https://youtu.be/PwDf6h0Om3c and https://youtu.be/SHk5IIOBRdA.

7.2 Recent Climbing Robot Platforms Developed by the ARA Lab

Before delving into the details of one particular practical climbing robot discussed in Section 7.3, this section highlights a selection of climbing robots developed by the ARA lab from 2014 to 2023, as depicted in Figure 7.3 [Pham et al., 2016, 2020; Pham and La, 2016; La et al., 2019; Nguyen and La, 2019a,b; Nguyen et al., 2022; and Otsuki et al., 2023]. Among these climbing robots, there are wheel-based models (Figure 7.3 (a, b, c)), adept at scaling flat steel surfaces. Notably, Figure 7.3 (b) showcases a robot with a flexible chassis designed for climbing on curving (cylindrical) surfaces [Pham et al., 2020]. These wheel-based climbing robots exhibit a commendable payload capacity of 3 kg and can operate for up to an hour.

Additionally, a tank-like robot (Figure 7.3 (d)) excels at traversing both flat and curving surfaces. However, due to its rigid tank chains, this robot may encounter difficulties navigating structural joints. Consequently, the ARA lab devised a solution in the form of a roller-chain-like robot (Figure 7.3 (e)) [Nguyen and La, 2019b]. This robot features a flexible body, enabling it to proficiently climb flat and curving

(a)   (b)   (c)

(d)   (e)   (f)

Figure 7.3 Climbing robot platforms developed by the ARA lab, University of Nevada, Reno.

surfaces. It boasts a maximum payload of 0.5 kg and can operate for up to 30 min. It is worth noting that controlling the orientation of this robot can be challenging.

The most recent advancement in climbing robots is the bicycle-like robot (Figure 7.3 (f)). This innovative robot tackles the challenges faced by its predecessors. Operating akin to a bicycle, it incorporates a freedom joint in the middle of its body, enabling it to ascend complex steel structures. For further insights into this bicycle-like robot, refer to Nguyen et al. [2022] and Ahmed et al. [2022].

7.3 Overall Design

The design concept of the hybrid climbing robot is illustrated in Figure 7.4, while its function is described in Figure 7.5. The robot is composed of two primary components: the feet and the body. Each foot is equipped with permanent magnets to generate adhesion force, enabling the robot to attach to steel surfaces without expending any energy. During mobile mode, the magnets hover at a 1mm distance from the steel surface in an untouched position. The ring-shaped magnets are arranged in an array to facilitate the smooth passage of nuts and bolts while maintaining full adhesion force. The torsion spring is an integral part of the foot design, as it enables each magnet cell to adjust individually to varying external

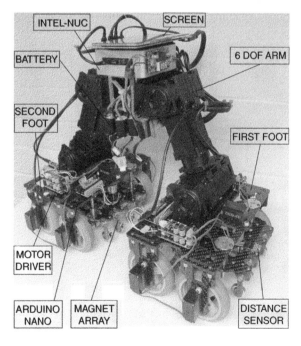

Figure 7.4 Overall design of robot.

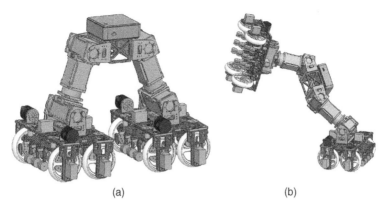

Figure 7.5 Robot function (a) Mobile mode; (b) Transforming mode, or worm mode.

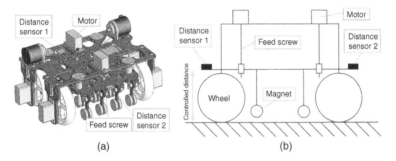

Figure 7.6 The robots foot with flexible magnet array.

stimuli and then return to its original structure. The magnets in the array are magnetized through their diameter, making them more suitable for their design purpose than if they had been magnetized through their thickness, as shown in Figure 7.6.

The distance between the magnet arrays and the surface is controllable, and each foot can function in both touched and untouched orientations. The control system utilizes two parallel feed screws with an actuator to modify the distance that each foot is kept from the surface. Feedback from a distance sensor ensures that the magnet arrays are kept at an optimal distance. Each foot has four wheels that provide stability when standing on one foot and support large moments while remaining lightweight. Rubber wheels are used to maximize the friction factor between the robot and the surfaces on which it adheres. The foot design enables the ARA robot to operate on various surface conditions, as described in Figure 7.7. The body of the robot has six degrees of freedom (DOF) and functions like a robot arm, as shown in Figure 7.8, when the robot is in transforming mode.

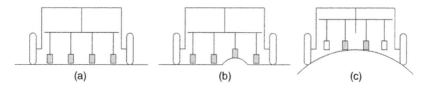

Figure 7.7   (a) The robot on flat surface, (b) The robot passing nuts, (c) The robot on a curving surface.

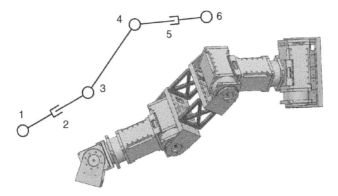

Figure 7.8   The robot body-6 DOF robot arm.

If the robot encounters difficulties traversing an area in mobile mode, it can alter the magnet orientation of one foot to touch the surface and shift into transforming mode, allowing it to find a new surface to travel along and complete its task. In transforming mode, one foot's magnet array will fully touch the surface to maximize its adhesion force, while the magnets on the second foot move up to release the adhesion force. The robot then functions as a 6-DOF robot arm. Once it touches a new surface, the process repeats on the opposite feet, allowing the robot to move the entire body to a new location.

7.3.1   Mechanical Design and Analysis

During the design process of the hybrid climbing robot, the ARA team conducted extensive static analyses to examine the mechanical behaviors underlying its construction. These analyses enabled the team to design and manufacture a robot that surpasses the capacities of previous steel bridge inspection robots such as climbing robot caterpillar (CROC) [Ward et al., 2015] and MINOAS [Eich and Vgele, 2011].

Transformation Analysis
One important analysis conducted by the team was the transformation analysis, which aimed to determine the maximum moment the robot would experience

under static conditions. This analysis allowed the team to calculate the moment the robot experiences when transforming and determining the minimum torque required from the servos to overcome the force of gravity.

Variable h_f represents the height of the feet, and m_f is the mass of the foot. L_1, L_2, L_3 are the lengths of links, one through three. m_e is the mass of equipment that the robot is carrying. m_{L1}, m_{L2}, m_{L3} are the masses of their respective links. g is for gravity.

The extended statics diagram shown in Figure 7.9 was used to calculate the torque output required from a servo at point A as a function of variables such as the height of the feet, the mass of the foot, and the lengths and masses of links. By using this equation, the ARA team could select a suitable servo for the robot's construction.

$$
\begin{aligned}
T_A = (\tfrac{1}{2}L_1)m_{L1}g + (L_1 + \tfrac{1}{2}L_2)(m_{L2} + m_e)g \\
+ (L_1 + L_2 + \tfrac{1}{2}L_3)m_{L3}g \\
+ (L_1 + L_2 + L_3 + \tfrac{1}{2}h_f)m_f g.
\end{aligned}
\tag{7.1}
$$

The moment the robot experiences when fully extended was determined in this analysis which informed the team about how large the moment acting upon the robot would be through Eq. (7.2). In the Turn-Over and Sliding Friction Analysis section later in this chapter, we take the results of this study and use them as input for the external moment that the foot is experiencing in the Turn-Over Analysis.

$$
\begin{aligned}
M_B = (\tfrac{1}{2}h_f)m_f g + (h_f + \tfrac{1}{2}L_1)m_{L1}g \\
+ (h_f + L_1 + \tfrac{1}{2}L_2)(m_{L2} - m_e)g \\
+ (h_f + L_1 + L_2 + \tfrac{1}{2}L_3)m_{L3}g \\
+ (\tfrac{3}{2}h_f + L_1 + L_2 + L_3)m_f g.
\end{aligned}
\tag{7.2}
$$

The mass of the feet plays a critical role in determining the acting moment and torque in both the Moment and Torque equations (7.1) and (7.2). Our analysis

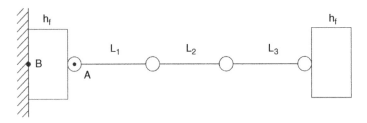

Figure 7.9 Extended statics diagram.

using aluminum as the material for the structural members of the robot revealed that our robot required a substantial amount of torque to overcome gravity, resulting in very low factors of safety for our servos. To address this issue, the ARA team decided to change the material used for the robot's structural members and chose carbon fiber instead of aluminum. This reduction in the total mass of the feet has significantly increased our robot's factor of safety, allowing it to operate more efficiently. The decrease in mass, which amounts to a reduction of 40%, has enabled our robot to traverse steel structures as its servos can now hoist the robot and hold it still or overpower the force of the moments acting on them. This has also enabled the ARA robot to carry more equipment than if it had been manufactured from aluminum.

Turn-Over and Sliding Friction Analysis

At this point in the design process, the ARA team needed to conclude whether our robot would be at risk of toppling over or slipping while in use. To do this, we performed two more statics analyses, one, on an adhered foot with an external moment acting on it and the other as a stationary foot adhered to a steel surface resisting sliding.

Our turn-over analysis (Figure 7.10a) builds off of the transformation analysis by using the proposed moment function Eq. (7.2) as an input to the external moment acting upon the foot in this analysis. In order to determine how much adhesion force was required by each foot, a moment was taken at point C, this produced equations (7.3) and (7.4).

Variables n_1, n_2, n_3, n_4 represent the number of magnets in each row. M_C is the moment at point C, and F_{mag} is the force created by each magnet. d_1, d_2, d_3 all

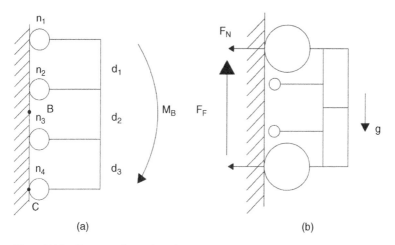

(a)　　　　　　　　　　　　　(b)

Figure 7.10 Turn-over/adhesion diagram.

represent distances, where $M_B = f(h_f, m_f, m_{L1}, m_e, m_{L2}, m_{L3}, L_1, L_2, L_3, g)$, is solved for in the Transformation Analysis subsection of this chapter.

$$M_c = -M_B + F_{mag}n_3d_3 + F_{mag}n_2(d_2 + d_3)$$
$$+ F_{mag}n_1(d_1 + d_2 + d_3). \tag{7.3}$$

Eq. (7.3) was then simplified and adjusted to emphasize the role magnet strength, and each distance would have upon the net moment, M_C, giving Eq. (7.4).

$$M_c = F_{mag}(d_3(n_1 + n_2 + n_3) + d_2(n_1 + n_2)$$
$$+ d_1n_1) - M_B = -M_B + F_{mag}n_3d_3$$
$$+ F_{mag}n_2(d_2 + d_3) + F_{mag}n_1(d_1 + d_2 + d_3). \tag{7.4}$$

The resulting equation Eq. (7.4) helped develop a better understanding of the importance of foot orientation, the location of magnets, and the power of the magnets in the design. This analysis was also instrumental to the design layout of each of the feet in our robot and helped to ensure that the robot would maintain its position on steel structures.

The sliding friction analysis (Figure 7.10b) assumes static conditions to determine what the required force of friction acting on the wheels is to prevent sliding. The main purpose of these calculations was to ensure that the robot's feet would not slide down the steel surfaces that they adhered to.

i represents the number of wheels that the normal force (F_N) is distributed between, and n is for the total number of permanent magnets on the foot. F_F is the friction force, and F_{mag} is the force generated by one magnet. μ represents the friction coefficient between steel and rubber. For the purposes of our study, we used a value of 0.7 [Coefficients of Friction for Steel, 2005].

This evaluation begins by defining the equation for the normal force acting perpendicular to the surface that the robot is adhered to; Eq. (7.5).

$$F_N = \frac{nF_{mag}}{i}. \tag{7.5}$$

Then, a base equation is written to solve for the force of friction acting against the motion, Eq. (7.6).

$$F_F = iF_N\mu. \tag{7.6}$$

When both are combined, yields Eq. (7.7).

$$F_F = nF_{mag}\mu. \tag{7.7}$$

Equation (7.7) models the force of friction as a function of the total number of magnets on the foot, the friction coefficient based on the materials in contact, and the force generated by one magnet.

7.4 Overall Control Architecture

In the previous section, the ARA robot was described by two transformation modes: mobile and worming. To enable the robot to adapt to different environmental conditions, the team integrated a switching control mechanism (as shown in Figure 7.11) into the ARA robot. This control allows the robot to automatically switch between the two configurations depending on the surface it encounters.

When the robot is on smooth steel surfaces, it activates the *Mobile transformation* mode (as shown in Figure 7.12(a)). In this configuration, the robot uses path planning control to navigate, differential wheels for movement, and performs visual inspections. The magnetic array is switched to the untouched mode in this mode.

However, when the robot encounters complex steel surfaces, it configures itself into the *Worming transformation* mode (as shown in Figure 7.12(b)). In this mode, the robot performs an inch-worm jump to the next surface as it cannot move on wheels on such structures. The magnetic array is switched into the touched mode during the jump. The switching control mechanism detects the environment type and sends the appropriate command to the executable nodes to control the robot's movement.

The control architecture of the ARA robot is composed of both low-level and high-level control structures, each performing various tasks. The low-level control structure, operated by an Arduino, manages several functions such as the conversion of velocity for robot wheel movement and the control of the magnetic array.

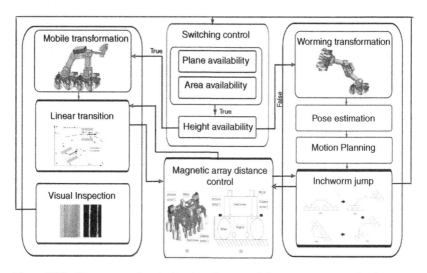

Figure 7.11 The proposed control system framework for autonomous navigation.

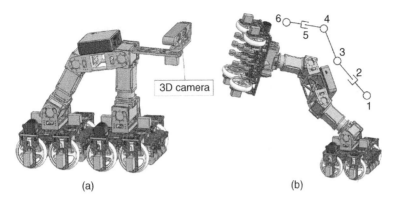

Figure 7.12 Steel bridge inspection robot in (a) mobile transformation, and (b) *Worming transformation.*

Additionally, the low-level control structure reads sensor measurements from the encoder and Inertial Measurement Unit (IMU) and relays this information to the high-level control. In contrast, the high-level control is embedded in an onboard processor (NUC) and performs tasks such as the switching control function, processing of 3D point cloud data, and the IK of the *Worming transformation*. To achieve the desired linear velocity and obtain data from advanced sensors such as the RGB camera and IMU, the high-level and low-level controls utilize a Kalman filter to fuse the sensor data. The arrangement of these controls is illustrated in Figure 7.13.

7.4.1 Control System Framework

The control system framework of the ARA robot is composed of four modules: switching control, *Mobile transformation*, *Worming transformation*, and magnetic array distance control. An overview of the overall framework is shown in Figure 7.11.

Switching Control
The ARA robot's ability to transform itself into two different modes (mobile and worming) is made possible by the switching control S. This control uses a switching function S, represented by Eq. 7.8, and takes in three boolean parameters: plane availability S_{pa}, area availability S_{am}, and height availability S_{hc}. These parameters are used to determine whether there is a still surface available and to estimate the area and height of the surface. A logical operation is then performed on these parameters using the function $f(.)$. To estimate the function parameters, 3D point cloud data of the steel surface is used.

$$S = f(S_{pa}, S_{am}, S_{hc}) = S_{pa}S_{am}S_{hc}. \tag{7.8}$$

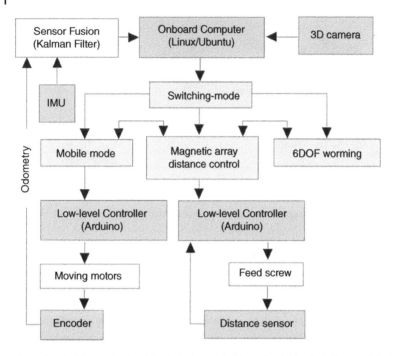

Figure 7.13 The control architecture integrated into the ARA robot. Source: Adapted from Nguyen et al. [2020].

The robot configures to *Mobile transformation* if the function returns a true value. The false value configures the robot into the *Worming transformation*.

Plane availability: The 3D point cloud of steel surface is processed using *pass-through* filtering, *downsampling*, and *plane detection* [Bui et al., 2020a]. The *plane detection* [Bui et al., 2020a] extracts the planar point cloud P_{cl} from the initial point cloud. The plane availability is checked using Eq.7.9:

$$\begin{cases} S_{pa} = \textit{False, if } P_{cl} = \varnothing \\ S_{pa} = \textit{True, otherwise} \end{cases} \tag{7.9}$$

Moreover, two functions, *get_centroid* and *get_normal_vector*, use the point cloud set to calculate the point cloud's centroid $C_{P_{cl}}$ and normal vector $\vec{N}_{P_{cl}}$.

Area availability: The ARA robot's leg movement requires an accurate estimation of the available planar surface area P_{cl} to ensure a successful transition. To achieve this, the team proposed **Algorithm 2** which checks the sufficiency of the available planar surface. However, the most crucial step of this algorithm is to extract the boundary points of the planar surface accurately. For this reason, the team also developed **Algorithm 1**, which outlines a boundary estimation procedure.

Algorithm 1 Boundary point estimation from 3D point cloud data of steel bridges.

1: **procedure** BOUNDARYESTIMATION(P_{cl}, α_s)

2:     $Planes = \{xy, yz, zx\}$

3:     $d_{min} = \forall_{i \in Planes}$ Point along minimum value of plane i

4:     $d_{max} = \forall_{i \in Planes}$ Point along maximum value of plane i

5:     Initialize $B_s = \{\}$

6:     **for** $p \in Planes$ **do**

7:         $i \rightarrow 1$

8:         **while** $sl_{p_i} < d_{max}$ **do**

9:             $sl_{p_i} = d_{min_p} + i * \alpha_s$

10:        $PS_{p_i} = $ Set of points in range $sl_{p_i} \pm \alpha_s/2$

11:        $P_{cl_A}, P_{cl_B} = \underset{\forall\{P_i, P_j\} \in PS_{p_i}}{\mathrm{argmax}} \{\|P_i - P_j\|\}$

12:        $B_s = B_s \cup \{P_{cl_A}, P_{cl_B}\}$

13:        $i = i + 1$

14: **return** B_s

In **Algorithm 1**, the boundary points are estimated using a window-based approach. The algorithm takes the 3D point cloud P_{cl} of the planar surface and a slicing parameter α_s as inputs. Firstly, the two furthest points d_{min} and d_{max} in P_{cl} along each plane are determined. The point cloud is then divided into multiple smaller slices along the three planes, and for each slice in a particular plane p, the slicing index sl_p, representing the center coordinate of the slice, is calculated (line 8 of Algorithm 1). Next, the point sets PS_p in the range $sl_p \pm \alpha_s/2$ are extracted from P_{cl}. This sliding factor is determined experimentally based on the size of the point cloud. For each set of points in PS_p, the two furthest points (P_{cl_A}, P_{cl_B}) are extracted, and these points are considered as the boundary points for that particular slice. The boundary points are added to the boundary point sets B_s. Figure 7.14 provides a visual representation of the boundary estimation algorithm.

Once the boundary points B_s are estimated, the team uses **Algorithm 2** to estimate the area availability parameter S_{am}. This algorithm takes as input the boundary points B_s, the point cloud centroid $C_{P_{cl}}$, the normal vector of the point cloud $\vec{n}_{P_{cl}}$, the length l and width w of the robot leg, and the wheel distance tolerance t.

First, the team calculates the n closest points (N_{clos}) from B_s to the point cloud centroid $C_{P_{cl}}$. For each point N_i in the set N_{clos}, the team estimates the corner points of a rectangle of width w and length l, which is also the robot foot width and length, respectively, that adheres to the robot wheels. The team estimates the rectangle edges parallel to the vectors $\vec{e}_{x_i}$ and $\vec{e}_{y_i}$, which are calculated for point N_i. The four corners R of the rectangle are estimated using these two vectors. To accommodate for the nonconvex shape of the steel surface and alleviate point

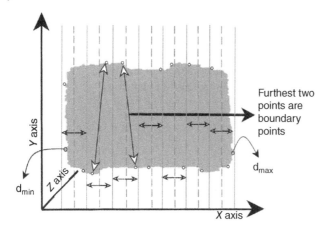

Figure 7.14 Boundary point estimation from 3D point cloud data.

cloud collection error, the team also includes four middle points of the estimated rectangle corners in R.

Next, the team finds the m closest points to R from B_s to measure if the points in R are inside the boundary. The team calculates the distance from point cloud centroid $C_{P_{cl}}$ to R and Q, respectively. The algorithm considers a point to be inside the boundary if its tolerance is less than t and the distance to the centroid is less than its neighbors. When the value of S_{am} is true for all the conditions, the team considers those sets of points as rectangular points.

Height availability: The height availability S_{hc} is crucial for the switching control. Based on this parameter, the switching control activates the robot transformations. At first the centroid of the point cloud $C_{P_{cl}}$ is calculated along the camera frame f_c. Then we transform the centroid coordinate to the robot base frame f_{rb} using Eq. 7.10.

$$P_{C_{f_{rb}}} = T_{f_{rb}f_c} P_{C_{f_c}} \tag{7.10}$$

where $p_{C_{f_{rb}}}, p_{C_{f_c}}$ is coordinate of the centroid C_c in the camera frame and the robot base frame, respectively. $T_{f_{rb}f_c}$ is the transformation matrix from the camera frame f_c to the robot base frame f_{rb}.

The plane height $z_{f_{rb}}$ coordinate is then compared with the robot base height. If they are equal, the returned result is *true* and the robot configures to *Mobile transformation*. Otherwise, it returns *false*, and the robot goes to *Worming transformation*. The height availability condition is shown in Eq.7.11.

$$\begin{cases} S_{hc} = True, \text{ if } z_{f_{rb}} = z_{robotbase} \\ S_{hc} = False, \text{ otherwise} \end{cases} \tag{7.11}$$

Algorithm 2 Area Checking from the plane surface boundary points and Pose Calculation.

1: **procedure** AREA($B_s, C_{P_{cl}}, \vec{n}_{P_{cl}}, w, l, t, S_{am}$)
2: $\quad N_{clos}$ = Find n closest points to $C_{P_{cl}}$ from B_s
3: $\quad$ **for** $N_i \in N_{clos}$ **do**
4: $\qquad R = \{\}$, //Estimated rectangle corner points,
5: $\qquad \vec{e}_{x_i} = N_i - C_{P_{cl}}$
6: $\qquad \vec{e}_{z_i} = \vec{n}_{P_{cl}}$
7: $\qquad \vec{e}_{y_i} = \vec{e}_{x_i} \times \vec{e}_{y_i}$
8: $\qquad k_w = \frac{w}{|\vec{e}_{x_i}|}\vec{e}_{y_i}$ and $k_l = \frac{b}{|\vec{e}_{y_i}|}\vec{e}_{x_i}$,
9: $\qquad \{R_1, R_2\} = \{N_i + k_w, N_i - k_w\}$
10: $\qquad R = R \cup \{R_1, R_2\}$
11: $\qquad R = R \cup \{R_1 + k_l, R_2 + k_l\}$
12: $\qquad M = \forall_{r_i \in R}\{\frac{r_i + r_{i+1}}{2}\}$
13: $\qquad R = R \cup M$
14: $\qquad S_{am} = True$
15: $\qquad$ **for** $r_i \in R$ **do**
16: $\qquad\quad Q_i$ = Find m closest points to r_i
17: $\qquad\quad d_{r_i} = \|d_{r_i}, C_{P_{cl}}\|$ and $d_{Q_i} = \|Q_i, C_{P_{cl}}\|$
18: $\qquad\quad S_i = (d_{r_i} < d_{Q_i}) \vee (\frac{d_{r_i} - d_{Q_i}}{d_{r_i}} < t$
19: $\qquad\quad S_{am} = S_{am} \wedge S_i$
20: $\qquad$ Pose = (Orientation, Position)
21: $\qquad$ **if** S_{am} == True **then**
22: $\qquad\quad R_c$ = Centroid of R
23: $\qquad\quad$ Orientation = $(\vec{e}_{x_i}, \vec{e}_{y_i}, \vec{e}_{z_i})$
24: $\qquad\quad$ Position = $\left(x_{R_c}, y_{R_c} - l/4, z_{R_c}\right)$
25: $\qquad\quad$ **return** Pose
26: $\quad$ **return** False

Magnetic Array Distance Control

The magnetic array of the ARA robot is equipped with distance control switches that allow for two modes: touched and untouched. The distance control framework, illustrated in Figure 7.15, is responsible for configuring the magnetic array into either mode. The "untouched" mode is used for creating a magnetic adhesion force at a distance of 1mm from the steel surface, while the "touched" mode fully adheres the robot to the surface for performing an inch-worm jump.

To adjust the adhesion force of the magnetic array, a PID controller is used, as depicted in Figure 7.15. This controller enables the magnetic array to move up and down, thereby adjusting the magnetic adhesion force to the steel surface.

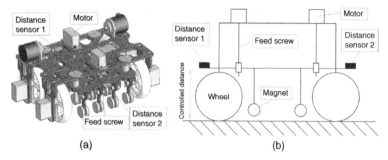

Figure 7.15 Distance control system of magnetic array, (a) 3D model, and (b) 2D diagram.

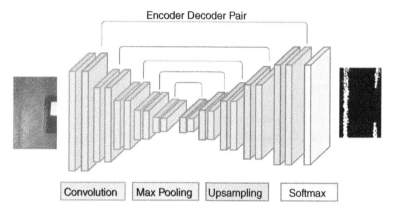

Figure 7.16 An encoder–decoder-based CNN architecture for steel bridge inspection.

Two high-resolution distance sensors (VL6180X) are mounted on the magnetic array for distance sensing, and the distance is transmitted to the controller via two motors and a parallel screw. The accuracy of the controller is crucial, as the magnetic array must maintain a 1mm distance to the steel surface. Any mechanical error could cause the magnetic array to transform into the "touched" mode, resulting in a high load on the two motors. Therefore, the design of the controller significantly improves its accuracy to avoid these difficulties.

The Inspection Framework

After performing the linear transition, the robot conducts the visual inspection. The onboard RGB camera module captures images from the steel surface and sends them to the NUC processor for defect identification.

The CNN architecture as shown in Figure 7.16 used for steel bridge defect identification is based on an encoder–decoder approach inspired by previous work in detecting defects on concrete structures. The architecture segments images into defective and healthy regions by passing the input images through five encoder layers followed by five decoder layers. Each encoder layer performs a 7×7 convolution operation, which is followed by a batch normalization operation and ReLU activation. The feature space is down-sampled using a 2×2 max-pooling unit and fed to the next encoder layer. Five such encoder layers are utilized for feature extraction. Once the last encoder layer is reached, the feature maps enter the decoder portion of the network, where they are up-sampled using bi-linear interpolation in each decoder layer. The up-sampling operations are again followed by a convolution operation, batch normalization, and ReLU activation. The CNN architecture is pretrained on 3,000 steel images containing severe defects, such as corrosion. For the hyperparameter optimization, the Adam optimizer with a learning rate of 0.0001 is used, and the network is trained for 150 epochs in this experiment.

Worming Transformation

The inch-worm jump mode of the robot allows it to move from one steel surface to another by performing a series of steps, as illustrated in Figure 7.17. Initially, the second permanent magnet on the second leg of the robot is activated, generating a strong adhesive force for the robot to adhere to the steel surface. The robot controller then manipulates the arm joints to move the first leg toward the target surface using a generated trajectory, as shown in Figure 7.17(b). Once the first leg touches the target surface, the second leg begins to detach from the starting surface, as depicted in Figure 7.17(c). Finally, in Figure 7.17(d), the second leg completely detaches from the starting surface and adheres to the target surface.

The conversion from Mobile configuration (Figure 7.12(a)) to the Worming transformation is challenging for the motion planner to create a trajectory. To improve performance, a convenient robot's pose P_{conv} is proposed as the starting point for the Worming transformation, as depicted in Figure 7.18-left in blue color. The motion planner generates a trajectory from P_{conv} to the destination in red color. An Inverse Kinematics (IK) solver processes the trajectory and generates a set of joint positions for the robot to move. The worming is completed by moving the second leg to the target surface and reforming the Mobile configuration, as shown in Figure 7.18(b). The first and third steps of the robot are done by point-to-point control. To perform the second step, the robot needs to determine the target plane and its pose, which are outputs of **Algorithm 2**.

The flexibility of the robot in worming is made of the six DOF arm. The revolute joints of the arm in Figure 7.12(b) can rotate along three different axes separately. For example, the joint 2 and joint 5 in Figure 7.12(b) are configured to rotate around *y-axis*. The rest of the joints are positioned to rotate around *z-axis*. This configuration was selected to maintain symmetry so that the manipulator can move efficiently in both worming and mobile modes. Our previous research states an in-detail elaboration of the robotic arm in Nguyen et al. [2020].

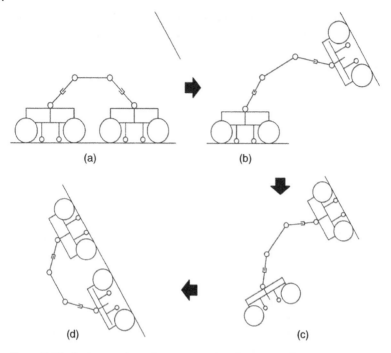

Figure 7.17 Inch-worm jump from one steel surface to another.

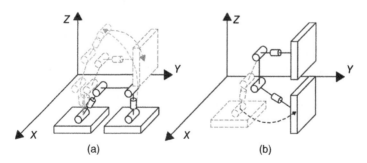

Figure 7.18 Inch-worm jump procedure.

7.5 Experiment Results

The team conducted experiments on the ARA robot, which was originally introduced by Nguyen et al. [2020], but with an additional camera module. An RGB-D camera, specifically the ASUS Xtion Pro Live, was attached to the robot to collect point clouds and perform visual inspections. To calibrate the camera,

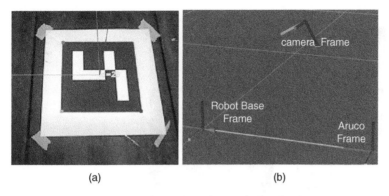

(a)   (b)

Figure 7.19 Aruco Marker and coordinate frames for localization.

the authors utilized a method implemented by Bui et al. [2020a] to integrate the necessary parameters. To localize the robot during the experiments, an aruco marker [Garrido-Jurado et al., 2014] was placed on a planar surface where the robot base could adhere. A geometric calculation was then performed to determine the position between the aruco marker and the robot base. To employ the robotic operating system (ROS) as well as the inspection framework, an Intel NUC 5 Business Kit (NUC5i5MYHE)-Core i5 vPro was incorporated. Due to the rectangular shape of the robot feet, a rectangle was drawn around them. The robot was then set to its *Mobile transformation* to move a distance d_c along the x-axis. To achieve an equivalent orientation, the marker was placed in the center of the previously drawn rectangle. The pose and position between the camera and the robot base were extracted using the *lookupTransform* function in the *tf* package. The aruco marker and required coordinates are depicted in Figure 7.19.

The team performed experiments on two individual steel slabs located perpendicularly from each other. The steel slabs are highly corroded to replicate steel defects. The following section describes the experiment and its results elaborately.

7.5.1 Switching Control

At the start, the RGB-D point cloud data of a bridge steel bar was collected from the robot camera. An example of the initial point cloud is shown in Figure 7.20(a). After performing some preprocessing operations such as *pass-through filtering* and *downsampling*, the data is sent to *plane detection* to extract the planar surface. The processed point cloud is shown in Figure 7.20(b). The coordinate frame is also shown in the figure with x-axis in red, y-axis in green, and z-axis in blue. After obtaining the planar surface the team extracts the boundary points of the surface and performs *Area availability* checking. For this purpose, **Algorithm 1** and **Algorithm 2** were employed on two different surfaces, one containing

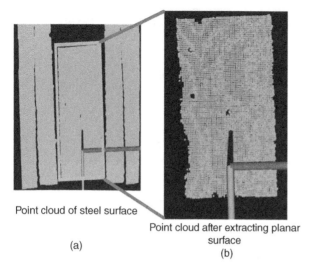

Point cloud of steel surface

(a)

Point cloud after extracting planar surface

(b)

Figure 7.20 Planar surface extraction from a 3D point cloud of the steel surface.

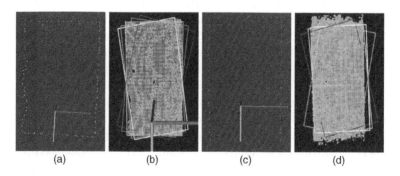

(a)   (b)   (c)   (d)

Figure 7.21 (a), (c) The Boundary set and (b), (d) Area rectangle set.

sufficient area for movement and the other without. Using **Algorithm 1**, the two boundary points of the two different point clouds are shown in Figure 7.21(a) and Figure 7.21(c).

The area availability **Algorithm 2** is employed using the boundary point estimated. The parameters of this algorithm are as follows: $n = 5, m = 3$, and $t = 0.02$. The estimate of five rectangles for robot feet with this algorithm is shown in Figure 7.21(b) and Figure 7.21(d) in red, yellow, blue, green, and purple color. In Figure 7.21(b), several corners of all the red, yellow, purple, and blue rectangles are outside of the point cloud area. It represents that the area of the rectangle is not sufficient enough for an inch-worm jump. Only the green rectangle is inside the point cloud, satisfying the area requirement. The selected rectangle is shown

Figure 7.22 (a) The selected area rectangle, and (b) Pose estimation.

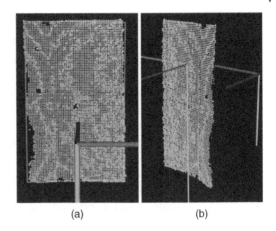

(a)   (b)

in Figure 7.22(a) (in red color). Since there is enough area for an inch-worm jump the variable S_{am} is set to true by the algorithm. After that, we calculate the pose of the planar surface in Figure 7.22(b). The pose is represented with three orientations shown in red, green, and blue color on the point cloud surface. In Figure 7.21(d) the steel surface area is not sufficient for an inch-worm jump. As a result, all the rectangle corners are outside the point cloud boundary.

After that, it is transformed into the robot base frame. The plane height (z-coordinate of the centroid of the robot base frame) is compared to the robot base height. When both the heights are equal the value of $S_{hc} = True$. The robot will configure itself into *Mobile transformation* in this case.

Figure 7.23(b) shows another scenario as the point cloud is from a surface that is $d = 7\,cm$ lower than the robot base. In this case, the heights are different, then the returned value of variable S_{hc} is *false*, and the robot will do the *Worming transformation* in the next step.

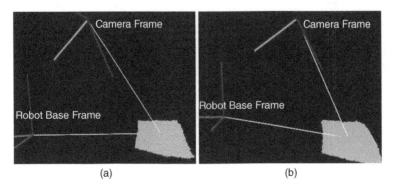

(a)   (b)

Figure 7.23 Surface height availability check: (a) Same height, and (b) Different height.

7.5.2 Robot Navigation in Mobile and *Worming Transformation*

In the *Mobile transformation* mode, the *ARA* robot can move simultaneously in both the x- and y-directions. The navigation process for *Mobile transformation* is depicted in Figure 7.24. During this mode, the *ARA* robot collects images of the steel surface and performs visual inspection using a CNN network. The network segments the images into either corroded or healthy regions, with a sample of the network output presented in Figure 7.24(a).

On the other hand, in the *Worming transformation* mode, the robot performs an inch-worm jump from point P_{conv} to the target plane, as illustrated in Figure 7.25. The team used the MoveIt package in ROS with a plane pose as input to generate a trajectory for the first robot foot to reach the destination. To accomplish this, they created a primitive model of the robot in *urdf* format, which had the exact dimensions, joint types, and limits of the *ARA* robot. The *Kinematics and Dynamics Library* (KDL) IK and *RRTConnect* motion planner were utilized to implement the task, and the resulting trajectory was a ROS topic consisting of the robot joint positions necessary to reach the target pose. In Figure 7.25(a), the robot's base frame (*base_link*) and the target pose (*target_pose_check*) are shown. After obtaining the trajectory, the robot moves toward the target pose, as displayed in Figure 7.25(b).

The robot obtained data from the trajectory topic and sent an array of robot joint positions to the ARA robot controller via ROS service. The worming performance of the robot is illustrated in Figure 7.26. Initially, the robot activates and lowers down the magnetic array to touch the steel surface, then it transitions from the mobile configuration to the convenient point P_{conv} by following a predefined trajectory shown in Figure 7.26(a) and Figure 7.26(b). Once it reaches point P_{conv},

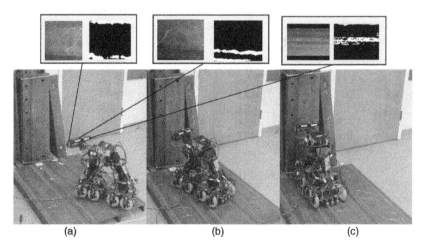

(a)　　　　　　　　　　(b)　　　　　　　　　　(c)

Figure 7.24 The movement of robot in *Mobile transformation* and visual inspection.

Figure 7.25 The motion planning for the ARA robot movement.

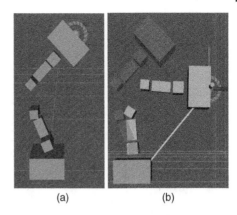

(a)   (b)

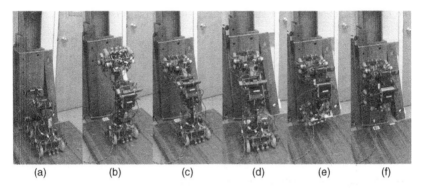

(a)   (b)   (c)   (d)   (e)   (f)

Figure 7.26 *Worming transformation*: (a) magnetic array of the second foot touches the base surface (b) first foot moves to convenient point (c) first-foot reaches target pose and touches the second surface (d) magnetic array of the second foot is released (e) and (f) second foot moves to target pose.

the robot follows the trajectory created by the RRTConnect motion planner toward the target surface. As it touches the surface, it activates both magnetic arrays, with the one in the first foot moving down to stick on the target steel surface and the one in the second leg moving up and detaching from the starting surface as depicted in Figure 7.26(c,d). The second foot of the robot then transforms into the target plane, as illustrated in Figure 7.26(e,f).

7.5.3 Robot Deployment

The purpose of the indoor testing was to evaluate the robots' climbing abilities (Figure 7.27) and examine how adhesion force varied with distance from the steel surface.

Figure 7.27 The robot is being tested on a steel structure indoors.

Figure 7.28 Deployment of the robot on a steel structure.

The ARA robot was able to successfully climb the lab's steel beams (Figure 7.27). With the ARA robot's climbing capabilities verified, the robot was then tested outside on more complex structures than what could be found in the lab.

Outdoor testing for the robot began at a local steel art statue on the university campus famous for having every commercial steel link type on it, as showcased in Fig 7.28.

The team examined the ARA Lab robot's physical ability to navigate complex geometries at an art statue, as shown in Figure 7.28. The results indicate that the robot is capable of traversing complex, jagged geometries like those found on the art statue, but there are still some improvements to be made. Specifically, the robot struggled to maintain its orientation when moving due to a lack of friction to resist twisting, especially in worming mode. Additionally, the rudimentary controller design resulted in sharp, sudden movements that need to be smoothed out for the robot to meet commercial demands for steel structure inspection. Overall, while the testing at the art statue was promising, further work is needed to ensure the robot's success in the steel bridge inspection deployments.

Figure 7.29 Mobile mode example on the bridge.

Additional evaluations were done at a bridge on campus, which is comprised of long cylindrical members. As shown in Figure 7.29, these members were useful for testing the robot's ability to adhere to round surfaces.

The link to the video demonstration of the robot deployments can be seen here: https://youtu.be/PwDf6h0Om3c.

The link to the video demonstration of the robot control and navigation framework can be seen here: https://youtu.be/SHk5IIOBRdA.

7.6 Conclusion and Future Work

In this chapter, the ARA robot is presented as a design that combines the best features of modern steel inspection robots, making it suitable for navigating complex geometries commonly found in steel structures like windmills, steel bridges, and buildings. The main challenge is ensuring that the robot functions as intended, given that it incorporates design parameters from various sources. Detailed design analyses are necessary to ensure that these diverse designs can work together seamlessly.

To enhance the flexibility of navigation and inspection, a switching control mechanism is proposed for the autonomous navigation of the ARA robot, allowing it to switch between two modes – inch-worm and mobile. The proposed algorithm estimates available surfaces based on their area, plane, and height, supporting navigation. In addition, a mobile control framework and a magnetic adherence distance controller are developed to facilitate robot navigation.

In the future, the ARA team plans to design a working arm for the robot, equipped with special equipment such as an eddy current sensor, to assist in performing detailed inspections of difficult-to-reach and dangerous areas on steel structures. The team also aims to improve the robot's resistance to twisting by increasing the friction generated with the steel surface during the robot's articulations in inch-worm mode. Furthermore, autonomous localization, navigation, and sensing functions will be further tested and validated to enable the robot to perform automated inspections, building upon previous work on localization, navigation, and sensing for bridge deck inspection robots La et al. [2013], La et al. [2014a], La et al. [2014b], La et al. [2015, 2017], Gibb et al. [2018].

Bibliography

BridgeBot, 2017. http://www.robotics.umd.edu/content/maryland-robotics-center-videos/.

Coefficients of Friction for Steel, 2005. https://hypertextbook.com/facts/2005/steel.shtml.

Crews inspect condition of golden gate bridge's towers, April 30, 2018. https://www.nbcbayarea.com/on-air/as-seen-on/crews-inspect-condition-of-golden-gate-bridge_s-towers_bay-area-2/61247/.

U.S Department of transportation highway administration, national bridge inventory data, 2019. http://www.fhwa.dot.gov/bridge/nbi.cfm.

Versatrax 100^{TM}, n.d. https://robotics.eddyfi.com/en/.

Washington bridge collapse, 2013. https://www.cnn.com/2013/05/24/us/gallery/skagit-river-bridge/index.html.

Habib Ahmed, Son Thanh Nguyen, Duc La, Chuong Phuoc Le, and Hung Manh La. Multi-directional bicycle robot for bridge inspection with steel defect detection system. In *2022 IEEE/RSJ International Conference on Intelligent Robots and Systems (IROS)*, pages 4617–4624, 2022. doi: 10.1109/IROS47612.2022.9981325.

T. Bandyopadhyay, R. Steindl, F. Talbot, N. Kottege, R. Dungavell, B. Wood, J. Barker, K. Hoehn, and A. Elfes. Magneto: A versatile multi-limbed inspection robot. In *2018 IEEE/RSJ International Conference on Intelligent Robots and Systems (IROS)*, pages 2253–2260, Oct 2018. doi: 10.1109/IROS.2018.8593891.

Hoang-Dung Bui, Hai Nguyen, Hung Manh La, and Shuai Li. A deep learning-based autonomous robot manipulator for sorting application. In *2020 Fourth IEEE International Conference on Robotic Computing (IRC)*, Taichung, Taiwan, 298–305. IEEE, 2020a. doi: 10.1109/IRC.2020.00055

Hoang Dung Bui, Son Thanh Nguyen, U.-H. Billah, Chuong Le, Alireza Tavakkoli, and Hung Manh La. Control framework for a hybrid-steel bridge inspection robot. In *2020 IEEE/RSJ International Conference on Intelligent Robots and Systems (IROS)*, pages 2585–2591. IEEE, 2020b.

M. Eich and T. Vögele. Design and control of a lightweight magnetic climbing robot for vessel inspection. In *the 19th Mediterranean Conference on Control Automation*, pages 1200–1205, June 2011. doi: 10.1109/MED.2011.5983075.

Sergio Garrido-Jurado, Rafael Mu noz-Salinas, Francisco José Madrid-Cuevas, and Manuel Jesús Marín-Jiménez. Automatic generation and detection of highly reliable fiducial markers under occlusion. *Pattern Recognition*, 47(6):2280–2292, 2014.

S. Gibb, H.M. La, T. Le, L. Nguyen, R. Schmid, and H. Pham. Nondestructive evaluation sensor fusion with autonomous robotic system for civil infrastructure inspection. *Journal of Field Robotics*, 35(6):988–1004, 2018.

Billah Umme Hafsa, Hung Manh La, and Alireza Tavakkoli. Deep learning-based feature silencing for accurate concrete crack detection. *Sensors*, 20, (16) 4403. doi: 10.3390/s201644032020.

H.M. La, R.S. Lim, B.B. Basily, N. Gucunski, J. Yi, A. Maher, F.A. Romero, and H. Parvardeh. Mechatronic systems design for an autonomous robotic system for high-efficiency bridge deck inspection and evaluation. *IEEE/ASME Transactions on Mechatronics*, 18(6):1655–1664, Dec 2013. ISSN 1083-4435. doi: 10.1109/TMECH.2013.2279751.

H.M. La, N. Gucunski, S.H. Kee, and L.V. Nguyen. Visual and acoustic data analysis for the bridge deck inspection robotic system. In *The 31st International Symposium on Automation and Robotics in Construction and Mining (ISARC)*, pages 50–57, July 2014a.

H.M. La, N. Gucunski, Seong-Hoon Kee, J. Yi, T. Senlet, and Luan Nguyen. Autonomous robotic system for bridge deck data collection and analysis. In *2014 IEEE/RSJ International Conference on Intelligent Robots and Systems*, pages 1950–1955, Sep 2014b. doi: 10.1109/IROS.2014.6942821.

H.M. La, N. Gucunski, S.-H. Kee, and L.V. Nguyen. Data analysis and visualization for the bridge deck inspection and evaluation robotic system. *Visualization in Engineering*, 3(1):1–16, 2015.

H.M. La, N. Gucunski, K. Dana, and S.-H. Kee. Development of an autonomous bridge deck inspection robotic system. *Journal of Field Robotics*, 34(8):1489–1504, 2017. doi: 10.1002/rob.21725.

H.M. La, T.H. Dinh, N.H. Pham, Q.P. Ha, and A.Q. Pham. Automated robotic monitoring and inspection of steel structures and bridges. *Robotica*, 37(5):947–967, May 2019. doi: 10.1017/S0263574717000601.

A. Leibbrandt, G. Caprari, U. Angst, R.Y. Siegwart, R.J. Flatt, and B. Elsener. Climbing robot for corrosion monitoring of reinforced concrete structures. In *The Second International Conference on Applied Robotics for the Power Industry (CARPI)*, pages 10–15, Sept 2012.

H. Leon-Rodriguez, S. Hussain, and T. Sattar. A compact wall-climbing and surface adaptation robot for non-destructive testing. In *2012 12th International Conference on Control, Automation and Systems (ICCAS)*, pages 404–409, Oct 2012.

A. Mazumdar and H.H. Asada. Mag-Foot: A steel bridge inspection robot. In *IEEE/RSJ International Conference on Intelligent Robots and Systems, 2009. IROS 2009*, pages 1691–1696, Oct 2009.

A. McCrea, D. Chamberlain, and R. Navon. Automated inspection and restoration of steel bridges – a critical review of methods and enabling technologies. *Automation in Construction*, 11(4):351–373, 2002. ISSN 0926-5805. doi: https://doi.org/10.1016/S0926-5805(01)00079-6. URL http://www.sciencedirect.com/science/article/pii/S0926580501000796.

Cadence Motley, Son Thanh Nguyen, and Hung Manh La. Design of a high strength multi-steering climbing robot for steel bridge inspection. In *2022 IEEE/SICE International Symposium on System Integration (SII)*, pages 323–328. IEEE, 2022.

S.T. Nguyen and H.M. La. Development of a steel bridge climbing robot. In *IEEE/RSJ International Conference on Intelligent Robots and Systems, 2019. IROS 2019*, Nov 2019a.

S.T. Nguyen and H.M. La. Roller chain-like robot for steel bridge inspection. In *The Ninth International Conference on Structural Health Monitoring of Intelligent Infrastructure (SHMII-9)*, Aug 2019b.

Son Nguyen and Hung La. A climbing robot for steel bridge inspection. *Journal of Intelligent & Robotic Systems*, 75:102, 2021.

Son T. Nguyen, Anh Q. Pham, Cadence Motley, and Hung M. La. A practical climbing robot for steel bridge inspection. In *2020 IEEE International Conference on Robotics and Automation (ICRA)*, pages 9322–9328. IEEE, 2020.

Son Thanh Nguyen, Hai Nguyen, Son Tien Bui, Van Anh Ho, Trung Dung Ngo, and Hung Manh La. An agile bicycle-like robot for complex steel structure inspection. In *2022 International Conference on Robotics and Automation (ICRA)*, pages 157–163, 2022. doi: 10.1109/ICRA46639.2022.9812153.

Yu Otsuki, Son Thanh Nguyen, Hung Manh La, and Yang Wang. Autonomous ultrasonic thickness measurement using a steel climbing mobile robot integrated with Martlet wireless sensing. In *2022. The 30th ASNT Research Symposium*, pages 9322–9328. IEEE, 2022.

Yu Otsuki, Son Thanh Nguyen, Hung Manh La, and Yang Wang. Autonomous ultrasonic thickness measurement of steel bridge members using a climbing bicycle robot. *Journal of Engineering Mechanics*, 149(8):04023051, 2023. doi: 10.1061/JENMDT.EMENG-7000.

N.H. Pham and H.M. La. Design and implementation of an autonomous robot for steel bridge inspection. In *54th Allerton Conference on Communication, Control, and Computing*, pages 556–562, Sept 2016. doi: 10.1109/ALLERTON.2016.7852280.

N.H. Pham, H.M. La, Q.P. Ha, S.N. Dang, A.H. Vo, and Q.H. Dinh. Visual and 3D mapping for steel bridge inspection using a climbing robot. In *The 33rd International Symposium on Automation and Robotics in Construction and Mining (ISARC)*, pages 1–8, July 2016.

Anh Q. Pham, Hung M. La, Kien T. La, and Minh T. Nguyen. A Magnetic Wheeled
Robot for Steel Bridge Inspection. In Kai-Uwe Sattler, Duy Cuong Nguyen, Ngoc Pi
Vu, Banh Tien Long, and Horst Puta, editors, *Advances in Engineering Research
and Application*, pages 11–17. Springer International Publishing, Cham, 2020.
ISBN 978-3-030-37497-6.

Anh Quyen Pham, Cadence Motley, Son Thanh Nguyen, and Hung Manh La.
A robust and reliable climbing robot for steel structure inspection. In *2022 IEEE/
SICE International Symposium on System Integration (SII)*, pages 336–343. IEEE,
2022.

B. Purna Chandra Rao. *Non-destructive Testing and Damage Detection*, pages 209–228.
Springer Singapore, Singapore, 2017. ISBN 978-981-10-2143-5.
doi: 10.1007/978-981-10-2143-5_11.

A. San-Millan. Design of a teleoperated wall climbing robot for oil tank inspection.
In *2015 23rd Mediterranean Conference on Control and Automation (MED)*, pages
255–261, June 2015.

W. Shen, J. Gu, and Y. Shen. Permanent magnetic system design for the wall-climbing
robot. In *IEEE International Conference on Mechatronics and Automation, 2005*,
volume 4, pages 2078–2083, July 2005. doi: 10.1109/ICMA.2005.1626883.

Y. Takada, S. Ito, and N. Imajo. Development of a bridge inspection robot capable of
traveling on splicing parts. *Inventions*, 2(3):22, 2017. ISSN 2411-5134.
doi: 10.3390/inventions2030022.

R. Wang and Y. Kawamura. A magnetic climbing robot for steel bridge inspection.
In *2014 11th World Congress on Intelligent Control and Automation (WCICA)*, pages
3303–3308, June 2014.

R. Wang and Y. Kawamura. Development of climbing robot for steel bridge
inspection. *Industrial Robot: An International Journal*, 43(4):429–447, 2016.
doi: 10.1108/IR-09-2015-0186.

P. Ward, P. Manamperi, P.R. Brooks, P. Mann, W. Kaluarachchi, L. Matkovic, G. Paul,
C.H. Yang, P. Quin, D. Pagano, et al. Climbing robot for steel bridge inspection:
Design challenges. In *Austroads Publications Online, ARRB Group*, 2015.

8

Underwater Robots for Cleaning and Inspection of Underwater Structures

Andrew Wing Keung To, Khoa Le, and Dikai Liu*

Robotics Institute, University of Technology Sydney, Broadway, NSW, Ultimo, Australia, 2007

8.1 Introduction to Maintenance of Underwater Structures

According to World Corrosion Organization (WCO) white paper [Schmitt, 2009], the estimated annual cost of damage due to corrosion across the world is approximately US$2.5 trillion, equivalent to 3–4% GDP of developed countries. Corrosion can cause expensive damage to civil infrastructures.

Routine inspection and maintenance of subsea structures, such as underwater pipes, bridge pylons, and piles, is crucial to maintain the integrity and prolong the lifespan of the structures and prevent potential damages caused by active corrosion and biological attack. Three levels of underwater inspection can be conducted to evaluate structural conditions [Kelly, 1999]:

- Level 1: general visual assessment without cleaning the structures. This level of inspection provides initial inspection data and detects substantial damage or deterioration.
- Level 2: This level requires partly removing marine growth from structures before or during inspections for greater details of data acquisition. With Level 2 inspection, inspectors can detect surface cracking, crumbling, and rust straining of underwater structures.
- Level 3: This level is employed to detect hidden and internal damages that result in progressive collapses and structural failure. Extensive cleaning of the structure is required before conducting inspection which may involve using destructive and invasive data sampling techniques.

*Email: dikai.liu@uts.edu.au

Infrastructure Robotics: Methodologies, Robotic Systems and Applications, First Edition.
Edited by Dikai Liu, Carlos Balaguer, Gamini Dissanayake, and Mirko Kovac.
© 2024 The Institute of Electrical and Electronics Engineers, Inc. Published 2024 by John Wiley & Sons, Inc.

Figure 8.1 Manual operation in removing marine growth using high-pressure water-jetting.

As described in levels 2 and 3 inspections, removing crustaceans and other marine growth that covers the structure is necessary for making observations and collecting high-quality data such as images. The current practice is usually for professional divers to manually remove the growth using hand scraping tools or high-pressure water-jet (hydro blasting). The marine growth removal operation, as shown in Figure 8.1, is an arduous task. Divers face poor working conditions, including (i) strong water currents, (ii) poor visibility, (iii) using awkward body poses for a long period, and (iv) dealing with large reaction force of up to 50 Newtons from the water-jetting nozzle. Statistically, as shown in Annual Fatality Rate (AFR) data, there are around 8.73 (6.85-10.96) deaths per 100,000 professional scuba divers for Australian residents [Lippmann et al., 2016], making diving-related tasks high risk.

Work-class Remotely Operated Vehicles (ROVs) equipped with a cleaning system [Curran et al., 2016] are an available technological offering to mitigate human risks associated with underwater structure cleaning and inspection. The current work-class ROVs usually operate in deep and open water environments. These ROVs tend to be large and quite expensive, with large and heavy hulls that prevent them from being deployed to work in narrow and complex underwater environments littered with obstacles, such as bridge pylons, wharf piles, etc. Furthermore, operating these commercial ROVs requires trained operators who have undergone intensive and costly training procedures [Bai and Bai, 2019].

There is a demand for small-scale intervention autonomous underwater robots capable of underwater structure cleaning and inspection. This demand has driven the research to develop a suitable underwater robot that potentially alleviates the challenges of underwater structure maintenance tasks.

8.2 Robot System Design

The Submersible Pile Inspection Robot (SPIR) (Figure 8.2) consists of 12 propulsion propellers (thrusters), four grasping arms, one 3-DOF arm for positioning a high-pressure water-jet (blasting arm), and a suite of cameras in an enclosure along with other sensors for navigation, localization, 3D map building and visual inspection. This robotic system has the functionalities required to conduct cleaning and inspection of underwater structures, which include:

- Maneuverability, Navigation and Stabilization: Water currents around underwater structures such as a bridge pylon are turbulent and can reach speeds of around 3 knots (1.5 m/s). The robot needs to be able to deal with this type of water condition and still maneuver smoothly in the space around the underwater structures (e.g. between the piles and pylons).
- Docking with the structure: Rigidly docking to establish a stable platform is an energy-efficient approach to maintaining the robot position for cleaning and inspection tasks. Human divers readily use a safety belt to secure themselves to pylons during cleaning and inspection work.
- Marine Growth removal: Surfaces of underwater structures are normally covered by marine growth, especially in the inter-tidal zone. The robot must be capable of removing the marine growth using appropriate tools.
- Underwater 2D and 3D map building: Sensing is a core requirement for autonomy, the robot will need both 2D and 3D maps to realize the functionalities described earlier.

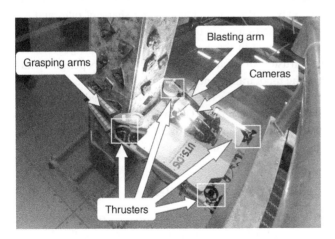

Figure 8.2 The underwater robot annotated with the main components of the system.

Figure 8.3 Computational fluid dynamics for evaluating hull designs.

8.2.1 Hull Design and Maneuvering System

In consultations with underwater inspection experts, it is a common practice to temporarily suspend cleaning and inspection operations when the current speed exceeds 3 knots (1.5 m/s), the safe limit for a diver to swim against. Therefore the underwater robot needs to be designed to operate at least 3 knots to supplement human operations.

For the robot to operate in high water currents, the underwater robot's hull must be carefully designed and optimized using the computational fluid dynamics (CFD) technique (Figure 8.3) and experimental approach to minimize the fluid drag force affecting the body. Furthermore, the size and shape of piles and the distance among piles are also considered in the designing process.

Twelve T200 thrusters vectored in the configuration shown in Figure 8.2 are installed in the robot: 8 thrusters in a transverse plane (2 thrusters per group and arranged at 4 corners of the robot frame) and 4 thrusters in the vertical direction. The arrangement of the thrusters makes the robot a fully-actuated vehicle capable of maneuvering in narrow environments. Given that each thruster provides 5 kgF at 300 W, the thruster system consumes nearly 3.5 kW at full throttle and can provide up to 20 kgF of thrust (up/down direction).

8.2.2 Robot Arms for Docking and Water-Jet Cleaning

Underwater structures are usually covered by a thick, hard layer of marine growth, sometimes reaching 20 cm in thickness. It is necessary to remove this layer of marine growth in order to inspect the structure. The high-pressure water-jet blasting method is selected due to its efficiency. The high-pressure water pump

(SCUD400) with the turbo-jet nozzle creates blasting pressure up to 4,000 psi and is proven to be efficient for removing marine growth from pile surfaces at a distance between 5-10 cm. However, the large reaction force from the water-jet, up to 50-70 N, causes challenges to design or select a lightweight and compact robot manipulator that can handle this force and ensure the stability of the overall robot during the cleaning operation.

As the payload of the underwater robot is only 6.5 kg in water, the arm designed for marine growth removal needs to be relatively lightweight and can still handle 50-70 N at the end-effector. However, no product on the commercial market meets this requirement [Sivčev et al., 2018]. From lab experiments and feedback from professional divers, it is determined that the water-jet cleaning performance is primarily affected by the blasting pressure and the distance from the nozzle to the surfaces, and to a lesser degree by the posture or the orientation of the nozzle. Hence, to ensure the weight of the robot arm is under the limited payload of 6.5 kg, a spherical-workspace robot manipulator is designed that can position the nozzle. The design solution is the combination of a high torque pan and tilt unit (SS109HT) with the linear actuator (UltraMotion), creating a 3-DOF spherical robot manipulator as shown in Figure 8.4.

The underwater robot body needs to be kept stable during the water-jet blasting operation to ensure cleaning quality and protection of piles from damage. Inspired by the safety belt approach used by divers, two pairs of claw-like grasping arms are designed for robot's docking mechanism (Figure 8.5). Actuated independently by four motors, the docking mechanism always provides four contacting points, hence being highly capable of grasping a wide range of sizes and shapes of bridge

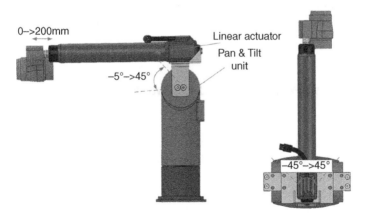

Figure 8.4 3-DOF arm for removing marine growth using water-jet blasting Le et al., [2020]/IEEE.

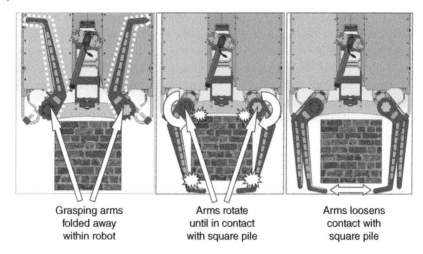

| Grasping arms folded away within robot | Arms rotate until in contact with square pile | Arms loosens contact with square pile |

Figure 8.5 Top-down view of different grasping arm configurations: (Left) folded away within the robot's body during navigation, (Middle) grasped onto a square pile to create a rigid dock for cleaning, (Right) loose-engaging to allow the robot to move around the pile. The lower pair of grasping arms that have no hook-tip is extended slightly wider for additional clearance.

pylons. The motor's current/torque control mode generates the grasping force in every pose.

The grasping arms can be set in a loose-engaging mode during sensing, 3D map building and inspection, where the tips of the arms assist the robot to maneuver around bridge piles without being drifted away by water current. The profile of the grasping arms is designed to be compatible with the most popular pile shapes, i.e., cylinder, square, hexagon, and octagon, with sizes ranging from 350 to 550 mm in diameter. The grasping arms are replaceable in the case of dealing with piles that are smaller than 350 mm in diameter or greater than 550 mm in diameter.

8.3 Sensing and Perception in Underwater Environments

Many sensors (Table 8.1) are used in this robot for navigation, 3D map building, marine growth identification, and environmental and situation awareness. Table 8.1 shows the list of sensors implemented on the robot.

For estimation of the robot state, sensors including inertial measurement unit (IMU), a pressure-based depth sensor, and a short baseline acoustic positioning system are installed in the robot. This combination of sensors is typical for underwater vehicles and performs well with certain limitations including acoustic dead

Table 8.1 Sensors installed in the robot.

Sensor type	Model name
9-DOF Inertial Measurement Unit (IMU)	Advanced Navigation Orientus
Pressure-based Depth Sensor	Bar30-BlueRobotics
Acoustic Positioning System	WaterLinked
Sonar Scanner	Tritech Micron
Stereo Camera	DuoM - Duo3D
Monocular Camera	Blackfly BFS-U3-31D4C-C with 2mm lens

zones created by the narrow structure of bridge piles and the magnetic readings affected by metal structure.

The robot's environmental awareness is achieved by using a sonar scanner to locate bridge piles and avoid obstacles. A stereo camera is used to create 3D maps of pile structures, and a monocular camera to capture surface images for condition assessment. The stereo and monocular cameras have been selected to have a wide field-of-view for capturing images at a close distance to the pile surfaces, which is necessary in turbid waters.

8.3.1 Underwater Simultaneous Localization and Mapping (SLAM) Around Bridge Piles

Vision-based SLAM is implemented to localize the underwater robot in an underwater environment with many bridge piles. Using SLAM, a 3D model of a pile can be created, and high-quality images of the pile can be geo-referenced to the 3D model. The 3D model provides a record of the marine growth on the pile, and the high-resolution images allow inspection for cracks or damage to the pile structure.

The sensors used for 3D mapping and high-quality image capturing are a gray-scale stereo camera and a high-resolution monocular camera. The cameras are mounted to the robot rigidly, and the extrinsic transform between the two cameras are known so that the high-resolution color images from the monocular camera can be registered to the 3D model generated from the stereo images.

The trajectory of the robot's movement around a pile is generated using ORB-SLAM2 on the stereo camera image feed [Leighton, 2019]. ORB-SLAM2 is a publicly available visual-SLAM algorithm that can automatically estimate the camera's movement from a temporal sequence of images. In the underwater scenario, image features of the pile and marine growth tend to be similar, which makes it challenging to match image features detected in overlapping images of the pile. ORB-SLAM2 estimates the camera's motion to predict the position of features in a new image and reduce the number of feature pairs compared. We believe it also improves the accuracy of feature matching for features with low uniqueness.

A dense 3D model of the pile is produced using the keyframe stereo images and camera pose. Individual viewpoint reconstruction is generated for each keyframe and is then merged using the pose of the keyframe to create the dense and complete model of the pile. Hence, a good quality 3D model can be achieved despite the poor quality of input images.

The monocular camera can be used to capture high-resolution images of the pile. The effect of floating materials in the water and sunlight flicker on image quality is reduced by stabilizing the camera by rigidly attaching the robot to the pile with the grasping arms. Multiple images over a few seconds are then captured and merged into a single median image. High-resolution image capture is done while ORB-SLAM2 is in operation so that the pose of the captured image relative to the full 3D model is known.

It is noted that there are several challenges for using vision sensors in an underwater environment, including:

- Limited visibility of 1 meter or less due to highly turbid water
- Varying floating materials
- Sunlight may illuminate the object or dominate the background

To overcome these challenges, a feasible approach is to keep the camera as close as possible to the pile surface and have strong artificial lighting that can overpower the sunlight. The artificial light sources need to be positioned such that the pile surface is well-illuminated and the light rays do not cross the turbid water in front of the camera to minimize back-scattering that affects imaging quality.

8.3.2 Marine Growth Identification

The underwater robot will need to identify marine growth on a pile for planning of cleaning. Marine growth identification is performed by comparing the 3D map created using the Vision-based SLAM with the pile dimensions provided by bridge construction drawings. Figure 8.6 shows a reconstructed 3D map of a pile registered using ICP (Iterative Closest Point) algorithm [Besl and McKay, 1992] onto the known dimension of the pile. After registration, any points on the 3D map outside the pile's dimension are considered marine growth and can be extracted based on the point-to-pile distance (refer to left graph in Figure 8.7). The marine growth points are further filtered based on cluster size being larger than a predetermined size (refer to right graph in Figure 8.7). Figure 8.8 shows a 3D map with marine growth identified (left), against the actual mapped pile (right).

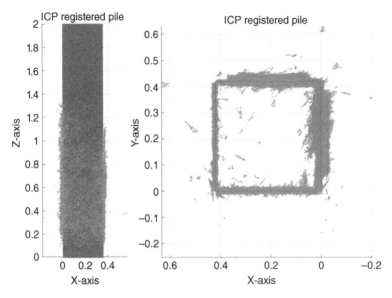

Figure 8.6 ICP-based registration of a 3D map onto a ground truth pile: Side view (left) top view (right).

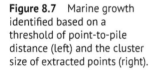

Figure 8.7 Marine growth identified based on a threshold of point-to-pile distance (left) and the cluster size of extracted points (right).

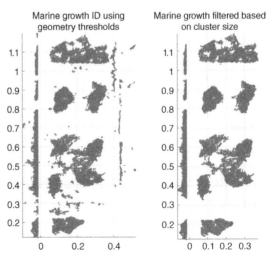

Marine growth regions on 3D map

Regions on actual pile structure

Figure 8.8 Marine growth identification results compared against actual pile structure.

8.4 Software Architecture

The software is developed using the Robot Operating System (ROS), which is open-source and widely adopted in the robotics community. The benefit of using ROS is the availability of hardware support for standard components (e.g. IMU, cameras, and motors) and access to state-of-the-art algorithms. Furthermore, ROS provides suitable hardware abstraction and messaging framework to allow new hardware to be easily integrated into the system.

The software developed for the underwater robot is a set of ROS packages that are custom developed and also taken from the ROS software repository. Figure 8.9 shows the packages in the system (rectangle boxes), the flow of information across the packages (thin arrows), and the external inputs/outputs signals (thick arrows). Each package provides a functionality for the robot, including hardware drivers, state estimation, motion planning, and more. The package "SPIR main" provides a state machine for the cleaning and inspection operations and converts human input from a joystick into relevant lower-level commands for other packages to execute.

8.5 Robot Navigation, Motion Planning and System Integration

8.5.1 Localization and Navigation in Open Water

A controller was developed for navigation of the underwater robot in an open water environment to move toward the target pile autonomously (Figure 8.10) while maintaining the desired depth and stabilizing the attitude (roll and pitch). Measurements from the IMU, depth sensor, magnetometer, and acoustic positioning system are combined to estimate the robot's state using [Le et al., 2015].

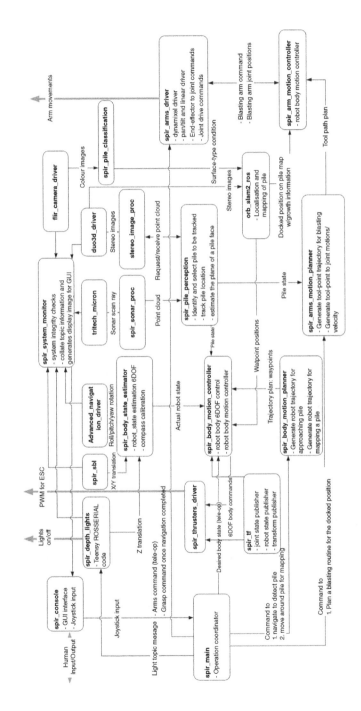

Figure 8.9 Software architecture.

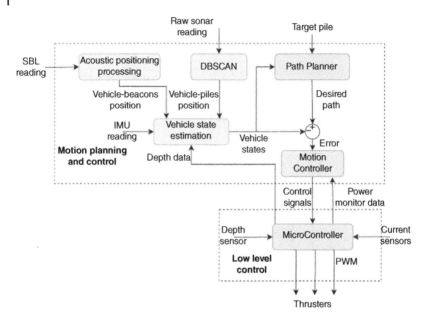

Figure 8.10 SPIR navigation in open water.

The location of underwater bridge piles relative to the robot's position is detected using a sweeping sonar sensor. The sonar's range and sweeping angle are set to 10 meter and 90°, respectively, to provide a view of the objects in front of the robot at a refresh rate of 2.5 seconds. The Density-based Spatial Clustering of Applications with Noise (DBSCAN) algorithm [Ester et al., 1996] is applied to recognize target bridge piles in the 2D sonar scans.

After the operator selects a target pile from the sonar scanning image, the path planner generates a feasible path for the robot to move to the target pile [Yu et al., 2019]. Defining **E** as the errors between the current robot states and the desired path toward the target pile, the goal of the robot motion controller is to generate the control signals to minimize **E** under the disturbance of environments. Several control algorithms such as anti-windup PID controller, reinforcement learning [Wang et al., n.d.], and adaptive controller [Lu and Liu, 2018] were studied and successfully designed.

The control signals generated by the motion controller are translation forces (longitudinal, lateral, and vertical directions) and moments (roll, pitch, and yaw). They are then converted to the thruster control signals using an allocation matrix. An electrical current sensor is used to monitor each thruster so that power overload, especially when working in high-disturbance environments, can be detected promptly.

When the robot is getting close to the target pile, the grasping arms are activated to attach the robot to the pile to form a rigid platform. Then, a cleaning or inspection routine can be started.

8.5.2 System Integration

The overall system of the underwater robot with auxiliary components, including a power supply with a generator, water pumps, and operator interfaces, is shown in Figure 8.11.

The underwater robot has a high-performance embedded computer, installed in the waterproof enclosure, for signal processing, vision processing, and closed-loop control. This embedded computer also interfaces and controls the arm motors, thrusters, and all onboard sensors. The robot is linked to a user interface via a tether cable that delivers power (400 V) and Ethernet. Video stream and signals are transmitted from the embedded computer over Ethernet and displayed on the top-side console monitor for supervisory purposes. The Ethernet connection is also used to transfer data, including sensory signals, robot states, and images logged in the robot's internal hard drive.

A petrol generator is selected over onboard batteries to power the robot to prolong operating time and minimize the robot's weight. As the power is transmitted over 100 m of tether cable, the voltage is boosted from 240 VAC to 400 VDC

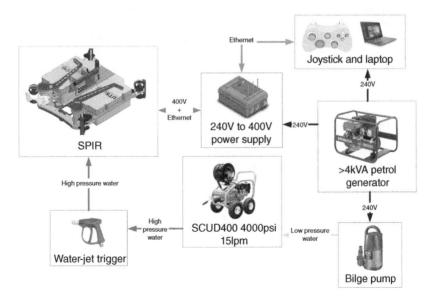

Figure 8.11 General view of SPIR system.

at the power supply box before transmitting to the robot to reduce the thermal dissipation. Here, the voltage is stepped down to 5, 12, and 24 V to be used by devices, including embedded computers, sensors, and actuators.

A high-pressure water pump (Model SCUD400) supplies a water-jet up to 4,000 psi with a 15 l/min flow rate. The high-pressure water hose is coupled to the power supply cable and connected to the blasting hose mounted at the tip of the blasting robot arm. A manual trigger deadman control valve is used to control the blasting process.

8.6 Testing in a Lab Setup and Trials in the Field

The underwater robot was tested in both a lab setup and real-world environments. The lab setup, as shown in Figure 8.12, consists of a $6m \times 4m \times 3m$ water tank and a mock pile structure placed in the tank. The lab setup provided a way to rapidly test new hardware configurations and algorithms, and to perform extensive and repeated testing. Once the robot had been tested to be reliable and robust in lab settings, field trials were conducted to validate the robot's capabilities in real-world settings and to discover new issues.

Field trials were performed on various bridges in New South Wales, Australia, including Peats Ferry Bridge, Mullet Bridge, Narrabeen Bridge, and Windang Bridge, with the support of Transport for NSW (TfNSW). Figure 8.13 shows the robot and research team on sites. Multiple tests were conducted to evaluate the operation procedure and the efficiency of the robot.

The following subsections will discuss (i) the operating procedure for cleaning and inspection, (ii) the trials of autonomous navigation, (iii) the trials of autonomous marine growth removal, and (iv) test results of inspection and marine growth identification

8.6.1 Operation Procedure

The operation procedure of the robot is summarized in the flowchart shown in Figure 8.14. The robot is first deployed into the water from the side of a boat.

Figure 8.12 Lab setup: water tank (left) and the mock bridge pile in the tank (right).

Figure 8.13 Field trial: the team was putting the robot into water next to a bridge.

Figure 8.14 Flowchart of the operation procedure (diagram Source: Le et al. [2020])/IEEE.

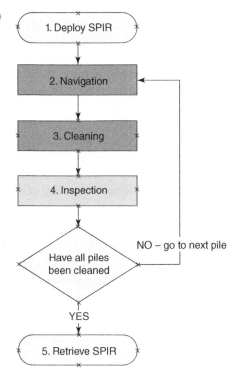

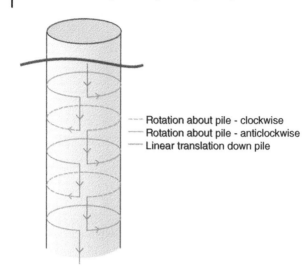

Figure 8.15 Diagram of the robot's spiral trajectory used to cover the pile structure (Diagram Source: Le et al. [2020]).

The robot will then activate the open water navigation mode, which uses the dynamic positioning controller (detailed in Section 8.5.1) to maintain the robot's desired poses against water disturbances. The human operator selects a bridge pile using the console monitor. Then, the robot navigates toward the selected pile and firmly engages it using the grasping arms. Next, the 3D model is created and marine growth removal process is started. To clean the entire piles without having the tether cable wraps around the pile, the robot follows a spiral trajectory, as shown in Figure 8.15, until reaching the preset depth. After completing the cleaning phase, inspection is conducted by following a similar spiral trajectory but in reverse, starting from the bottom and returning to the initial position. Data, such as the robot's trajectory and inspection images of the pile surfaces, is stored for assessment. After the robot returns to the initial position, the operator can select another bridge pile, and the operation procedure is repeated again.

8.6.2 Autonomous Navigation in Narrow Environments

Tests of autonomous navigation were conducted on the Narrabeen Bridge, which is built upon piles covered by pile caps. This environment is complicated as the vertical bridge piles are un-observable from above water due to being covered by the pile caps. At high tide, the vertical bridge pile structures are only visible after submerging 1.5 meter below the water line. This type of underwater structure is difficult for tele-operation and is ideal to use autonomous navigation. Furthermore, the distance between the vertical piles is approximately 2 meter, creating a narrow working environment for the robot.

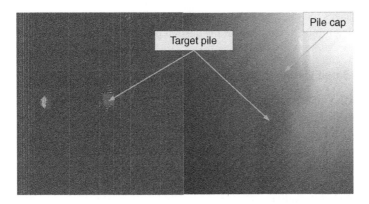

Figure 8.16 Bridge pile detection using sonar scanning.

In this test, the acoustic positioning receivers and the robot were deployed into the water from the side of the boat using a crane (Figure 8.13). Sensor calibration, presented in Section 8.5, was also conducted. Then the dynamic positioning of robot was activated to maintain the position of the robot under the disturbances of water current measured around 0.2-0.4 m/s, while the sonar sensor was used to identify the target pile. Figure 8.16 shows the result from the sonar sensor, showing that a pile was located around 2.5 meter ahead of the robot, which is impossible to be detected using a live video feed due to the turbidity of the water.

Figure 8.17 presents the results showing the navigation performance. It include three phases: finding the target pile, navigating toward the pile, and grasping pile. It can be seen that the robot position fluctuated in the range of ±0.3 m around the desired position when finding the target pile (Phase 1). When the robot

Figure 8.17 Navigation toward the target pile.

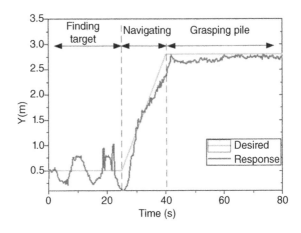

successfully engages the pile using the grasping arm, the robot body is shown to be stable, showing minor fluctuations at the desired pose.

8.6.3 Vision-Based Marine Growth Removing Process

Tests of removing marine growth were conducted at the Windang Bridge (Wollongong, Australia) where the marine growth thickness is more than 20 cm (Figure 8.18). As the growths flourishing on the surface are stiff and condense, the water pressure of 4,000 psi and the big nozzle were used for this scenario. In addition, the desired distance from the nozzle to the pile surface, and the moving speed of the nozzle were set to 5 cm and 20 cm/s, respectively, to maximize the cleaning capability of the robot.

To efficiently remove marine growth from the bridge piles, the distance between the water-jet nozzle and the surface needs to be maintained between 5 and 10 cm. If the distance is too small, the high-pressure water stream could damage the pile structure; vice versa, if it is excessively large, it is not powerful enough to remove hard marine growth such as barnacles, corals, etc. For this reason, the profile of the marine growth on pile surface needs to be identified in order to generate proper trajectory for the nozzle that is mounted to the end-effector of the blasting arm. A stereo camera was used as a cost-effective and lightweight sensor to provide the robot with the 3D surface profile.

In the trials, after the robot firmly grasps the pile, the blasting manipulator which carries the stereo camera is controlled to sweep-scan the pile surface in front to collect depth images. The depth images are then registered to the center of the robot's body and aggregate to form the 3D view of the surface, including marine growth. To prevent damage to the structure and avoid frequent or aggressive deceleration and acceleration due to sharp turns of the cleaning nozzle, a Deformable Spiral Coverage Path Planning (DSCPP) algorithm is utilized to generate a spiral path based on the 3D surface profile constructed from the stereo camera [Hassan and Liu, 2018]. The DSCPP algorithm appropriately generate

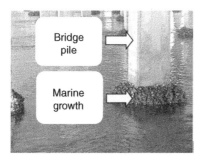

Figure 8.18 Thick marine growth on the Windang Bridge piles.

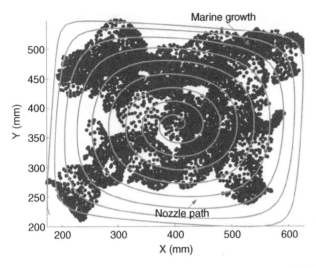

Figure 8.19 Deformable spiral path (diagram source Le et al. [2020])/IEEE.

a smooth spiral path inside a bounding rectangle, hence providing a path with smooth turns and nozzle speeds, while minimizing the path length.

As an example, Figure 8.19 shows a deformable spiral trajectory generated by the DSCPP algorithm for removing the marine growth on the pile. The performance, shown in Figure 8.20, was assessed by a professional diver. The IMU measurement shows that the fluctuation of the robot poses is small (below 0.5°), demonstrating the effectiveness of the grasping arm in maintaining the robot's stability against the reaction force from the water stream and the water current. After cleaning a surface area, the robot will move to the next position and repeat the process.

8.6.4 Inspection and Marine Growth Identification

The inspection and marine growth identification routine were extensively tested on piles at several bridges. The key challenge faced during the site trials was low

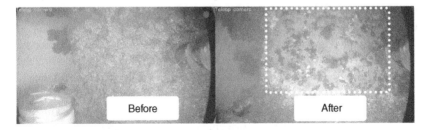

Figure 8.20 A surface before and after cleaning.

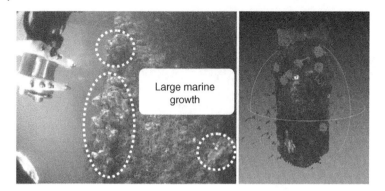

Figure 8.21 (Left) circled regions showing large marine growth on bridge pile, (right) 3D map of the bridge pile and the identified large marine growth (thicker than 50 mm).

visibility (about 50 cm) due to high turbidity caused by a period of rain and vortex water current around the pile that stirs particles around the pile. The robot design and software were thoroughly demonstrated to be capable of addressing this challenge, including the robustness of the stereo and monocular cameras with wide field-of-view to capture images at close viewing distances, the 3D mapping procedure using SLAM described in Section 8.3.1, and the high-resolution image capture procedure is also described in Section 8.3.1.

Figure 8.21 shows the 3D map of a pile and marine growth on the pile at the Narrabeen bridge. The 3D map of the pile is shown in Figure 8.21, and the significant marine growth (thicker than 50 mm) is illustrated in red on the 3D map. The ICP registration result of the 3D map with the ground truth model of the pile that is used to identify the marine growth is also shown in Figure 8.21

8.7   Reflection and Lessons Learned

The robot is designed so that it can be quickly deployed and recovered by two operators with the light duty crane. Extra time, from 20 to 30 min, is required to setup the beacons of the SBL system for the underwater localization.

The robot body is designed to achieve the speed of 4 knots (2 m/s). However, site trial results show that the drag force from the umbilical cables, including the power cable and the high-pressure water hose, creates a significant resistance to the robot motion, hence reducing the maximum speed of the vehicle. The drag force from the cable is highly nonlinear and challenging to be accurately estimated. Therefore, a more effective cable management system is required.

The site trials had to be paused for safety reason when the tidal stream was over 3 knots, as the power supply can be overloaded when all thrusters operate at the full speed for a prolonged period of time to maintain the position against the water current. To improve its capability against higher speed currents, two solutions can be considered: increasing the thruster power supply system or optimizing the robot hull to be more streamline. The former solution requires the use of more powerful thrusters and higher power supply capacity. As the robot speed has the quadratic relationship to the fluid drag force, it is required to triple the thruster power to gain the desired max speed of 4 knots. Optimizing the hull design can reduce water resistance. Several prototypes have been tested in a lab-environment, showing the potential of this solution.

The combination of the short baseline acoustic positioning system with the sonar sensor provides the robot with the ability to autonomously navigate toward a target pile. However, dealing with complex underwater structures such as pile caps, where the vertical piles are hidden beneath a horizontal thick base (cap), is challenging. In this case, the sonar sensor can only detect the cap as a large cluster due to its coarse resolution. To address this issue, robot operators have to manually control the robot to dive to a proper underwater depth where the vertical piles can be detected by the sonar.

The results from the site trials show that the 3-DOF blasting robot arm is relatively effective in removing the marine growth. The photos collected after the water-jet cleaning process are examined by experts, confirming that the core of the pile structure is exposed after being cleaned. The image quality is sufficient for assessing the conditions of the pile structure.

The robot is able to clean a surface area of 40 x 40 cm in about 2-3 minutes on average. The performance rate is slower than those of professional divers. In the long term operation, the robot performance is comparable to the human divers as the robot is able to operate continuously without a rest and recovering time.

8.8 Conclusion and Future Work

In this chapter, we presented the world-first autonomous robotic system for underwater bridge pile cleaning and inspection. The robot hardware, software and the algorithms are proven capable of conducting operations in hash underwater environments. The robot is successfully tested in various field trials, demonstrating that the system is feasible to clean underwater bridge piles and collect high-quality data for condition assessment. The field trial results also show that there is still room for improvement to enhance the robot's capability and performance.

In the future, we hope to see more adoption of underwater robots for underwater structure maintenance, where one operator can supervise multiple robots simultaneously, increasing the operation's efficiency [Hassan et al., 2015].

Acknowledgments

This work was supported in part by the Australian Research Council (ARC) Linkage Project (LP150100935), the Transport for NSW, Australia, and the Robotics Institute at the University of Technology Sydney.

The authors thank Prof. Shoudong Huang, Prof. Kenneth Waldron, Mr Craig Borrows, Dr Wenjie Lu, Dr Marc Carmichael, and Assoc. Prof. Jonghyuk Kim for their contributions to the work.

Bibliography

Yong Bai and Qiang Bai. Chapter 27 - ROV Intervention and Interface. pages 805–833. Gulf Professional Publishing, Boston, MA, 2019. ISBN 978-0-12-812622-6. doi: 10.1016/B978-0-12-812622-6.00027-0. URL http://www.sciencedirect.com/science/article/pii/B9780128126226000270.

P.J. Besl and N.D. McKay. A method for registration of 3-D shapes. *IEEE Transactions on Pattern Analysis and Machine Intelligence*, 14(2):239–256, Feb 1992. ISSN 1939-3539. doi: 10.1109/34.121791.

Andrew Curran, Evan King, Carolyn Lowe, and Brendan O'Connor. Identified hull cleaning robots. Technical report, Worcester Polytechnic Institute, 2016.

Martin Ester, Hans-Peter Kriegel, Jiirg Sander, and Xiaowei Xu. A density-based algorithm for discovering clusters in large spatial databases with noise. Proceedings of the Second International Conference on Knowledge Discovery and Data Mining. pp. 226-231, 1996. URL https://dl.acm.org/doi/10.5555/3001460.3001507.

M. Hassan and D. Liu. A deformable spiral based algorithm to smooth coverage path planning for marine growth removal. In *2018 IEEE/RSJ International Conference on Intelligent Robots and Systems (IROS)*, pages 1913–1918, 2018. ISBN 2153-0858. doi: 10.1109/IROS.2018.8593563.

M. Hassan, D. Liu, G. Paul, and S. Huang. An approach to base placement for effective collaboration of multiple autonomous industrial robots. In *2015 IEEE International Conference on Robotics and Automation (ICRA)*, pages 3286–3291, 2015. ISBN 1050-4729. doi: 10.1109/ICRA.2015.7139652.

S.W. Kelly. Underwater inspection criteria. Technical report, California State Lands Commission, 1999.

Khoa Duy Le, Hung Duc Nguyen, Dev Ranmuthugala, and Alexander Forrest. A heading observer for ROVs under roll and pitch oscillations and acceleration disturbances using low-cost sensors. *Ocean Engineering*, 110:152–162, 2015. ISSN 0029-8018. doi: 10.1016/j.oceaneng.2015.10.020. URL http://www.sciencedirect.com/science/article/pii/S0029801815005570.

Khoa Le, Andrew To, Brenton Leighton, Mahdi Hassan, and Dikai Liu. The SPIR: An autonomous underwater robot for bridge pile cleaning and condition assessment. In *2020 IEEE/RSJ International Conference on Intelligent Robots and Systems (IROS)*, pages 1725–1731, 2020. doi: 10.1109/IROS45743.2020.9341687.

Brenton Lee Leighton. Accurate 3D reconstruction of underwater infrastructure using stereo vision. Master's thesis. University of Technology Sydney, 2019.

J. Lippmann, C. Stevenson, D.McD. Taylor, and J. Williams. Estimating the risk of a scuba diving fatality in Australia. *Diving and Hyperbaric Medicine*, 46(4):241–247, 2016.

W. Lu and D. Liu. A frequency-limited adaptive controller for underwater vehicle-manipulator systems under large wave disturbances. In *2018 13th World Congress on Intelligent Control and Automation (WCICA)*, pages 246–251, 2018. doi: 10.1109/WCICA.2018.8630712.

Günter Schmitt. Global needs for knowledge dissemination, research, and development in materials deterioration and corrosion control. *The World Corrosion Organization*, 2009. URL https://corrosion.org/Corrosion+Resources/Publications/_/whitepaper.pdf.

Satja Sivčev, Joseph Coleman, Edin Omerdić, Gerard Dooly, and Daniel Toal. Underwater manipulators: A review. *Ocean Engineering*, 163:431–450, Sep 2018. ISSN 0029-8018. doi: 10.1016/J.OCEANENG.2018.06.018. URL https://www.sciencedirect.com/science/article/pii/S0029801818310308.

Tianming Wang, Wenjie Lu, and Dikai Liu. Excessive disturbance rejection control of autonomous underwater vehicle using reinforcement learning. Proceedings of the 2018 Australasian Conference on Robotics and Automation (ACRA), New Zealand, 4-6 December 2018, 10 pages.

H. Yu, W. Lu, and D. Liu. A unified closed-loop motion planning approach for an I-AUV in cluttered environment with localization uncertainty. In *2019 International Conference on Robotics and Automation (ICRA)*, pages 4646–4652, 2019. ISBN 1050-4729 doi: 10.1109/ICRA.2019.8794300.

9

Tunnel Structural Inspection and Assessment Using an Autonomous Robotic System

Juan G. Victores*, E. Menendez, and C. Balaguer

Robotics Lab research group within the Department of Systems Engineering and Automation, Universidad Carlos III de Madrid, Spain, 28911

9.1 Introduction

One of the biggest challenges that engineers face nowadays is the inspection of civil infrastructure to ensure that bridges, tunnels, and roads remain in safe condition and continue to provide reliable levels of service. These infrastructures deteriorate with time as a result of natural and artificial factors, the effect of aging, inadequate, or poor maintenance and changes in load criteria. Many tunnels in the United States were constructed over 50 years ago, and some of them have exceeded their intended design service life [Bergeson and Ernst, 2015]. Tunnels of all kinds, including water supply, metro, railway, and road, have increased in both total length and number, and will globally go on. Only in Europe in 2002 did the overall length of operational transportation tunnels had exceeded 15,000 km [Haack, 2002].

Tunnel linings require routine inspections, preferably through nondestructive evaluation (NDE) [Wimsatt et al., 2014] techniques. Several NDE techniques are currently used in tunnel inspection, including visual inspection, impact-echo testing, and ultrasonic methods. These inspection procedures are generally performed by human operators in environments characterized by humidity, absence of natural light, dust, and the existence of toxic substances among others. Manual inspection processes must rely on costly expert trained operators, and are time-consuming in general. Additionally, the human factor combined with the unfriendly environment can lead to lack of guarantee regarding quality control [Balaguer et al., 2014].

*Corresponding Author: Juan G. Victores; jcgvicto@ing.uc3m.es

Infrastructure Robotics: Methodologies, Robotic Systems and Applications, First Edition.
Edited by Dikai Liu, Carlos Balaguer, Gamini Dissanayake, and Mirko Kovac.

In recent years, robotics and automation technologies have gained attention for tunnel inspection and maintenance. Automated systems are expected to provide exhaustive and cost-effective inspections, increasing productivity and safety. Mobile robot-based systems have been developed for off-line detection of defects on the tunnel lining. A mobile robot for detection of deformation by fusing camera and ultrasonic sensor (US) data is reported in Yao et al. [2003]. Yu et al. [2007] present a similar system to detect cracks via computer vision algorithms. A system composed of a truck and a robotic arm equipped with a multi-hammer that generates impact sounds to find cavities and exfoliation areas has additionally been developed [Suda et al., 2004]. Fujita et al. [2012] present a mobile vehicle to perform a hammer-like inspection to detect inner defects in transportation tunnels. The TUNCONSTRUCT teleoperated system [Victores et al., 2011] performs robotic maintenance tasks on the tunnel lining: superficial preparation, fissure injection, and fiber reinforced polymer (FRP) composite adhesion to the tunnel lining.

The previous systems do not allow automated inspection and need to be partially or completely teleoperated. Different from all of the previously mentioned works, this chapter focuses on the development of the final integrated ROBO-SPECT system to perform automated tunnel inspection using NDE methods and data collection [Montero et al., 2015a]. The system includes a vehicle with autonomous navigation, an extended crane, a high-precision positioning robotic arm system and relevant sensors to provide useful inspection data in tunnel environments.

The next sections describe the main hardware and software components of the integrated system. This chapter also presents the mobile vehicle localization and navigation methods, followed by the inspection procedure and the results of experiments performed in real conditions inside a working road tunnel with actual traffic along the lanes. Finally, conclusions are presented.

9.2 ROBO-SPECT Project

The ROBO-SPECT research project was part of the European Commission 7th Framework Program (FP7) under ICT-2013.2.2 on robotics use cases and accompanying measures, that ran between October 2013 and October 2016. This project was conducted by a Consortium of 12 partners, and coordinated by the Institute of Communication and Computer Systems. The main objective of ROBO-SPECT is to provide robotized, reliable and faster tunnel inspection of concrete cracks, and other defects [Montero et al., 2015b]. The solution proposed in this work can perform inspection and detailed structural assessment in one pass and minimally interfering with tunnel traffic. The use of systems such as ROBO-SPECT will decrease the use of scarce tunnel inspectors, while improving the working conditions of such personnel, reduce inspection and assessment costs, and minimize tunnel total of partial closures for inspection.

The ROBO-SPECT robotic system includes different sensors and subsystems to autonomously perform the inspection of the tunnel lining. The system is also supported by three main software components. The first one is the Intelligent Global Controller (IGC), which manages the correct execution of the different tasks of a complete inspection and is located inside the controller PC of the robotic system. The second is the Ground Control Station (GCS), where the operator prepares and monitors the inspection mission. The third one is Structural Assessment Tool (SAT), where the data collected by the robotic system is processed and evaluated. The GCS and the SAT are placed off-board the robotic system.

9.2.1 Robotic System

The ROBO-SPECT robotic system is composed of a mobile vehicle that navigates along the tunnel, an articulated crane capable of reaching up the transportation tunnel's height, and a robotic arm for positioning an US robotic tool to measure the width and depth of detected cracks inside the tunnel. A vision system is used for detecting cracks and other defects on the tunnel lining, and a 3D laser profiler provides data to detect deformations of the tunnel through the SAT. Figure 9.1 depicts

Figure 9.1 The ROBO-SPECT robotic system is composed of a mobile vehicle, an automated crane, a robotic arm, a US tool, and a vision system.

the actual design of the robotic system and its components. The mechanical design of the system is based on the one used in the TUNCONSTRUCT European FP6 project, which used a similar vehicle, crane and robotic arm configuration.

Mobile Vehicle and Crane

The mobile robotic platform chosen is the Genie Z30/20N, an industrial lift vehicle with an articulated hydraulic crane that can reach up to 10 meter in height. This commercial platform was dismantled in order to integrate encoders on the steering and driving wheels and the crane joints, a complete control rack on the back of the vehicle, batteries, cameras for monitoring, and laser sensors for autonomous navigation.

Due to the darkness and similarity that characterize tunnels, the autonomous navigation is based on reflective beacons tracking via a NAV200 navigation laser located at the front of the vehicle. Before the inspection, a set of reflective beacons are placed inside the tunnel in an asymmetric pattern to locate the robot inside the tunnel with enough precision and simultaneously build a map of the navigated section. The majority of tunnel operator's companies can provide a map of the tunnel section to be inspected that can be used by the mobile vehicle. One SICK S3000 laser sensor is placed at the front, and one at the rear part of the vehicle, to provide obstacle avoidance capabilities. In addition, a gyroscope sensor to increase odometry precision is also installed. The selected crane is a 6 degree-of-freedom (DOF) system with independent motion of every joint. Nevertheless, the hydraulic compressor pressure limit defines constraints on the crane motion. Only a single joint can be moved at a time, so a crane trajectory is composed of a sequence of independent joint motions.

The mission of the articulated crane is to position the platform with the robotic arm equipped with the US tool, the 3D vision system, and the 3D laser profiler to collect inspection data. The angular and linear encoders installed on the joints monitor their state for kinematic operations to control the platform position and orientation. Furthermore, these joints are adapted through the installation of custom brakes to reduce the oscillation of the platform. An additional laser sensor is placed in the front of the platform, to perform 3D scans of the tunnel ceiling and avoid collisions during the crane movements.

Robotic Arm

An industrial high-precision robotic arm is placed on the crane platform to position the US tool in the detected cracks with the accuracy needed to take measurements. Figure 9.2 depicts the selected Mitsubishi PA-10 robotic arm, which has 7 DOF to avoid obstacles typically found in tunnels, such as pipes or cables. The workspace of the manipulator covers 1 meter from the base manipulator to the end-effector. A 2D Hokuyo range laser sensor is attached to one of the final links

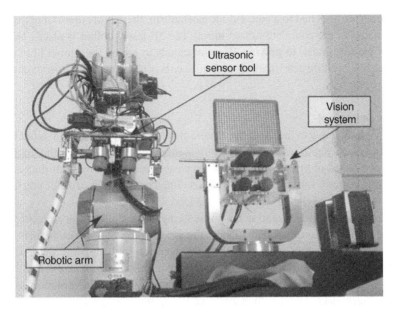

Figure 9.2 The robotic arm with ultrasonic sensor tool and vision system are located on the crane platform.

of the arm. The robotic arm performs a sweep motion to obtain a 3D point cloud of the surroundings of the crack before approaching the wall. A collision-free trajectory to the crack is then computed. Additionally, an IP camera is attached to the tip of the robotic arm to monitor the procedure and provide visual feedback along with the vehicle camera in case teleoperated inspection is used.

Vision System

The robotic inspection is based on a computer vision system composed of two pairs of high-resolution cameras, a lighting system, a pan and tilt mechanism, and a 3D laser profiler. Figure 9.2 depicts these components along with the robotic arm. The first pair of cameras are designed to identify defects commonly found in tunnels like cracks, spalling, and efflorescence. These defects are detected using Convolutional Neuronal Networks trained with data extracted from real images of tunnel defects [Makantasis et al., 2015]. 3D position and orientation of cracks are provided to measure them with the US tool attached to the robotic arm. The other defects are detected at the end of the inspection. The mission of the second pair of cameras is to take stereo images of the crack to be analyzed later.

The lighting system is placed behind the cameras outside of the field of view of the cameras. It provides strong and diffused illumination to take images that are not overexposed or highly shadowed. The cameras and the lighting system behind them are attached to a pan and tilt mechanism, that points these cameras to take images of the complete tunnel lining. The 3D laser profiler is placed on the crane platform next, to provide 3D point clouds to model the tunnel lining and detect small profile deformations with an accuracy of 2 mm in the assessment software.

Ultrasonic Sensor Tool

The US tool is attached to the tip of the robotic arm to be positioned on the tunnel wall during crack depth and width measurements. Figure 9.3 depicts this tool which is composed of three transducers: two piezo-electric ceramic transducers (pulser and receiver) for crack depth, and a sliding contact tip sensor for crack width.

For depth measurements, the crack must be between the two piezoelectric ceramic transducers, and perpendicular to the straight line that connects them. The ceramic pulser sends a US signal to the receiver, and the time delay is measured [Pinto et al., 2007]. These transducers are installed on a small internal 2 DOF XY mechanism to perform several measurements at controlled distances to the crack. The collected data is then analyzed using the US propagation velocity on concrete to calculate the crack depth. Regarding the width measurements, the sliding contact tip sensor is placed on an additional XY positioning system to perform scans of the tunnel surface.

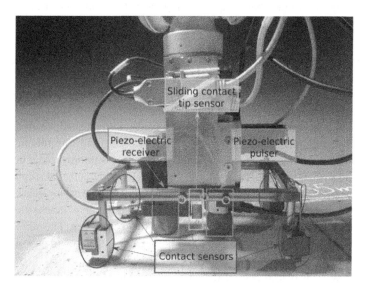

Figure 9.3 US tool positioned on a detected crack to take width and depth measurements.

Due to the contact requirement, the US sensors are developed to be robust to resist collision with the tunnel lining. An aluminum rectangular frame with dedicated contact sensors on each corner is attached to the tip of the robotic arm, because of a stable position during the US measurements is needed. These contact sensors allow for the detection of the tunnel wall during the approach of the US tool.

9.2.2 Intelligent Global Controller

The IGC receives data from the different components of the robotic system (Figure 9.4) and commands the tasks required in the inspection procedure. It controls the mobile vehicle navigation on the road, identifies the detection of a crack, and commands the crane and arm movements to place the US tool on the crack to perform the measurements.

The IGC receives the inspection missions from the GCS and controls the complete system in order to autonomously perform the requested inspection mission. The IGC updates the GCS with the state of the mission and stores the inspection data in a shared folder. This shared folder contains all the data gathered during the inspection, consisting of images of the tunnel lining, laser scanners, robot position, and orientation inside the tunnel, crack measurements, and stereo images of the cracks. The IGC communicates simultaneously with the different components through a local network. YARP [Metta et al., 2006] is used to manage the message exchange with the vision system and the US tool. The mobile vehicle, the crane, the robotic arm, and the GCS communicate with the IGC using a mixed solution of ROS [Quigley et al., 2009] and YARP.

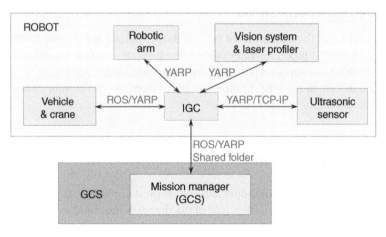

Figure 9.4 The IGC communicates with the different subsystems of the robotic system to perform the inspection.

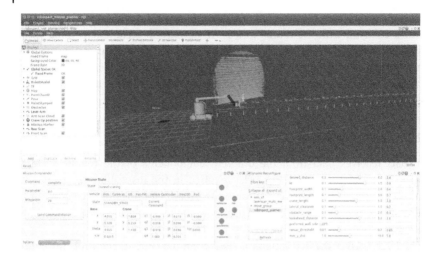

Figure 9.5 The GCS allows the operator to prepare and monitor the state of the inspection mission.

9.2.3 Ground Control Station

The GCS provides the end-user with a graphical interface to prepare the inspection mission: distance to inspect, autonomous or teleoperated inspection, etc. As shown in (Figure 9.5), this interface also monitors the complete state of the robot, displaying the robot model, camera streams and sensor readings on the screen.

9.2.4 Structural Assessment Tool

The SAT receives all the data gathered by the robotic system once the inspection mission has ended. It processes and evaluates the inspection data to provide the end-users with an assessment report. This report includes details and locations of the detected defects in the tunnel lining, images of them, as well as information on tunnel deformation.

9.3 Inspection Procedure

The autonomous inspection procedure is based on the division in to segments of half of the tunnel lining. A complete half of the tunnel is inspected in one pass, and the other half of the tunnel is inspected on a returning pass. The robotic system inspects each segment entirely, and when finished, it advances to the next. The width of each segment depends on the tunnel geometry and the defined user mission. The segments are designated by a user in the GCS, so the system can inspect all the tunnel, or just a few significant segments. During the tunnel inspection, the vehicle advances through the tunnel while maintaining a distance to the wall, using the method described in detail in Section 9.5. The vehicle stops when it arrives at a new segment that needs to be inspected. Afterward, the mobile vehicle will remain static while the inspection of the tunnel segment is performed. On each segment, the crane platform moves to different locations to take images in order to cover the tunnel segment. On each crane location, the pan and tilt mechanism orients the cameras to be perpendicular to the wall before taking images.

The locations computed for a specific installment are shown in (Figure 9.6). The mission is defined as inspecting one-half of the tunnel lining, allowing vehicles drive along the other half. Computation results in two locations to position the crane platform, with four images taken on the first location, and five on the other one. A total of nine images are taken in order to cover all the tunnel lining in this setup.

If a crack is detected in one of the images, the vision system computes the distance from the camera to the crack. The automated crane positions the platform near the crack, with horizontal orientation, and close enough for the robotic arm to reach the wall, while avoiding possible obstacles. While the automated crane is in this configuration, the robotic arm initializes the approaching process described in the flowchart of Figure (9.7).

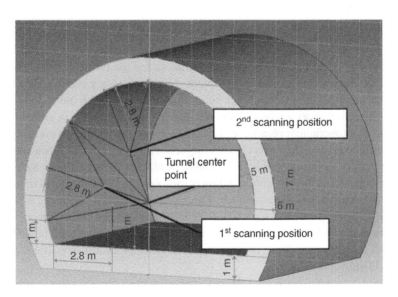

Figure 9.6 Crane platform locations to take images of each segment of the Metsovo tunnel.

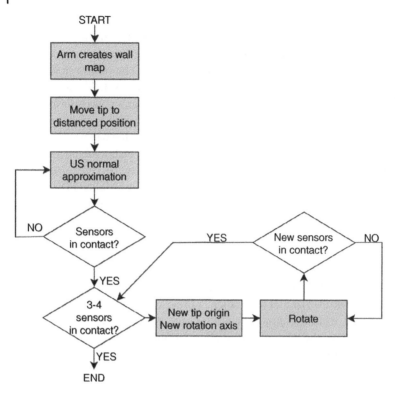

Figure 9.7 General scheme of approaching process of the US Inspection Tool on a crack.

First, the robotic arm performs a sweep motion facing the tunnel lining with the 2D range laser sensor to obtain a 3D point cloud of the crack surroundings. The normal to tunnel lining in the crack point is computed using this point cloud. A collision-free trajectory to position the US tool at a distance of 10 cm away in the normal is autonomously planned and executed. Then, the robotic arm uses the contact sensors to detect the correct positioning of the US tool. The robot advances slowly, and reorients the tip until 3 or 4 contact sensors are in contact with the wall. Figure (9.8) illustrates the final configuration of the robotic arm with the tip in contact with the concrete wall over the crack. Although this process is performed in a complete autonomous mode, it is possible to define the mission for the movements of the robotic arm to be teleoperated by a worker using the on-board camera and the contact sensors as feedback. This was included in the procedure to guarantee safety in difficult cases.

Once the US tool has performed the required measurements, the robotic arm retrieves the sensors from the wall, and the crane returns to the position where the crack was detected by the vision system. From this position, the cameras continue

Figure 9.8 The robotic arm positions the ultrasonic sensor tool to perform crack measurements while in contact with the tunnel lining.

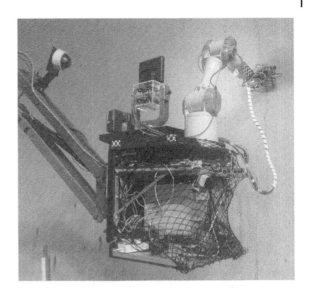

to inspect the rest of the tunnel segment and, when finished, the robot can continue to the next one. Before moving to the next segment, if a crack has been detected, the 3D laser profiler takes a complete scan of the tunnel segment. The mission is finished when every segment has been inspected. The robot returns to the starting point, and the data is then passed to the tunnel assessment tool to be analyzed and stored.

9.4 Extended Kalman Filter for Mobile Vehicle Localization

This section presents the mobile vehicle localization. The SICK NAV200 laser navigation sensor provides position and orientation estimates based on trilateration of fixed reflective beacons. While these beacons have been placed asymmetrically along the tunnel (Figure 9.9), the sensor requires at least three beacons to be detected. Similar to Herrero-Pérez et al. [2013], an Extended Kalman Filter (EKF, as introduced in Chapter 3) that fuses the laser navigation data with robot wheel odometry is used for persistent state estimation.

The EKF filter is a discrete-time state estimator, which linearizes about the mean and covariance of the state to be estimated. It estimates the state of a process modeled by a nonlinear stochastic differential equation $\mathbf{f}$, where $\mathbf{x}_k$ and $\mathbf{u}_k$ are the system state vector and the control vector at time step k, and $\mathbf{w}_k$ is a random variable that represents a zero-mean process noise (Eq. 3.1). It additionally counts on measurement noise, modeled by the functions $\mathbf{h}$ and $\mathbf{v}$ as in Eq. 3.2.

Figure 9.9 The mobile vehicle is able to estimate its position with the reflective beacons placed on both sides of the road tunnel.

The EKF is a recursive cycle with two steps. The first is a **Predict** step, where in this setup the state vector $\bar{\mathbf{x}}_k$ and the error covariance $\bar{\Sigma}_k$ at time step k are estimated using Equations 3.2 and 9.1, respectively.

$$\bar{\Sigma}_k = F_k \bar{\Sigma}_{k-1} F_k^T + R_k \tag{9.1}$$

where F_k corresponds to the linearization of $\mathbf{f}$ in Eq. 3.1, and R_k is the covariance of $\mathbf{w}_k$. The second step is the **Update** step, where $\bar{\mathbf{x}}_k$ is corrected using measurement $\mathbf{z}_k$. The Kalman filter gain K_k in this setup is defined by the Eq. 9.2.

$$K_k = \bar{\Sigma}_k H_k^T (H_k \bar{\Sigma} H_k^T + Q_k)^{-1} \tag{9.2}$$

where H_k is the linearization of function $\mathbf{h}$ of Eq. 3.2, and Q_k is the covariance of $\mathbf{v}_k$. Eqs. 9.3 and 9.4 show the measurement update for $\mathbf{x}_k$ and Σ_k.

$$\mathbf{x}_k = \bar{\mathbf{x}}_k + K_k(\mathbf{z}_k - \mathbf{h}(\bar{\mathbf{x}}_k)) \tag{9.3}$$

$$\Sigma_k = (I - K_k H_k)\bar{\Sigma}_k \tag{9.4}$$

In this mobile vehicle localization case, the system state vector is the current pose $\mathbf{x}_k = [x_k, y_k, \theta_k]^T$, where x_k and y_k represent the position, and θ_k is the heading angle of the vehicle in the reference frame. The current pose is estimated using the previous pose $\mathbf{x}_{k-1}$ and the odometry information in the motion model. An incremental encoder is installed to compute the forward velocity $(\dot{x}_k)$, and an absolute encoder is used to directly measure the steering angle (β). Equation 9.5 is the velocity motion model based on the geometry of motion of the Ackerman vehicle

[LaValle, 2006]. This model is simplified to a two-wheel bicycle model, where L is the distance between rear and front wheels.

$$\begin{bmatrix} x_k \\ y_k \\ \theta_k \end{bmatrix} = \begin{bmatrix} x_{k-1} \\ y_{k-1} \\ \theta_{k-1} \end{bmatrix} + \begin{bmatrix} \dot{x}_k \delta t_k cos(\theta_{k-1}) \\ \dot{x}_k \delta t_k sin(\theta_{k-1}) \\ \dot{x}_k \delta t_k tan(\beta)/L \end{bmatrix} \tag{9.5}$$

Since the navigation range laser SICK NAV200 provides the state to estimate directly, the function $\mathbf{f}$ that relates the estimated state $\mathbf{f}(\overline{\mathbf{x}_k})$ with the estimated measurement $\overline{z_k}$ is simplified to $\mathbf{f}(\overline{\mathbf{x}_k}) = \overline{\mathbf{x}_k} + \mathbf{v}_k$.

9.5 Mobile Vehicle Navigation

The mobile vehicle navigates following the tunnel lining at a constant distance using a frontal SICK S3000 safety laser sensor. A 2D point cloud is obtained with the laser sensor scans. A line is fit to the point cloud via Random Sample Consensus (RANSAC) [Fischler and Bolles, 1981], and a path parallel to this line at a d_w distance is computed.

The Pure-Pursuit method (Figure 9.10) [Coulter, 1992] computes the curvature γ that will drive the vehicle to the point (x_g, y_g) in local coordinates that is a look-ahead distance d_l in the path from the vehicle following Eq. 9.6.

$$\gamma = \frac{-2x_g}{d_l^2} \tag{9.6}$$

Figure 9.10 Schematics of the pure pursuit method for an Ackermann vehicle.

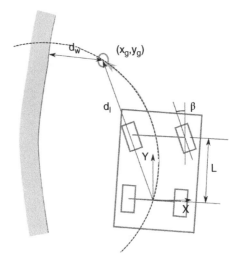

Equation 9.7 is used to compute the correspondent steering angle β_{ref}, where L is the distance between rear and front wheels in X axes.

$$\beta_{ref} = atan\left(\frac{L}{1/\gamma}\right) \tag{9.7}$$

The look-ahead distance is scaled with the longitudinal velocity of the vehicle, and saturated at a minimum and maximum value.

9.6 Field Experimental Results

The field experiments were performed in the Metsovo tunnel in Greece, managed by Egnatia Odos. This tunnel has a 7 meter height and a 10 meter diameter. Figure 9.11 depicts the ROBO-SPECT robotic system inspecting one-half of the tunnel, with ongoing traffic by its side.

First, each of the subsystems were tested individually. The navigation of the vehicle was verified first, with reflective beacons placed inside the tunnel at both sides of the road. Next, the crane movements and the camera acquisition of images were tested, by performing the mission without the US tool measuring enabled. Finally, the robotic arm approaching process was tested in autonomous and teleoperated mode. Timing and accuracy of all of the different processes were annotated from 10 complete autonomous inspections performed on the tunnel during

Figure 9.11 Autonomous inspection with ongoing traffic in the Metsovo tunnel.

Table 9.1 Inspection process timing.

Inspection Process Stages	Time [s]
Vehicle motion to next 1.4 m segment	10
Crane motion to first scanning position	21
Crane stabilization	4
Crack detection in first position	150
Crane motion to second scanning position	35
Crane stabilization	4
Crack detection in second position	120
Crane go to home position	26
Inspect one-half-segment without cracks	**370**
Crane approach to the wall	7
Robotic arm tip touch the wall	15
US Measurements	80
Laser profiler scanning	120
Inspect one-half-segment with one crack	**592**

these final trials. The top part of Table 9.1 summarizes the timing of each inspection process stage of a tunnel segment, where no cracks are present (more usual). The bottom part of the table adds the needed process stages when one crack is detected and measured with the US tool. These results show that the complete inspection with one crack measured can be performed in approximately 10 min per half-tunnel segment. If no cracks are detected, the time spent is about 6 min per segment. Using these numbers and taking into account that cracks are not present on every segment of the tunnel, a full pass (both sides) of a 20 meter section of the tunnel was performed in 2 h and 10 min.

A manual inspection test was performed by two trained engineers during the same tests, in order to compare these results with the traditional manual inspection. The inspection was carried out with manual electronic and ultrasonic equipment. This manual inspection of less than a half-segment (only the lower part of the tunnel wall, 1.5 meters high) takes about 15 min to be completed. Comparing these results with the autonomous robot, it is clear that the ROBO-SPECT system can provide quality inspection at higher locations (up to 10 meter), and with many sensor outputs such as laser readings, location of cracks, images and ultrasonic measurements, all with a timing not higher than the manual inspection.

Finally, the accuracy of the inspection was determined by taking into account the accuracies and precisions of all the different components of the system. Tables 9.2, 9.3, and 9.4 describe the accuracies of each of the ROBO-SPECT components.

Table 9.2 Robotic System accuracy: average precision.

Robotic System Component	Position [mm]	Orientation [°]
Vehicle	25	0.06
Crane	30	0.22
Vision system	10	0.35
Robot laser sensor	2	0.01
Robotic arm	3	0.01
Ultrasonic Sensors	1	0.05

Table 9.3 Robotic System accuracy: component maximum error.

Robotic System Component	Position [mm]	Orientation [°]
Vehicle	38	0.10
Crane	47	0.34
Vision system	16	0.55
Robot laser sensor	3	0.02
Robotic arm	4	0.02
Ultrasonic sensors	2	0.07

Table 9.4 Robotic System accuracy: chain maximum error.

Robotic System Chain	Position [mm]	Orientation [°]
Vehicle	38	0.10
Vehicle + Crane	85	0.44
Vehicle + Crane + Vision	101	0.99
Vehicle + Crane + Vision + Laser	104	1.01
Vehicle + Crane + Vision + Laser + Robotic arm	108	1.03
Vehicle + Crane + Vision + Laser + Robotic arm + Ultrasonic sensors	**110**	**1.10**

The overall absolute precision of the ROBO-SPECT system is described in terms of tunnel global coordinates, which is approximately 11 cm in the worst case. This value refers to the location error of the US on a given position in global coordinates, reflecting the cumulative error from the chain, starting with the vehicle position, crane encoders, vision, laser and arm components. In practice, the US is correctly placed on the cracks due to the autonomous approximation phase in local reference frames.

9.7 Conclusion

In this chapter, a highly complex system to perform transportation tunnel inspection autonomously with operators only monitoring the process has been described. This system was developed within the framework of the EU FP7 ROBO-SPECT project.

The presented solution uses cameras, laser profilers, and an US tool to evaluate cross-section deformations, detect, and measure cracks online, and perform the detection of other defects when the inspection mission has finished. The modular design of the robotic system allows the addition of further NDE inspection methods, as well as the improvement or replacement of system components.

The system has robustly performed a comprehensive inspection during the numerous field tests. The mobile vehicle has demonstrated its capability to autonomously navigate through the real tunnel using its reflective beacons system and the dedicated laser sensors, while avoiding obstacles and maintaining a constant parametrized distance to the wall. The ability of the automated crane to perform trajectories with collision avoidance was tested successfully. The robotic arm has positively demonstrated the placement of the US tool with the required accuracy on detected cracks. The ROBO-SPECT robotic system provides accurate, fast and reliable tunnel lining inspection and assessment with ongoing traffic in safe working conditions.

Bibliography

Carlos Balaguer, Roberto Montero, Juan G. Victores, Santiago Martínez, and Alberto Jardón. Towards fully automated tunnel inspection: A survey and future trends. In *Proceedings of the 31st International Symposium on Automation and Robotics in Construction and Mining (ISARC)*, pages 19–33, Sydney, Australia, 2014. University of Technology, Sydney.

W. Bergeson and S. Ernst. Tunnel operations, maintenance, inspection and evaluation (TOMIE) manual. *Sage Journals*, 2592(1), 2015. doi: 10.3141/2592-18.

R. Craig Coulter. Implementation of the pure pursuit path tracking algorithm. Technical report, Robotics Institute at Carnegie Mellon University, 1992.

Martin A. Fischler and Robert C. Bolles. Random sample consensus: A paradigm for model fitting with applications to image analysis and automated cartography. *Communications of the ACM*, 24(6):381–395, 1981.

Masayuki Fujita, Oleg Kotyaev, and Yoshinori Shimada. Non-destructive remote inspection for heavy constructions. In *CLEO: Applications and Technology*, pages ATu2G–3. Optical Society of America, 2012.

A. Haack. Current safety issues in traffic tunnels. *Tunnelling and Underground Space Technology*, 17(2):117–127, 2002. doi: 10.1016/S0886-7798(02)00013-5.

David Herrero-Pérez, Juan José Alcaraz-Jiménez, and Humberto Martínez-Barberá. An accurate and robust flexible guidance system for indoor industrial environments. *International Journal of Advanced Robotic Systems*, 10(7):292, 2013.

Steven M. LaValle. *Planning algorithms*. Cambridge University Press, 2006.

Konstantinos Makantasis, Eftychios Protopapadakis, Anastasios Doulamis, Nikolaos Doulamis, and Constantinos Loupos. Deep convolutional neural networks for efficient vision based tunnel inspection. In *IEEE International Conference on Intelligent Computer Communication and Processing (ICCP)*, pages 335–342. IEEE, 2015.

Giorgio Metta, Paul Fitzpatrick, and Lorenzo Natale. YARP: yet another robot platform. *International Journal on Advanced Robotics Systems*, 3(1):43–48, 2006.

Roberto Montero, Juan G. Victores, Santiago Martínez, Alberto Jardón, and Carlos Balaguer. Past, present and future of robotic tunnel inspection. *Automation in Construction*, 59:99–112, 2015a. doi: 10.1016/j.autcon.2015.02.003.

Roberto Montero, Juan G. Victores, Elisabeth Menéndez, and Carlos Balaguer. The robot-spect EU project: Autonomous robotic tunnel inspection. In *Robocity2030 13th Workshop EU Robotic Projects Results*, pages 91–100. Leganés, 2015b.

Roberto C.A. Pinto, Arthur Medeiros, I.J. Padaratz, and Patrícia B. Andrade. Use of ultrasound to estimate depth of surface opening cracks in concrete structures. *The Open Access NDT Database*, 2007.

Morgan Quigley, Ken Conley, Brian Gerkey, Josh Faust, Tully Foote, Jeremy Leibs, Eric Berger, Rob Wheeler, and Andrew Ng. ROS: An open-source Robot Operating System. In *ICRA Workshop on Open Source Software*, volume 3, page 5. Kobe, Japan, 2009.

Takeshi Suda, Atsushi Tabata, Jun Kawakami, and Takatsugu Suzuki. Development of an impact sound diagnosis system for tunnel concrete lining. In *Tunneling and Underground Space Technology. Underground Space for Sustainable Urban Development. Proceedings of the 30th ITA-AITES World Tunnel Congress Singapore, 22–27 May 2004*, volume 19, 2004. doi: 10.1016/j.tust.2004.01.026.

Juan G. Victores, Santiago Martinez, Alberto Jardón, and Carlos Balaguer. Robot-aided tunnel inspection and maintenance system by vision and proximity sensor integration. *Automation in Construction*, 20(5):629–636, 2011. ISSN 09265805. doi: 10.1016/j.autcon.2010.12.005.

Andrew Wimsatt, Joshua White, Chin Leung, Tom Scullion, Stefan Hurlebaus, Dan Zollinger, Zachary Grasley, Soheil Nazarian; Hoda Azari, Deren Yuan, et al. Mapping voids, debonding, delaminations, moisture, and other defects behind or within tunnel linings. Number SHRP 2 Report S2-R06G-RR-1. 2014. doi: 10.17226/22609.

Fenghui Yao, Guifeng Shao, Ryoichi Takaue, and Akikazu Tamaki. Automatic concrete tunnel inspection robot system. *Advanced Robotics*, 17(4):319–337, 2003. doi: 10.1163/156855303765203029.

Seung-Nam Yu, Jae-Ho Jang, and Chang-Soo Han. Auto inspection system using a mobile robot for detecting concrete cracks in a tunnel. *Automation in Construction*, 16(3):255–261, 2007. ISSN 0926-5805. doi: 10.1016/j.autcon.2006.05.003.

10

BADGER: Intelligent Robotic System for Underground Construction

Santiago Martínez[1]*, Marcos Marín[1], Elisabeth Menéndez[1], Panagiotis Vartholomeos[2], Dimitrios Giakoumis[3], Alessandro Simi[4], and Carlos Balaguer[1]

[1]*Department of Systems Engineering and Automation, Carlos III University of Madrid, Avd. Universidad, 30, Madrid, Leganés, Spain, 28911*
[2]*Depterment of Computer Science and Biomedical Informatics, University of Thessaly, Papasiopoulou, 2-4, Lamia, Greece, 35131*
[3]*Centre of Research and Technology Hellas, Information Technologies Institute, 6th Km Charilaou-Thermi Road, Thermi-Thessaloniki, Greece, 57001*
[4]*Research & Development, IDS Georadar s.r.l., Via E. Calabresi, 24, Pisa, Italy, 56121*

10.1 Introduction

The development and growth of cities drive the demand to implement new underground service networks in the public and private underground, mainly, in areas with a high density of existing ones. In this way, this need favors creating new systems that can go underground without altering the current service lines.

The traditional approach to access the underground was through open-cut excavations, which require the excavations of large areas, causing damage to infrastructures, traffic issues, or noise among other disturbances. Moreover, it is extremely difficult to be applied in areas with densely buried utilities, a very common situation in the urban environment.

However, particularly in urban regions, underground works favor the need for new boring technologies with extended capabilities and big trenches of ground. This has led to the development of many trenchless excavations. These include micro-tunneling or horizontal directional drilling (HDD) techniques. The improvement of this technology would significantly lower construction costs and significantly reduce time and movement. Many studies and research has proved these savings over time.

*Corresponding Author: scasa@ing.uc3m.es

Infrastructure Robotics: Methodologies, Robotic Systems and Applications, First Edition.
Edited by Dikai Liu, Carlos Balaguer, Gamini Dissanayake, and Mirko Kovac.

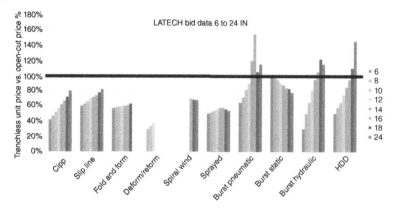

Figure 10.1 Trenchless methods costs vs. open-cut costs. Source: Adapted from LATECH [2003].

The Louisiana Technical Institute (LATECH) has produced reports that present all of the data from bids (successful and unsuccessful) on U.S. municipal pipelines, by plotting each bid against the diameter of the pipe, for each type of technology. Figure 10.1 represents the ratio of the unit cost from various trenchless methods vs. the cost of open cut. The data presented would indicate that in 2003 trenchless technology installation and rehabilitation methods provided significant savings over open cut, of up to 75%, particularly in the smaller pipe diameter ranges. This value is currently applicable. In terms of the surface civil infrastructure, approximately 70% of construction or **direct costs** of open-cut excavations are excavation and refill materials. These costs would be **reduced by 75%**, regarding the study case [Hashemi et al., 2008] because these technologies are less invasive, nondestructive, and minimize interference with current subterranean utilities. The damage in natural environments is significantly less than in the open-cut procedure. Social costs such as the visual impact on the environment have a smaller impact reducing both disruption and stress on the lifestyle [Apeldoorn, 2010]. Low levels of noise, dirt, and vibrations protect residents, nature, and the environment. The disruption and congestion of traffic are minimized by reducing carbon emissions. **Annual traffic congestion** in EU15 in 2010 was estimated at €80B where street works cause 10–20% of congestion costs [ESWRAC, 2004]. **Trenchless methods can save up to 10%** to these congestion and social costs only in Europe. **Green house gas emissions** from trenchless construction methods resulted in **78% or 100% lower** than emissions from open-cut methods [Pemberton, 2009]. Reduction in greenhouse emissions by use of HDD was around 97% according to the study from Ariaratnam and Sihabuddin [2009]. The comparison of emissions between HDD and open-cut works (in pounds (lb) and States Tons (S/T)) is shown in Figure 10.2.

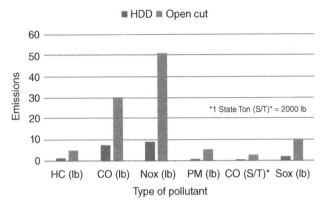

Figure 10.2 Comparison of emissions from the equipment used in both open-cut and HDD construction methods from Ariaratnam and Sihabuddin [2009]. Source: Adapted from Ariaratnam and Sihabuddin [2009].

Population growth and aging underground utility systems force emerging technologies to assist in providing sustainable solutions to address this situation. It is projected that 483,000 km of underground utilities are constructed annually worldwide with a market valuation of $35B [Sterling, 2000]. Trenchless techniques, currently, are used for less than 5% of street works despite their significant benefits over traditional open-cut excavations. These data reinforce the growing need for advanced robotic solutions, such as BADGER, capable of operating in a large variety of trenchless underground applications.

This chapter presents the novel BADGER robotic system for HDD operations. This system is the result of the European Project BADGER: Autonomous unDerGround trenchless opERations, mapping, and navigation, and this chapter presents it as a whole, not done in previous publications. The main part of the BADGER system is the BADGER robot that performs the drilling task. The biomimetic inspiration of an earthworm is reflected in the design of the modules that make up the robot, enabling underground crawling motion. Before describing the BADGER system, an overview of trenchless boring methods is presented. As well, the most similar robotic systems for drilling are presented in relation to BADGER robot design principles.

10.2 Boring Systems and Methods

10.2.1 Directional Drilling Methods

The building or replacement of existing underground infrastructure requires a series of techniques, materials, and equipment that causes minimal disruption to surface traffic, business, and other operations. This set of activities can be defined as trenchless technologies, which are mostly applied to urban areas. HDD, pipe bursting (PB), and pipe ramming (PR) are the most often used trenchless methods for underground service construction in urban regions.

Horizontal Directional Drilling This directional drilling technique is the most widely used trenchless method for installing underground infrastructures such as pipes, conduits, or cable installation. This drilling technique is referred to as directional since it offers the possibility to drill through a relatively shallow radius predetermined route. Usually, the process is divided into three steps. The first stage, called pilot drilling, produces a pilot bore following the desired path (Figure 10.3a).

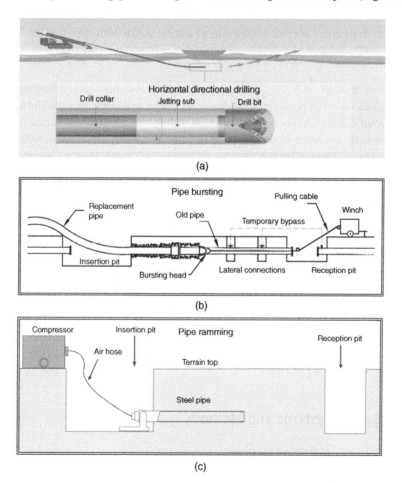

(a)

(b)

(c)

Figure 10.3 (a) HDD method, (b) pipe bursting, and (c) pipe ramming.

The main parameter regarding the path design is the bending radius, which is usually considered 1000 times the pipe diameter. The second stage upsizes the hole by passing a larger cutting tool, the back reamer, whose diameter depends on the size of the pipe to be pulled back through the borehole. The last stage is the pulling process of the pipe, conduit, or set of cables [Hair, 1995].

Pipe Bursting   The repairing of deteriorated and undersized pipes is mainly done through PB procedures, where a bursting head tool is used to both fragment the old pipe and reduce friction in the pulling pipe (Figure 10.3b). The strategy is to fracture the old tube while simultaneously installing the new tube. PB techniques have nevertheless certain weaknesses when handling curved pipes or vast soils that can collapse the conduit hole, but it is one of the most cost-effective approaches when there are few lateral connections and additional room is required [Simicevic and Sterling, 2001a].

Pipe Ramming   PR is a trenchless operation that uses a pneumatic drill to hammer a new pipe or a casing into the ground while removing waste ground to the surface (Figure 10.3c). If the element installed has been a casing, new pipes of other kinds are then placed for underground facilities. The greatest weakness of this process is that it is not steerable [Simicevic and Sterling, 2001b].

10.2.2   Drilling Robotic Systems

Boring technologies are far from being robotized and fully automated. Their operation is usually is semi-automated and still relies on the human experience. There are only a few approaches to developing a fully robotized tunneling machine. Almost all come from research projects to use these machines in spatial environments, such as the excavation of Mars or the Moon to study the subsoil for scientific purposes. Four examples of embedded robotic systems are presented in this section. They all use biomimetic propulsion for vertical drilling but none of them are able to perform HDD, which is one of the main achievements in the BADGER robot presented.

The underground boring robot designed in the Ground Mole Demonstrator European Space Agency (ESA) project in 2005 is the first example, initially intended for space purposes [Campaci et al., 2005]. Figure 10.4a shows the robotic system. The combination of rotative and percussive drilling techniques allowed the application of the system in both soft and tough soils. The propulsion was carried out by a nut screw, and the robot had steering capabilities. The robot was equipped with a localization system, however, does not have perceptional sensors for the detection of obstacles in its surroundings.

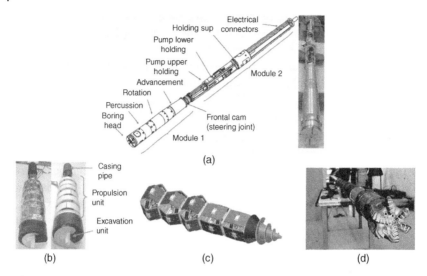

Figure 10.4 (a) Ground Mole robot. Source: [Campaci et al., 2005]/European Space Agency. (b) JAXA drilling robot. Source: [Mizushina et al., 2015]/Taylor & Francis. (c) NASA drilling robot. Source: [Kuhlman et al., 2010]/Johns Hopkins University. (d) Badger-Explorer drilling robot [Helgesen, 2010].

The lunar surface excavation robot with the peristaltic crawling system developed by Chuo, in Japan, and the Japanese Space Agency (JAXA) is the second example coming from the domain of space technologies [Mizushina et al., 2015], shown in Figure 10.4b. This system has no steering capabilities; therefore, no directional boring can be carried out.

The robotic boring worm shown in Figure 10.4(c) is the third example of space technology. It is able to operate in a hazardous environment, using biomimetic peristaltic propulsion created by NASA in 2010 [Johnson et al., 2011]. The boring system is composed of a conical arch with a piezoelectric ultrasonic drill, being the first approach in which ultrasonic drilling was applied to pulverize any hard rock or compact minerals. It has no sensors or steering capabilities.

The fourth case, not for the space environment, is shown in Figure 10.4d by the Badger-Explorer petrol drilling business in Norway [Helgesen, 2010] to drill the bottom of the ocean. The concept is a drilling robot that drills itself down into the ground without the use of a drilling rig, using peristaltic crawling.

10.3 Main Drawbacks

Currently, these technologies require almost obstacle-free space and very good prior knowledge of the subsurface. The human user is typically manually guiding the boring machines by means of a teleoperation user interface. Furthermore, existing trenchless technologies are limited to the realization of only straight-line paths (or performing curves of small curvature over long distances), therefore, their maneuverability and workspace are severely constrained and their applicability is limited to the installation of straight conduits or slightly curved ones, practically taking advantage of only one (1D) of the three possible dimensions of the underground workspace. Consequently, the effect and usefulness of underground technology are actually limited to a very small fraction of the total applications that may possibly be carried out in underground space, and even in these cases, there is much room for development. The fundamental barriers preventing the widespread use of trenchless technologies can be summarized:

No Capabilities for Advanced Maneuverability Existing trenchless technologies lack maneuverability and thus allow drilling paths that are straight or at most can describe a shallow arc of a large radius of curvature over long distances (useful only for digging long curved distances under rivers, etc.), practically offering technology for 1D workspace. Unless curved paths can be drilled, no large-scale topographically complex underground infrastructure can be built. Moreover, unless trenchless technology becomes maneuverable, no autonomous navigation can be executed because the device cannot implement the required motion in order to track the desired path.

No Capabilities for Environment Perception and Mapping Dense obstacle distributions, such as existing buried objects and utilities, require a tunneling machine to get around, go over, or push through diverse terrain conditions and obstacles. Existing devices lack real-time localization and mapping of buried objects, and thus they cannot be navigated unless the environment is fully mapped a-priory and the conditions during drilling are fully predictable (hardly ever the case). As a result, a time/resource-consuming preparation has to be carried out by experts with (manual) radar equipment before the initiation of each drilling procedure, in addition to the (handheld) radar scanning during the procedure.

No Capability for Motion Autonomy and Perception Existing trenchless technologies lack feedback control and trajectory tracking, therefore, in most cases, they operate in an open-loop mode where any feedback and motion compensation involves the human operator, i.e. the operator observes the path covered and corrects by manually sending commands to the underground system, which in turn increases effort, cost, and time.

No Capability for Curved Segment Pipe Installation Today, all installed pipes (even small diameter ones) have to be jacked from the entrance of the bore either by jacking or by hydraulic ramming. Jacking prefabricated linear segments of pipes into a bore allows the construction of only straight pipes and cannot be used for building more complex curved networks. Jacking or hydraulic ramming involves costly infrastructure with a relatively big footprint, which renders this technology inappropriate for low-cost small-bore diameter installations.

Although some of the technological components for overcoming these limitations may exist, an integrated functional, smart robotic system is not there yet; to varying extents, this has long been an obstacle to the realization of the strategic vision of extended underground utilization.

10.4 BADGER System and Components

BADGER stands for **"roBot for Autonomous unDerGround trenchless opER-ations, mapping and navigation"**. The BADGER European Project in the HORI-ZON2020 frame program has been carried out in the period 2017–2020, and its main goal has been the design and development of an "autonomous robotic system that will be able to drill, maneuver, localize, map and navigate in the underground space, and which will be equipped with tools for constructing horizontal and vertical networks of stable bores and pipelines." The BADGER operation concept is depicted in Figure 10.5a–c, and comprises **three main subsystems**. The first one is the **underground robot** that autonomously performs drilling tasks in the subsurface space. The second one is the **surface rover** devoted to subsurface mapping, thanks to the use of Ground Penetrating Radar (GPR) technology. The last one is a monitoring and control station.

The main achievement of this project has been the BADGER underground robot that incorporates novel mechatronic ideas and methods for sensing, mapping, and interpreting the surrounding subterranean environment. The operation of the robotic system is controlled by intelligent functions, based on robot perceptions, which determine the task execution, path, and motion planning. The huge information collected during its operation is applied to constantly enhance perception while offering human users the possibility of storing, processing, and analyzing this information. BADGER underground robot can be defined as the first autonomous robotic prototype to perform low radius curvature HDD with advanced perception capabilities from the sensor data.

10.4.1 Main Systems Description

The Underground Robotic System The BADGER robot is a mole-type tunneling machine, in which the entire mechanical structure is under the ground during operation, having only an umbilical connection to the surface for power supply, data communication, and cuttings extraction.

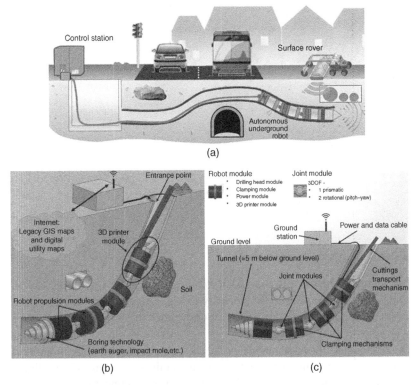

Figure 10.5 BADGER concept. (a) Overview of the concept and drilling goal, (b) and (c) different BADGER system components.

The robot mechatronics is accomplished by adopting a modular hardware architecture that comprises building blocks, three modules, the drill head, the clamping modules, and the joint modules. All these modules are serially connected as shown in Figure 10.6. The modular design is expandable and could allow more joint/clamping modules to be connected serially if required by the application.

The drill head is mainly responsible for cutting the soil. This procedure is performed using a rotating scraper disk. The resulting cutting goes inside the drill head and is extracted thanks to the use of a vacuum system through a 60 mm diameter pipe. In addition, it hosts three GPR components mounted at angles of 120° on the drill-head periphery and whose antennas are inclined toward the front of the robot [Simi et al., 2019]. They compose a partial view at the front with a range of 1.5 m. The shape and the range of this perception are enough to detect pipes underground.

The clamping module is also called the service module. Internally, this module accommodates pneumatic, hydraulic and electrical actuation systems, control electronics, and communication interfaces. Externally, it is equipped with an

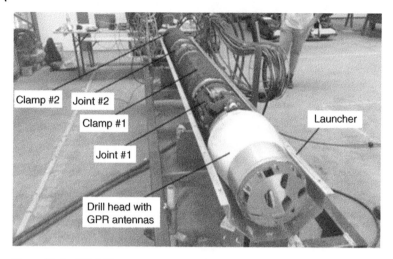

Figure 10.6 BADGER underground system.

airbag to attach the robot to the underground once inflated. Due to the lack of a solid connection to the surface (contrary to, e.g. a classic HDD machine), the force need to counteract the drilling reactions against the bore-hall wall must be provided by the clamping system.

The last module, the joint module, produces the motion of the robot (propulsion and steering). Each joint is composed of two hydraulic subsystems with six linear actuators. Three linear hydraulic cylinders moving at the same time provides one prismatic degree of freedom (DoF) along the principal axis of the robot. The other three hydraulic cylinders can be controlled to provide two steering DoFs for the pitch and yaw angles.

The Surface Rover A commercially available surface radar installed in a rover is employed to facilitate the initial subsurface mapping process, either in an automatic way or manually (see Figure 10.7). The surface rover creates an initial

Figure 10.7 BADGER surface rover.

surface and subsurface map to be used in the tunneling path planning process. The perception of the robot will expand and refine this underground map during the boring process. The surface rover could also be applied in underground robot localization.

The Control System A central computer is used to plan, control, and present the information of a drilling task. The information collected from the surface robotic system is used for planning the trajectory. Once the drilling task is started, underground simultaneous localization and mapping (uSLAM) [Menendez et al., 2019] algorithms perform the required high-level control strategies to follow the path planned. This computing station records all data from the drilling task and provides the human–machine interface (HMI) consisting of a web-based interface for the operator who controls the operation of the robots.

10.4.2 BADGER Operation

The BADGER system operation procedure comprises mainly a sequence of three tasks in order to perform the proper drilling process: surface mapping, path planning, and the drilling process.

Surface Mapping The operator has to select the desired area where the drilling will be performed. With the help of the graphical user interface (GUI) in the Central PC, the area is marked based on 4 selected points under the OpenStreet mapping database [OpenStreetMap Contributors, 2017]. The selected area 10.8 (a) is scanned with the surface rover and the trolley transporting a GPR device. The surface rover is capable to navigate to the designated area and follow a scanning path generated automatically (Figure 10.8b).

The surface rover system is able to produce a 3D volumetric map coupled with underground utility detections with mapping overall error significantly below 10 cm (Figure 10.8c). Hyperbola detection algorithm [Kouros et al., 2018] was able to detect buried utilities and yet in situations where the subsurface humidity was severally increased occasionally failures were also detected (Figure 10.8c). Yet in the overall pipe's registration, such failures were compensated due to point cloud filtering, outliers rejection, and the exploitation of the structural knowledge of the pipe. Most of the hyperbolic signatures are detected and successfully tracked.

Trajectory Generation and Path Planning Having the area scanned and the underground utilities detected, the operator is then able to select the entry and exit points for the underground robot to perform the drilling (Figure 10.9).

Underground global path planning methods that accommodate adaptations to the previously unregistered regions in the metric map. The mapped space has

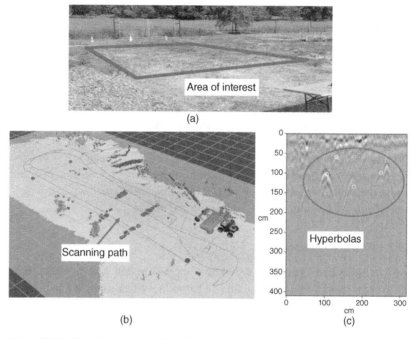

Figure 10.8 Underground mapping: (a) area of interest, (b) automatic path planning, (c) obstacle identification (hyperbolas).

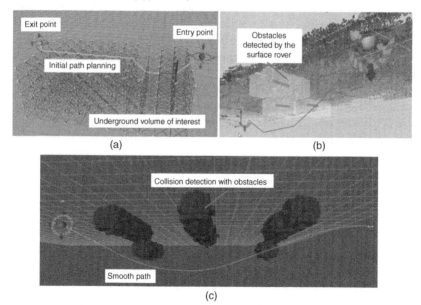

Figure 10.9 BADGER underground robot path planning method. (a) A* planning algorithm of the trajectory, (b) obstacles detected including uncertainty, and (c) Bezier smoothed trajectory and uncertainty boxes for collision avoidance.

been converted into a graph, and the searching algorithm has been implemented to compute the obstacle-free map respecting the robot's kinematics constraints. For path planning the A* (A star algorithm) has been implemented since it is very accurate, fast with high performance, and supports replanning (Figure 10.9a and b). As a final refining step, the path undergoes a filtering step in accordance with which cubic Bezier curves are calculated in order to minimize the curvature of the trajectory (Figure 10.9c).

Drilling Process With the robot placed right in front of the soil, the operator has the choice to perform either manual or autonomous control drilling. Then the operation can start. Once the operation is activated, the robot moves toward the front of the pit (Figure 10.10a and b) to perform the boring task (Figure 10.10c).

Using manual control, the operator can select through the GUI the desired forward and radial velocities, as well as the rotational velocity at the bore head. The operation starts with the user requests execution, followed by the system initialization (both clamps inflated and both joints retracted). The control system automatically generates the gait sequence and executes it following the programmed gait sequence [Vartholomeus et al., 2020].

BADGER motion is earthworm biomimetically inspired, resulting in a net linear propelling motion within the soil as reproduced in BADGER Gait Motion. The operator is able to visualize both numerically and graphically the operation:

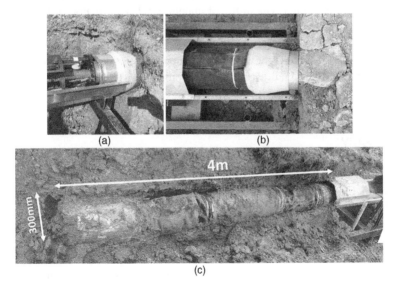

(a)   (b)

4m

300mm

(c)

Figure 10.10 Drilling operation. (a) and (b) Entry points and start drilling, (c) result of a 4-meter drilling test.

The clamps and joints status as well as the advance from the starting point. In addition, a 3D representation of the system is available in the GUI. Robot navigation and localization through the underground are based on the uSLAM technique [Menendez et al., 2019]. The robot-embedded GPR antennas will detect the buried utilities and compare them to the previous scanned subsurface map to locate the robot and avoid new utilities previously undetected following the path planning change as in BADGER trajectory re-planning. This information will also appear in the 3D representation of the system available in the GUI. All the data is recorded and stored in a database as shown in the video BADGER GUI operation.

10.5 Future Trends

PBADGER has been presented as the most advanced robotic drilling solution unique in its class. No other earthworm-like drilling system exists with similar structure, self-motion, sensing, and navigation capabilities.

The earthworm biomimetic design has shown its capabilities to perform directional drilling similar to current HDD techniques. Installation, replacement, or repair of underground utilities or conduits is limited to the realization of only straight-line paths (or performing curves of slight curvature over long distances). Therefore, their maneuverability and workspace are severely constrained, and their applicability is limited. The future of underground construction in Europe will be devoted to the development of the ecological and digital transitions declared in the EU Urban Agenda [European Commission, 2023a,c]. The ecological transition pursues the decontamination and the decarbonization of the continent, and the digital transition is devoted to developing Smart Cities strategies [European Commission, 2023b].

Bibliography

Steve Apeldoorn. Comparing the costs-trenchless versus traditional methods. In *International Society for Trenchless Technology Conference*, Australasian Society for Trenchless Technology, Sidney, Aug 2010.

Samuel T. Ariaratnam and Shaik S. Sihabuddin. Comparison of emitted emissions between trenchless pipe replacement and open cut utility construction. *Journal of Green Building*, 4(2):126–140, 2009.

R. Campaci, Stefano Debei, Roberto Finotello, Giorgio Parzianello, C. Bettanini, M. Giacometti, G. Rossi, Gianfranco Visentin, and Mirco Zaccariotto. Design and optimization of a terrestrial guided mole for deep subsoil exploration - boring performance experimental analysis. In *European Space Agency, (Special Publication) ESA SP*, Aug 2005.

ESWRAC. European Street Works Research Advisory Council, 2004.

European Commission. European digital transition. https://ec.europa.eu/reform-support/what-we-do/digital-transition.en, accessed Jan 2023a.

European Commission. Smart Cities — European Commission. https://ec.europa.eu/info/eu-regional-and-urban-development/topics/cities-and-urban-development/city-initiatives/smart-cities.en, accessed Jan 2023b.

European Commission. Urban Agenda for the EU — Futurium — European Commission. https://ec.europa.eu/futurium/en/urban-agenda/, accessed Jan 2023c.

J.D. Hair. *Installation of pipelines by horizontal directional drilling: An engineering design guide*. American Gas Association, 1995.

Behnam Hashemi, Mohammad Najafi, and Rayman Mohamed. Cost of underground infrastructure renewal: A comparison of open-cut and trenchless methods. In *Pipelines 2008: Pipeline Asset Management: Maximizing Performance of our Pipeline Infrastructure*, pages 1–11. 2008.

Ole K. Helgesen. Badger Explorer has dug down; Badger Explorer har gravd seg ned, Jul 2010.

J.A. Johnson, B. Sanders, and C.L. Carmen. Design and development of a ground based robotic tunnelling worm for operation in harsh environments. In *62nd International Astronautical Congress, Cape Town, ZA*, volume 2, pages 1702–1712, Jan 2011.

Georgios Kouros, Ioannis Kotavelis, Evangelos Skartados, Dimitrios Giakoumis, Dimitrios Tzovaras, Alessandro Simi, and Guido Manacorda. 3D underground mapping with a mobile robot and a GPR antenna. In *2018 IEEE/RSJ International Conference on Intelligent Robots and Systems (IROS)*, pages 3218–3224. IEEE, 2018.

Michael Kuhlman, Blaze Sanders, Lafe Zabowski, and J. Gaskin. Robotic tunneling worm for operation in harsh environments. *Hopkins Undergrad Research Journal*, (12), 50–53, 2010.

LATECH. Louisiana Technical Institute, 2003.

Elisabeth Menendez, Santiago Martínez De La Casa, Marcos Marín, and Carlos Balaguer. uSLAM implementation for autonomous underground robot. In *2019 IEEE SmartWorld, Ubiquitous Intelligence & Computing, Advanced & Trusted Computing, Scalable Computing & Communications, Cloud & Big Data Computing, Internet of People and Smart City Innovation (SmartWorld/SCALCOM/UIC/ATC/CBDCom/IOP/SCI)*, pages 237–241. IEEE, 2019.

Asuka Mizushina, Hayato Omori, Hiroyuki Kitamoto, Taro Nakamura, Hisashi Osumi, and Takashi Kubota. Study on geotechnical tests with a lunar subsurface explorer robot using a peristaltic crawling mechanism. *SICE Journal of Control, Measurement, and System Integration*, 8(4):242–249, 2015.

OpenStreetMap Contributors. Planet dump retrieved from https://planet.osm.org. https://www.openstreetmap.org, 2017.

Kate Pemberton. Trenchless: A greener solution. *Trenchless Australasia*, page 52, 2009.

Alessandro Simi, Davide Pasculli, and Guido Manacorda. Badger project: GPR system design on board on a underground drilling robot. In *10th International Workshop on Advanced Ground Penetrating Radar*, pages 1–9. European Association of Geoscientists & Engineers, 2019.

Jadranka Simicevic and Raymond L. Sterling. Guidelines for pipe bursting. *US Army Corps of Engineers, Vicksburg, Miss. TTC technical report*, 2001a.

Jadranka Simicevic and Raymond L. Sterling. Guidelines for pipe ramming. *Technical Rep.*, pages 1–21, 2001b.

Raymond L. Sterling. The expanding role of trenchless technology in underground construction. *No-Dig*, page 1, 2000.

Panagiotis Vartholomeus, Panos Maratos, George Karras, Elisabeth Menendez, Marcos Marin, Santiago Martínez de la Casa, and Carlos Balaguer. Modeling, gait sequence design and control architecture of BADGER underground robot. *Robotics and Automation Letters with ICRA*, 6(2):1160–1167, 2020.

11

Robots for Underground Pipe Condition Assessment

Jaime Valls Miro

Robotics Institute, University of Technology Sydney, NSW, Australia
AZTI Foundation, Bizkaia, Spain

11.1   Introduction to Ferro-Magnetic Pipeline Maintenance

Nondestructive testing (NDT) or evaluation (NDE) is extensively employed by the energy and water industry to assess the integrity of their network assets. In particular, their larger and most critical conduits (generally referred to as those larger than 350 mm in diameter). This is a decision-making process which informs their renewal/repair/rehabilitation programs. The key advantage of NDT/NDE is that the structure of the asset is not compromised in estimating its condition.

The sensing modality to use is strongly influenced by the material of the asset. Gray cast iron (CI) pipelines remain the bulk of the buried critical water infrastructure in the developed world as that was the material of choice for mass production with the advent of the Industrial Revolution in the middle of the 18th century (alongside its less brittle relative of DI since the 1950s), until carbon steel, asbestos cement, or plastic pipelines (PVC) among other materials made them redundant over the years. The nonhomogeneity of the CI produce means that sensing techniques widely employed in the (mild) carbon steel networks in the energy pipeline sector, such as ultrasonics or electromagnetic acoustic transducers (EMAT), are inadequate for CI, and the underlying techniques of most commercial propositions for CI are instead based on either magnetics (e.g. magnetic flux leakage (MFL), pulsed Eddy currents (PEC), and remote field Eddy currents (RFEC)), or the study of the propagation of pressure waves in the pipeline and/or fluid.

*Email: jaime.vallsmiro@uts.edu.au, jvalls@azti.esThis paper is contribution n° 1191 from AZTI,

Marine Research, Basque Research and Technology Alliance (BRTA)

Infrastructure Robotics: Methodologies, Robotic Systems and Applications, First Edition.
Edited by Dikai Liu, Carlos Balaguer, Gamini Dissanayake, and Mirko Kovac.

11.1.1 NDT Inspection Taxonomy

NDT techniques produce results that tend to be a trade-off between deployment costs and information gain. Local inspection techniques (i.e. 1–3 meters) can provide dense measurements but are time consuming and generally costly per unit length as significant preparatory civil works are required (excavations, supply rerouting, traffic control, etc.). Moreover, inspections can only be undertaken at locations which are accessible from the surface. An example of these tools can be seen in Figure 11.1a.

On the other hand, the taxonomy of long-coverage tools can be broadly split into (a) techniques that provide average pipe wall measurements over longer distances

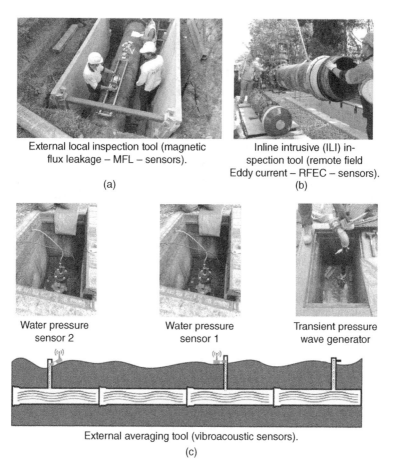

External local inspection tool (magnetic flux leakage – MFL – sensors).

(a)

Inline intrusive (ILI) inspection tool (remote field Eddy current – RFEC – sensors).

(b)

Water pressure sensor 2

Water pressure sensor 1

Transient pressure wave generator

External averaging tool (vibroacoustic sensors).

(c)

Figure 11.1 Example of various configurations of nondestructive testing/evaluation tools.

(generally from a few to 100's of meters, even kilometers). And (b) in-line intrusive (ILI) devices ("smart pigs") deployed inside the pipeline to inspect in higher detail over longer distances (generally 100's of meters to kilometers to make it more cost effective), while propelled by the operating pressure of the fluid.

The former are generally deployed by accessing the external pipe wall or water column at a few access points spread over the length of the pipeline, either through small key-hole excavations or through external access points such as valves or hydrants. As such they tend to have low or no impact in the continuing operation of the pipeline and are more affordable alternatives for condition assessment (CA). An example can be seen in Figure 11.1c. However given the averaging nature of their results, these tools are aimed at providing an initial screening of the condition of an asset and lack the ability to provide the type of detailed geometry information needed to ascertain likelihood of pipe failure.

Flow-driven ILI tools, on the other hand, are inserted into the charged water column either through standard large appurtenances present in critical mains or more often than not via dedicated launch and retrieval mechanisms, as depicted in Figure 11.1b. While these tools are able to provide direct measurements related to the pipe wall condition over long distances, they do so at the expense of higher disruption to the utilities and combined costs from the substantial civil engineering support from the utility prior, during, and post inspection.

ILI tools represent an opportunity given the extended coverage, yet they also exhibit important shortcomings in attaining an accurate depictions of the condition of a pipe wall:

- they are at the mercy of the pressure of the fluid driving them (both in the tethered and free-flowing case).
- should the tools be operated in dewatered conditions, they necessitate complicated winch mechanisms between the entry and exit points.
- operating parameters need to be closely controlled (e.g. tool velocity), meaning that discriminating flow controls need to be in place, not necessarily an easy feat to achieve in a complex interconnected network.
- they lack the ability to do fine control and adjustments for mapping, e.g., ensuring tight tolerances in sensor lift-off, repeatability, or rectifying missed measurements.

11.2 Inspection Robots

The emerging automated field inspection paradigm is driven by internal NDT inspection robotic vehicles, as they are able to:

1. undertake localized, controlled inspections.
2. generate dense mapping suitable for CA and failure prediction.

3. tightly control inherent lift-off during sensing, as induced by the common presence of nonmagnetic cement lining and pipeline wall irregularities.
4. access arbitrary (within tether range) pipeline spools from a single point of entry, hence reducing costs to utilities and allowing inspection of inaccessible sections from the surface (e.g. under a rail pass) and minimizing disruption to customer (e.g. a pipeline under a driveway).

While this solution requires pipes to be dewatered for deployment, this serves a clear mandate from the utility sector that necessitates a robotic NDT inspection vehicle that can be deployed in an opportunistic manner to ascertain the condition of a particular pipeline, specifically when a mains break occurs, or on the back of a valve inspection or repair program when pipelines are inevitably taken offline. Moreover, while time is always at essence in any maintenance and inspection routines, this is particularly the case for critical assets that need to be put back online as soon as feasibly possible. To that effect, an efficient robotic inspection solution with the ability to produce detailed dense maps fitting for pipe failure analysis, even from incomplete inspection data, is highly desirable to minimize collection time and information losses against the original wall thickness maps.

11.2.1 Robot Kinematics and Locomotion

Two broad robotic locomotion mechanism designs have been proven more suitable to align the robotic unit inside a pipeline while placing the sensors along the inner wall for dense CA measurements. The first is based on omnidirectional wheels, while variations about standard rotating wheels conform the other. Some examples are studied next.

Omnidirectional Design

Mecanum wheels enable holonomic robot motion in planar applications as they allow control in all three degrees-of-freedom (DoF) available to the robot [Xie et al., 2015]. In a pipeline context, it is only necessary to control two degrees-of-freedom, longitudinal and circumferential motion. Figure 11.2 depicts the robot concept in more detail (a) and in an open-cut pipeline in the field (b). By applying a nonstandard wheel configuration, it is possible to exploit the unique geometry of the operating environment to passively align with the central pipe axis, automatically tracking the pipe should minor changes in direction occur. Figure 11.3a and Figure 11.3b demonstrate the layout designed to achieve these requirements. In this configuration, the axis of rotation of the pipe contacting rollers all pass through a single point allowing the robot to rotate freely about this point in response to an external force. When resting on a cylindrical surface such as a pipe wall, an external restoring force is generated in response to angular disturbances which acts to return the robot to the aligned position.

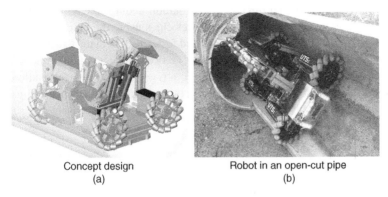

Concept design
(a)

Robot in an open-cut pipe
(b)

Figure 11.2 An omnidirectional NDT inspection robot. Design with two arms mounted (a). Real robot with a single arm (b).

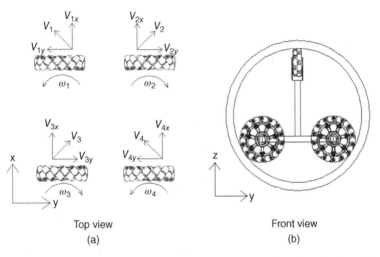

Top view
(a)

Front view
(b)

Figure 11.3 Self-alignment Mecanum wheel layout. Longitudinal wheel spacing along the x-axis prevents angular motion about the z-axis, therefore maintaining a heading along the longitudinal pipeline direction. The pipe wall stops the robot translating horizontally along the y-axis—instead resulting in the full circumferential rotation desired for the pipe inspection task given the additional support of the deployed vertical stability arm as shown in Figure 11.3b, which also allows for consistent surface contact of all the Mecanum rollers (at the expense of increased friction with the pipe walls).

Control in the longitudinal direction and rotation about the circumferential direction are achieved by controlling wheel velocities using the kinematic relations derived in Eq. 11.1, which follow standard forwards kinematic equations in simplified form [Taheri et al., 2015], where $v_x(t)$ reflects the longitudinal velocity (m/s), $v_y(t)$ is the circumferential velocity (m/s), ω_i ($i = 1...4$) is the wheel rotation speed (rad/s), $\omega_z(t)$ denotes angular velocity on the x/y plane, r is the wheel radius (m), and l_x, l_y indicate the wheel separation and body length, respectively.

$$v_x(t) = (\omega_1 + \omega_2 + \omega_3 + \omega_4) \times \frac{r}{4}$$

$$v_y(t) = (-\omega_1 + \omega_2 + \omega_3 - \omega_4) \times \frac{r}{4} \tag{11.1}$$

$$\omega_z(t) = (-\omega_1 + \omega_2 - \omega_3 + \omega_4) \times \frac{r}{4 \times (l_x + l_y)}$$

It is evident from Eq. (11.1) that maintaining zero angular velocity (and therefore a constant heading) is a wheel speed control task. As demonstrated in Figure 11.3b, wheel spacing prevents angular motion about the z axis, therefore maintaining a heading along the longitudinal direction. The pipe wall stops the robot translating horizontally–instead resulting in the circumferential rotation that is desired for this task. It is, however, essential that the angular velocities of diagonally opposite wheels are matched to prevent excessive motor loads precisely given the robot is constrained in the z axis. Driving each pair of diagonally opposite wheels with a single motor would achieve this requirement; however, the required drivetrain is complex and a design with a separate motor per wheel is preferable.

To maintain stability during circumferential rotations, a set of free-wheeling omni-wheels are mounted on a parallel four bar linkage shown in Fig. 11.4a. This is linked to a pair of gas struts to provide a consistent opposing force to the Mecanum wheels surface contact point. The applied force cancels out the gravity vector as the robot rotates, allowing consistent continuous surface contact for each of the wheels rollers. To achieve this, the assembly is pressed against the pipe wall with a preload of approximately twice the robot weight, maintaining control authority regardless of orientation while simultaneously compensating for variation in pipe diameter. A linear actuator is included to retract the omni-wheels from the pipe surface during insertion. This actuator features a spline so that it does not affect the self-correcting behavior of the parallel linkage during normal operation. Figure 11.4b demonstrates the applied force during the robots rotation along the vector F_act.

NDT sensors are coupled to actuated lever arms using a stiff rubber joint. This joint allows the sensor to conform to the pipe surface in the presence of minor irregularities while maintaining a precise placement and contact. The actuators

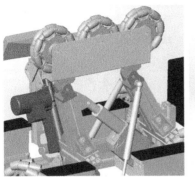

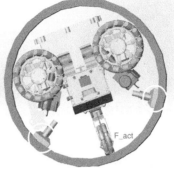

Arm design

Rotated stabilizing arm, with NDT sensors
shown extended in contact with the
pipe wall (detailed in white circle)

(a)

(b)

Figure 11.4 Stabilizing arm.

Concept design

Real robot

(a)

(b)

Figure 11.5 A cart with umbrella sensor arrangement NDT inspection robot.

drive until a stall condition is detected, allowing the sensor to be reliably placed on the pipe surface regardless of pipe variations or actuator drift. In the example presented in Figure 11.4b, two such sensors are mounted on this robot configuration.

Cart with Umbrella Sensor Design

An alternative approach would follow a traditional cart model, with standard rotational wheels on independent motors and non-holonomic constraints. This is a more effective arrangement to allow continuous measurements along the length

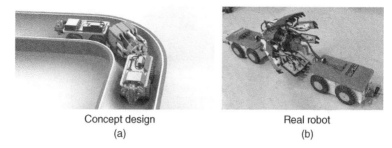

Concept design	Real robot
(a)	(b)

Figure 11.6 A multicart with umbrella sensor arrangement NDT inspection robot. The real robot (b) is shown with the umbrella sensor head expanded.

of the pipeline. The sensors would be mounted on an "umbrella" mechanism to accommodate measurements of pipelines within a given range of diameters. While an Ackerman car-like model could be adopted, a simpler skid-steering non-holonomic model is sufficient as only longitudinal motion along the pipeline is required with this design, with no circumferential motion. An example is shown in Fig 11.5. For larger mains, pipelines are laid fundamentally straight, with only small turns. In that regard, a passive spring-loaded self-turning sensor head can be incorporated in the design. This is the case of the robot depicted in Figure 11.5. For smaller mains, on the other hand, sharper turns up to 90° can be expected in the pipelines, and an alternative design is necessary to accommodate for such constraints. An example following the same cart concept is shown in Figure 11.6, where a dual cart with the sensing unit in the middle is adopted instead, hence permitting taking pronounced turns more easily, such as the 90° bent shown in the synthetic illustration on the left.

11.3   PEC Sensing for Ferromagnetic Wall Thickness Mapping

Recent research in the space of stress analysis and failure prediction of critical CI water mains has revealed that over and above pit depths, as traditionally provided during CA of a critical asset, there is a need to ascertain the presence and geometries of large corrosion patches in the pipe walls [Ji et al., 2017; Kodikara et al., 2016], such as those depicted in Figure 11.12d. There exist a wide range of NDT technologies developed for the purpose of material characterization for CI [Liu and Kleiner, 2013], yet the provision to build dense 2.5D maps of remaining wall geometries for lined water mains has driven the need to design an internal inspection tool around pulsed Eddy current (PEC) sensing technology, as a proven

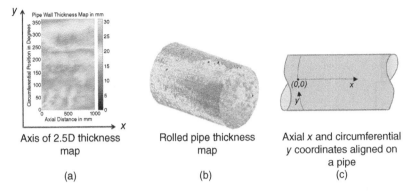

Axis of 2.5D thickness map	Rolled pipe thickness map	Axial x and circumferential y coordinates aligned on a pipe
(a)	(b)	(c)

Figure 11.7 Axial and circumferential coordinates of a 2.5D pipe thickness map.

technique typically used in the NDT sector for ferromagnetic material *thickness* estimation [Huang et al., 2010, 2011; Xu et al., 2012], resilient to sensor lift-off. It is noteworthy to emphasize that while there are a myriad of commercial NDT tools available, they are mostly aimed at visual inspection of an asset, or target carbon steel with the ultimate aim to pinpoint single pitting deficiencies reliably, as that has been the main drive in the energy sector (oil and gas industries), where CA pigging tools and localized inspection devices are more widely used.

Figure 11.7 enables interpreting a typical 2.5D maps of remaining wall thickness as will be depicted later in this chapter, and the conventions shown hold for all thickness maps presented herein. The axial location indicates the distance along the pipe's longitudinal axis, while the circumferential location represents rotational degrees around the pipeline. It should be noted that despite the visual representation of the circumferential dimensions in degrees throughout the chapter to aid the reader's intuitive understanding of cylindrical measurements around a pipeline, in mapping terms they are treated as 2D length measurements in mm in the Euclidean space, for which a single thickness measurement is obtained via the PEC sensor (hence 2.5D maps). The color bar to the right of the thickness maps is a legend representing thickness in mm, between black (0 mm, or a through-hole) and light gray (maximum thickness, 30 mm in the illustration).

A typical PEC sensing system for ferromagnetic materials consists of an exciter coil, a detector coil, a voltage pulse generator for excitation, and an amplifier for the detected signal decay. A representative block diagram of a PEC sensing set up is shown in Figure 11.8a. Amplified decaying signals captured by the detector coil for different CI thicknesses are shown in Figure 11.8b. As reported in the literature, features can be extracted from such signals, which can be directly linked to material thickness [Huang and Wu, 2015; Ulapane et al., 2014, 2017].

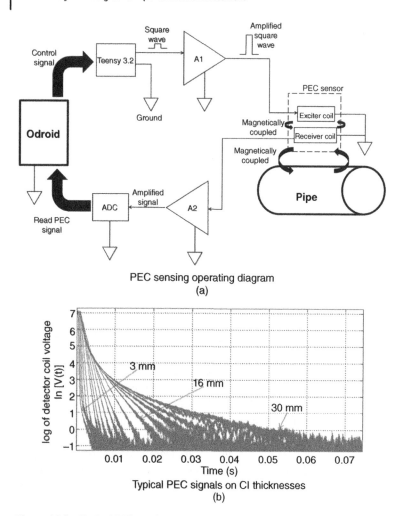

Figure 11.8 Typical PEC sensing setup (a) and signals (b).

11.3.1 Hardware and Software System Architecture

As a way of providing a systemic overview of the architecture of a pipeline CA robotic platform, the specific example of the omnidirectional design presented in Section "Omnidirectional Design" will be described next. The system uses two computers, one onboard the robot for data acquisition and actuator control and one outside the pipe for the user interface. The entire system runs from a generator on the surface with power delivered to the robot with a power over ethernet (PoE) connection. The user interface can receive data and issue control commands

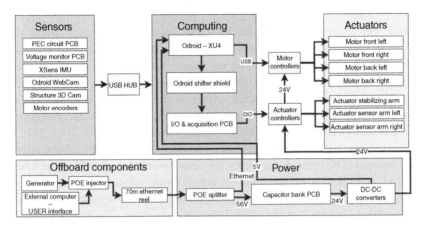

Figure 11.9 System block diagram depicting the four onboard major components on the robotic system (power, sensors, computing, and actuators) and the off-board components linking the robot to a user-driven computer via a 70 meter power over ethernet tether.

Table 11.1 Core component specifications.

COMPUTING	Odroid XU4–Arm based single board computer
SENSORS	Xsens Mti-10 IMU w/ Gyro 450°/s, acc 50 m/s²
	Odroid USB-Cam 30FPS, FOV: 68°
	3D Structure Sensor w/ HFOV: 58° , VFOV: 45°
CONTROL	Maxon Motor DCX26L, gear ratio 231:1, sensor 500 counts/turn
	Sensor Actuator–linear actuator, max force: 1,000 N
	Stability Actuator–linear actuator, max force: 2,500 N
POWER	PoE injector/splitter, 60 W, 70 meter cable reel

back to the operator in real time over the local area network (LAN) connection provided by the same ethernet tether. Figure 11.9 demonstrates the overall hardware system layout, while Table 11.1 lists details about each component.

The long deployment duration precludes battery operation for this particular platform, hence the choice of tethered PoE is to provide significantly longer operation times. In contrast, the traditional cart platform described in Section "Cart with Umbrella Sensor Design" runs on battery power given the faster deployment capabilities of the robot. The POE injector provides 60 W of power, the maximum supplied by readily available off-the-shelf equipment, so an ultracapacitor bank and bespoke charger was developed to supply bursts of high power while ensuring that the PoE equipment maintains an optimal power delivery rate.

While deploying the sensors and taking a reading, the average system power is 40 W. This increases dramatically to 100 W when the motors are driven to reposition the robot. Since the time spent driving the motors is relatively low in comparison to sensing acquisition, the overall average power requirement is less than the 60 W supplied by the PoE system. Thus, the chosen setup provides a steady power supply for the overall system on the condition that high power maneuvers are not sustained for extended periods, as is the case for the inspection of critical water mains which lay flat and straight in the ground.

The onboard ODROID computer system, running Linux, and the robotic operating system (ROS) receives data from the sensor suite through a powered USB hub and controls onboard actuators via digital input/outputs pins. Each sensor has its own monitoring node to manage incoming data and publish to the communication layer. Custom task allocation/behavior nodes then subscribe to the data streams, processing, and publishing control commands as required to the motor and actuator nodes. System control is accomplished using a state machine, which allows both user and autonomous control modes for consistent data retrieval and safe user override. When switched into automatic scanning mode, the circumferential angle and longitudinal position are managed using independent set-point control loops. This simplifies both the kinematics and the algorithms required for control. Controlling the circumferential angle is achieved using the onboard IMU and a standard PID control algorithm. The IMU publishes attitude data to ROS at a fixed rate of 100 Hz. As each data packet is received, the attitude data is transformed into the local coordinate frame to maintain consistency even when the pipe is not leveled. Similarly, longitudinal control is achieved using odometry calculated using encoder readings published at a 100 Hz and filtered to detect wheel stalls. In addition, an overriding human in the loop (HITL) input allows direct control of the longitudinal position. This is used to recover when odometry fails due to motor stalls or excessive wheel slip during the ring-to-ring transitions. A laser distance sensor is utilized to confirm longitudinal position when conditions are safe to deploy in the excavation pits, and there is sensor line-of-sight from the entry point.

11.4 Gaussian Processes for Spatial Regression from Sampled Inspection Data

Whiles accurate and suitable for older ferrous pipes, PEC inspection is in general a slow mapping process technique given the decaying response of the sensing technique. Live deployment of robotic inspection on critical assets such as water infrastructure is necessarily curtailed by reinstating water flows to customers as quickly as possible in a rupture event, or when taken offline for scheduled maintenance.

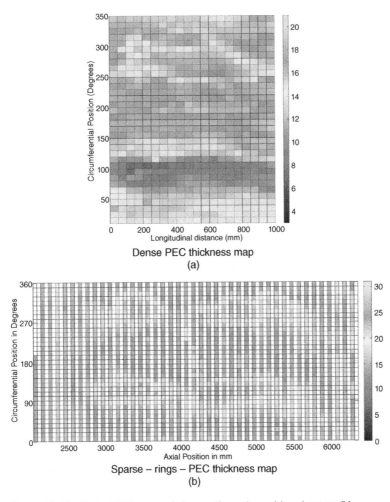

Figure 11.10 Typical PEC scans of pipe sections; dense (a) and sparse (b).

Moreover, service to customers cannot be interrupted for the case of sewerage. It is thus desirable to speed-up the inspection of assets by taking sparse measurements and then regress missing sensor data incurred by the sampling strategy adopted. Two examples of field data collected with the omnidirectional PEC robot described earlier are shown in Figure 11.10, with deliberate gaps in the rings when scanning the pipe spool depicted on the right. In robotics, probabilistic tools such as Gaussian processes (GPs) have been employed in terrain and surface modeling [O'Callaghan and Ramos, 2012; Smith et al., 2010; Vasudevan et al., 2009]. Similarly, tailoring GPs for thickness mapping and fusion to regress missing data has also been proven effective in the domain of pipeline CA [Sun et al., 2015; Vidal-Calleja et al., 2014].

11.4.1 Gaussian Processes

Modeling spatial dependencies for 2.5D data has been largely studied in the past. Mathematically, the problem can be described as a random field, which is a collection of random variables of the form $\{y_x, x \in \mathbf{R}^d\}$, where y_x is the quantity measured at the position $\mathbf{x}$ [Lord et al., 2014]. Random fields are also known as spatial processes, for instance, univariate [Kroese and Botev, 2015] or multivariate [Schlather et al., 2015] processes, that are defined for modeling spatially arranged measurements and patterns. Random fields can be statistically specified by mean and covariance [Lord et al., 2014; Kroese and Botev, 2015]. When the mean is a constant, depending on the covariances there are stationary random fields whose covariances are invariant under translations, isotropic stationary random fields whose covariances are invariant under both translations and rotations, and anisotropic stationary random fields whose covariances are directionally dependent [Lord et al., 2014; Kroese and Botev, 2015]. A more specific type of random field being studied extensively is Gaussian random fields [Davies and Bryant, 2013], which is also known as Gaussian spatial processes [Kroese and Botev, 2015] or Gaussian Process (GPs) [Rasmussen and Williams, 2006; Bishop, 2006].

GPs define the probability distribution over functions, any finite number of which have consistent joint Gaussian distribution. Consider n thickness-location pairs D defined as

$$D = \left\{ (y_1, \mathbf{x}_1), (y_2, \mathbf{x}_2), ..., (y_n, \mathbf{x}_n) \right\}, \tag{11.2}$$

where $\mathbf{x}_i \in X$ is the position in $\mathbf{R}^d$ ($d = 2$ in the case of 2.5D data) where the thickness measurements $y_i \in Y$ were taken. The dataset D is assumed to be drawn from a noisy process

$$y_i = f(\mathbf{x}_i) + \epsilon_i, \text{ where } \epsilon_i \sim \mathcal{N}(0, \sigma_n^2), \tag{11.3}$$

where noise ϵ_i follows independent, identically distributed zero-mean Gaussian with variance σ_n^2. GPs are used to learn the distribution $p(f|X, D)$ from D and have the capability of inferring $p(f|X^*, D)$ for arbitrary location X^*.

Having specified the mean and covariance functions[1] and identified the hyperparameter set θ, parameter estimation can be conducted through optimization by maximizing the likelihood function as described in equation 11.4.

$$\log p(\mathbf{y}|X) = -\frac{1}{2}(\mathbf{y} - m(X))^\top K_y^{-1}(\mathbf{y} - m(X)) - \frac{1}{2} \log |K_y| - \frac{n}{2} \log 2\pi, \tag{11.4}$$

---

1 The terms covariance and kernel function are used indistinctively.

where $\boldsymbol{m}$ and K are the mean and covariance functions, respectively, and $K_y = K(X, X) + \sigma_n^2 I$ denotes the joint prior distribution covariance of the function at positions X. The variance of the noise σ_n^2 constitutes another parameter to be learned together with θ.

Inference at a finite set of query locations X^* can be performed by calculating the predicted mean μ_P and covariance Σ_P:

$$\mu_P = \boldsymbol{m}(X^*) + K(X^*, X) K_y^{-1} (\mathbf{y} - \boldsymbol{m}(X)) \tag{11.5}$$

$$\Sigma_P = K(X^*, X^*) - K(X^*, X) K_y^{-1} K(X^*, X)^{\mathsf{T}} \tag{11.6}$$

The covariance matrix $K(X^*, X)$, obtained from a given covariance function K, is indicative of the cross-correlation between the function at X^* and the training inputs X.

GPs are thus completely specified by the choice of mean and covariance functions. The mean function can be usually set to be a constant value, while the covariance function controls the smoothness of the process, and its parameters govern the effective range of correlation and the variability observed in the data. There is no single covariance function that fits all modeling tasks. Depending on the purpose at hand and any insights that might be available from the underlying physical phenomenon described by the data, modified or composite covariance functions may allow more flexibility in the model. Indeed, the usage of prior knowledge in choosing appropriate covariance functions is encouraged in the literature [Tesch et al., 2011], e.g., using periodic covariance functions in the analysis of seasonal variation and physical phenomena [Rasmussen and Williams, 2006; Tartakovsky and Xiu, 2006].

In pursuit of effectively modeling the wall thickness of buried pipelines, a number of covariance functions have been considered which include characteristics related to the physical properties of the target to be modeled. Commonly used stationary kernel functions – mattern and square exponential, compounded with additional characteristics revealed by the data in terms of directionality and periodicity have been tested [Miro et al., 2018]. These were incorporated in the form of a 2D anisotropic composite covariance function with a periodical wrapping construction. The period in the circumferential direction was clamped to guarantee the 2π periodic property of a pipe wall thickness map. An example with a 2D Matern kernel (v = 3/2) is given by

$$K(X, X^*) = K(r) = \left(1 + \sqrt{3}r\right) \exp\left(-\sqrt{3}r\right) \tag{11.7}$$

where the input distance r is defined by

$$r = \frac{1}{l}\sqrt{(X - X^*)^T (X - X^*)} \tag{11.8}$$

for an isotropic kernel, with l being the length scale, and by

$$r = \sqrt{(X - X^*)^T \wedge^{-2} (X - X^*)} \tag{11.9}$$

for an anisotropic kernel, where $\wedge$ is a diagonal matrix with characteristic length scales l_1 and l_2 on the main diagonal.

In the case of modeling periodical data, an established approach is warping. This is generally done by mapping each one-dimensional input variable x to 2D input variable

$$\mathbf{u}(x) = [sin(x_p), cos(x_p)] \tag{11.10}$$

where $x_p = 2\pi x_p$, and p is the period parameter, hence constructing a covariance matrix $K(\mathbf{u}(X), \mathbf{u}(X^*))$ to turn an anisotropic kernel periodic [Rasmussen and Williams, 2006].

The study on PEC pipeline inspection data show that Matern v3/2 with a combination of anisotropic and periodic composite covariance model add-ons provides the best performance for cylindrical structures such as buried pipes, which corresponds to the observation that pipe wall thickness correlations in the extracted data appear differently in circumferential and axial directions, and the correlation in circumferential direction is 2π periodic. For further details on the GP model selection and analysis of the stability of kernel choices, the reader is referred to Miro et al. [2018] and Shi and Miro [2017].

11.5 Field Robotic CA Inspection Results

Some examples of PEC robotic inspections of CI buried infrastructure are provided. In the first instance, examples with the omnidirectional kinematic architecture described in Section "Omnidirectional Design" are given, followed by the cart with umbrella setup described in Section "Cart with Umbrella Sensor Design". Both have been extensively deployed in the water and pressure sewer mains of a utility in Sydney, Australia [Miro et al., 2014].

The omnidirectional robot was deployed in a 1 km live CI cement lined (CICL) pipeline. An example of an inspection plan is shown in Figure 11.11. Pipe sections

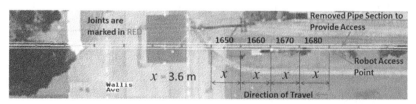

Figure 11.11 A typical inspection plan supplied to the utility partner for the deployment of the NDT inspection robot.

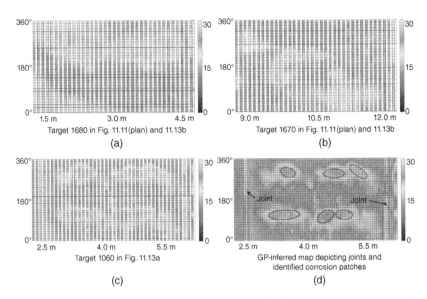

Figure 11.12 Various examples of remaining pipe wall thickness maps as measured by the omnidirectional robotic wall inspection during field deployment on a buried critical water main. Distances shown are with respect to the edge of closest access point.

between 3 and 4 meter in length were targeted for scanning by inserting the inspection robot through a removed pipe section, be that a previously replaced section, as shown in Figure 11.1b, or a new cutout. After reaching the target spool, circumferential and longitudinal ring inspections were undertaken as described in Section 11.2.1 to generate maps such as those depicted in Figure 11.12. Following the inspection pattern ascertained in Shi and Miro [2017], circumferential rings 100 mm apart in axial distance were evaluated with the robotic platform, which given the 50 mm sensor footprint effectively meant skipping every other ring, thus doubling the inspection rate and corresponding time savings. Examples of these are shown in Fig 11.12a, 11.12b and 11.12c. Adopting the GP spatial statistics learning scheme described in Section 11.4, a dense map for each inspected spool was inferred on the missing rings. An example of the final outcome achieved is shown in Figure 11.12d, where measurements indicative of the lead joints are also shown. The spread of the reduction is clearly evident where wall loss is present and can therefore be identified and measured. Such patches are modeled as ellipsoids (also depicted in Figure 11.12d overlaid over the reconstructed map), and their defining parameters can then be incorporated for stress calculation and remaining life prediction of the asset [Ji et al., 2017].

Additional thickness maps gathered by the NDT robot during an extensive period of deployment between 2016 and 2018 are depicted in Figure 11.13,

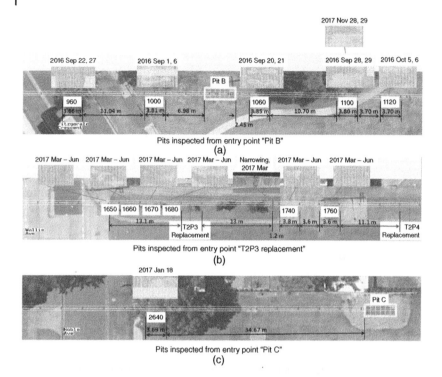

Figure 11.13 Final 2.5D spool thickness maps attained from the field deployment of the NDT robotic inspection robot on a buried critical water main in Sydney Three sections closer to the deployment entry points are shown. In the middle figure, two access points (T2P3 and T2P4) were required given an unexpected narrowing found in the pipeline.

referenced on an aerial picture showing the location where the pipeline is buried. The layout of the pipeline spool structure is also shown in yellow, where white lines are illustrative of the pipeline diameter, and red segments identify spool joints, with the spool number inspected with the robot labeled accordingly. The date of inspection has also been added for context.

CA inspections deployments of the umbrella design robot in the utility partner network are ongoing, covering various kilometers so far in the span of 3 years. Some examples of the maps recovered are shown in Figure 11.14. As described, the mechanical design allows for continuous measurements to take place, as illustrated by the data which in this case depicts two sections of around 25 meter each – part of the overall traversed pipelines. It can be seen how the gaps left by the spread of the umbrella sensor head occur in the longitudinal direction. The corresponding dense GP-inferred maps are depicted alongside.

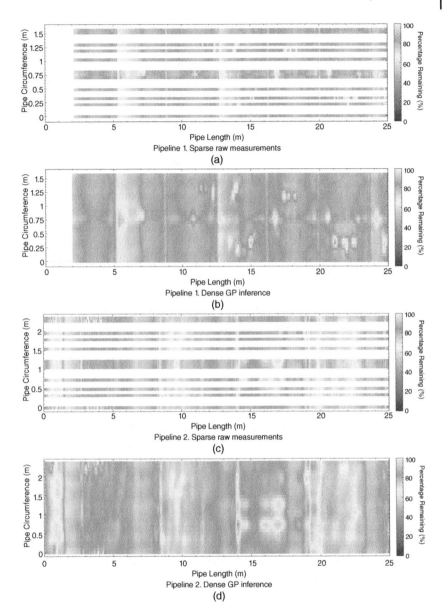

Figure 11.14 Sparse and dense remaining pipe wall thickness data maps at two inspection sites, Pipeline 1 (a, b) and Pipeline 2 (c, d). Measured by the cart with umbrella PEC sensor head robotic wall inspection during field deployment on a buried critical water main. As observed, measurements are taken continuously while the robot traverses the pipeline longitudinally.

11.6 Concluding Remarks

Maintenance of buried infrastructure is a necessary yet expensive and hazardous task. Utilities around the world face the same challenges, made worse in aging cities where infrastructure is made up of older ferromagnetic materials such as CI. It is not uncommon for a large percentage of the asset inventory to be run to failure given the lack of affordable and pragmatic solutions. But for the most critical infrastructure subset, this is not an option. Novel robotic solutions for pipeline CA are emerging to address a utility sector need for automatic, reliable NDT inspections that can report dense pipe wall thickness discrimination as prescribed by failure prediction analysis models. Manageable tools that can be deployed in an opportunistic manner – e.g., when a mains break occurs, or during valve inspection or repair programs when pipelines are discharged and access made available.

An embedded NDT sensing solution based on PEC data measurements – proven unsusceptible to a extended sensor lift-off bound, as typically found in cement lined water and force sewerage mains – has been proven a suitable sensing solution for collecting remaining thickness data from inside pipelines, unlike traditional manual operations carried out externally at spot excavations. Incorporation of these type of sensors into kinematic locomotion robot designs that optimize mobility in such tubular environment has been presented as an attractive and feasible proposition to automate CA. This allows both planned maintenance to occur, as well as rapid, opportunistic deployments after a failure. Time is of essence in the latter instance in particular, as utility services to customers need to be restored promptly, so robot kinematic designs such as the cart arrangement with umbrella sensor head presented are emerging as the preferred choice by utilities. Moreover, a Gaussian processes framework tailored for pipeline geometries to regress missing sensor data constitutes a practical proposition to further speed up the inspection process. Evaluations of the novel robotic CA tools during field deployments have demonstrated the feasibility of these robotic and sensing configurations to provide meaningful 2.5D geometric maps. The data gathered represents not only a visual understanding of the condition of the pipe for asset managers but also constitutes a quantitative measure of the current remaining wall thickness of an asset, a critical input for remaining life calculations that define the likelihood of a pipeline to be left in situ, or be scheduled for renewal or repair instead.

Bibliography

Christopher M. Bishop. *Pattern recognition and machine learning*, chapter Kernel Methods, pages 291–324. Information Science and Statistics. Springer-Verlag, New York, 2006.

Tilman M. Davies and David Bryant. On circulant embedding for Gaussian random fields in R. *Journal of Statistical Software*, 55(9):1–21, 2013.

Chen Huang and Xinjun Wu. An improved ferromagnetic material pulsed eddy current testing signal processing method based on numerical cumulative integration. *NDT & E International*, 69:35–39, 2015. doi: 10.1016/j.ndteint.2014. 09.006.

Chen Huang, Wu Xinjun, Xu Zhiyuan, and Yihua Kang. Pulsed eddy current signal processing method for signal denoising in ferromagnetic plate testing. *NDT & E International*, 43(7):648–653, 2010.

Chen Huang, Xinjun Wu, Zhiyuan Xu, and Yihua Kang. Ferromagnetic material pulsed eddy current testing signal modeling by equivalent multiple-coil-coupling approach. *NDT & E International*, 44(2):163–168, 2011.

Jian Ji, D.J. Robert, Chunshun Zhang, David Zhang, and Jayantha Kodikara. Probabilistic physical modelling of corroded cast iron pipes for lifetime prediction. *Structural Safety*, 64:62–75, 2017.

Jayantha Kodikara, Jaime Valls Miro, and Robert Melchers. Failure prediction of critical cast iron pipes. *Advances in Water Research*, 26(3):6–11, 2016.

Dirk P. Kroese and Zdravko I. Botev. *Spatial process simulation*. Stochastic Geometry, Spatial Statistics and Random Fields. Lecture Notes in Mathematics, vol. 2120. Springer, Cham, Switzerland, 2015.

Zheng Liu and Yehuda Kleiner. State of the art review of inspection technologies for condition assessment of water pipes. *Measurement*, 46(1):1–15, 2013.

Gabriel J. Lord, Catherine E. Powell, and Tony Shardlow. *An introduction to computational stochastic PDEs*, chapter Random Fields, pages 257–310. Cambridge University Press, 2014.

Jaime Valls Miro, Jeya Rajalingam, Teresa Vidal-Calleja, Freek de Bruijn, Roger Wood, Dammika Vitanage, Nalika Ulapane, Buddhi Wijerathna, and Daoblige Su. A live test-bed for the advancement of condition assessment and failure prediction research on critical pipes. *Water Asset Management International*, 10(2):3–8, 2014.

Jaime Valls Miro, Nalika Ulapane, Shi Lei, Dave Hunt, and Michael Behrens. Robotic pipeline wall thickness evaluation for dense nondestructive testing inspection. *Journal of Field Robotics*, 35:1293–1310, 2018. doi: 10.1002/rob.21828.

Simon T. O'Callaghan and Fabio T. Ramos. Gaussian process occupancy maps. *International Journal of Robotic Research*, 31(1):42–62, 2012.

Carl E. Rasmussen and Christopher K.I. Williams. *Gaussian process for machine learning*, chapter Regression, pages 7–32. MIT Press, Cambridge, 2006.

Martin Schlather, Alexander Malinowski, Peter J. Menck, Marco Oesting, and Kirstin Strokorb. Analysis, simulation and prediction of multivariate random fields with package randomfields. *Journal of Statistical Software*, 63(8):1–25, 2015.

Lei Shi and Jaime Valls Miro. Towards optimised and reconstructable sampling inspection of pipe integrity for improved efficiency of non-destructive testing. *Water Science and Technology: Water Supply*, 18(2):515–523, 2017.

Mike Smith, Ingmar Posner, and Paul Newman. Efficient non-parametric surface representations using active sampling for push broom laser data. In *Robotics: Science and Systems*, 2010.

Liye Sun, Teresa Vidal-Calleja, and Jaime Valls Miro. Bayesian fusion using conditionally independent submaps for high resolution 2.5 D mapping. In *IEEE International Conference on Robotic and Automation*, pages 3394–3400, 2015.

Hamid Taheri, Bing Qiao, and Nurallah Ghaeminezhad. Kinematic model of a four Mecanum wheeled mobile robot. *International Journal of Computer Applications*, 113:6–9, 2015.

Daniel M. Tartakovsky and Dongbin Xiu. Stochastic analysis of transport in tubes with rough walls. *Journal of Computational Physics*, 217(1):248–259, 2006.

Matthew Tesch, Jeff Schneider, and Howie Choset. Using response surfaces and expected improvement to optimize snake robot gait parameters. In *IEEE/RSJ International Conference on Intelligent Robots and Systems*, pages 1069–1074, 2011.

Nalika Ulapane, Alen Alempijevic, Teresa Vidal-Calleja, Jaime Valls Miro, Jeremy Rudd, and Martin Roubal. Gaussian process for interpreting pulsed eddy current signals for ferromagnetic pipe profiling. In *IEEE Conference on Industrial Electronics and Applications*, pages 1762–1767, 2014. doi: 10.1109/ICIEA.2014.6931453.

Nalika Ulapane, Linh Nguyen, Jaime Valls Miro, and Gamini Dissanayake. Designing a pulsed eddy current sensing setup for cast iron thickness assessment. In *IEEE Conference on Industrial Electronics and Applications*, pages 892–897, 2017.

Shrihari Vasudevan, Fabio T. Ramos, Eric Nettleton, Hugh Durrant-Whyte, and Allan Blair. Gaussian process modeling of large scale terrain. In *IEEE International Conference on Robotic and Automation*, pages 1047–1053, 2009.

Teresa Vidal-Calleja, Daobilige Su, Freek De Bruijn, and Jaime Valls Miro. Learning spatial correlations for Bayesian fusion in pipe thickness mapping. In *IEEE International Conference on Robotics and Automation*, pages 683–690, 2014. doi: 10.1109/ICRA.2014.6906928.

Li Xie, Christian Scheifele, Weiliang Xu, and Karl A. Stol. Heavy-duty omni-directional Mecanum-wheeled robot for autonomous navigation: System development and simulation realization. In *International Conference on Mechatronics*, pages 256–261, 2015.

Zhiyuan Xu, Xinjun Wu, Jian Li, and Yihua Kang. Assessment of wall thinning in insulated ferromagnetic pipes using the time-to-peak of differential pulsed eddy-current testing signals. *NDT & E International*, 51:24–29, 2012. doi: 10.1016/j.ndteint.2012.07.004.

12

Robotics and Sensing for Condition Assessment of Wastewater Pipes

Sarath Kodagoda[1]*, Vinoth Kumar Viswanathan[1], Karthick Thiyagarajan[1], Antony Tran[1], Sathira Wickramanayake[1], Steve Barclay[2], and Dammika Vitanage[2]

[1] *iPipes Lab, Robotics Institute, Faculty of Engineering and Information Technology, University of Technology Sydney, 15 Broadway, New South Wales, Sydney, Australia, 2007*
[2] *Sydney Water Corporation, 1 Smith Street, New South Wales, Parramatta, Australia, 2150*

12.1 Introduction

In today's civilized world, a functional network of underground wastewater pipes is an essential part of the infrastructure for both developed and developing nations. The concrete wastewater pipelines that transport discharges from residential and industrial regions are old and degrading in many nations, including Australia. This is primarily due to microbial-induced corrosion [Raju et al., 2022], a century-old problem for water utilities worldwide. According to Wu et al. [2020], microbial-induced corrosion is a four-stage process. Sulfates in the waste stream are converted to aqueous hydrogen sulfide in Stage 1 by sulfate-reducing bacteria living in biofilms beneath the waterline. In Stage 2, aqueous hydrogen sulfide is released into the wastewater pipe atmosphere as gaseous hydrogen sulfide due to turbulent wastewater flow inside the pipes. Aerobic sulfur-oxidizing bacteria convert gaseous hydrogen sulfide into sulfuric acid moisture on wastewater pipe walls during Stage 3. In Stage 4, the produced sulfuric acid penetrates the pipe's cement material and reacts with the concrete matrix, causing progressive disintegration of the material and a chemical attack on the reinforcing bars. Failure to restrict this microbial-induced corrosion process at the appropriate time can result in catastrophic pipe failures, with severe economic consequences, environmental hazards, and negative societal impacts.

*Corresponding Author: Sarath.Kodagoda@uts.edu.au

Infrastructure Robotics: Methodologies, Robotic Systems and Applications, First Edition.
Edited by Dikai Liu, Carlos Balaguer, Gamini Dissanayake, and Mirko Kovac.
© 2024 The Institute of Electrical and Electronics Engineers, Inc. Published 2024 by John Wiley & Sons, Inc.

Millions of kilometers of concrete wastewater pipe systems exist throughout the world [Wu et al., 2020]. The wastewater pipe infrastructure assets are currently valued at over $1 trillion and $100 billion, respectively, in the United States and Australia [Thiyagarajan et al., 2018b]. According to Romanova et al. [2014], the cost of the damages to the wastewater pipe assets, mostly due to corrosion, is projected to be £104 billion annually in the United Kingdom, £4 million per year in Belgium, US $36 billion per year in the United States, and 100 million euros annually in Germany. Sydney Water Corporation alone manages more than 300 km of concrete wastewater pipes spanning from Greater Metropolitan Sydney, the Illawarra, and the Blue Mountains. They spend around $30 million a year on repairs for the concrete wastewater pipes that are most affected by microbial-induced corrosion. For water utilities like Sydney Water, periodic inspection and condition assessments of the corrosion state of concrete pipes are essential to prevent catastrophic pipe breakdown events and save millions of dollars through targeted pipe renewals.

The majority of water utilities around the world, including Sydney Water Corporation, have adopted the destructive process of concrete wastewater pipe core sampling [Taheri et al., 2018] in order to evaluate the state of concrete pipe corrosion via human-based inspections in larger wastewater pipelines. Asset inspectors are confronted with a situation where they face an increased risk of occupational health and safety hazards as a direct consequence of this inspection. The water utilities conduct those inspections despite the risks they pose to the field workers for a greater cause in order to safeguard the public's health and safety from the detrimental consequences of wastewater pipe failure. The iPipes research team has developed a drill resistance sensing system [Giovanangeli et al., 2019] based on a mechatronics approach to measure corrosion conditions. Despite the fact that this method is destructive, it offers an in situ corrosion evaluation process that takes less time, which is helpful for field inspectors. A predictive model for predicting wastewater pipe corrosion throughout the network was developed using long-term environmental conditions [Thiyagarajan et al., 2018b, 2020; Montazeri et al., 2018; Rente et al., 2021] of the wastewater pipe as data inputs for accurate corrosion prediction. However, due to adverse environmental conditions within wastewater pipes, sensors that monitor critical environmental parameters on a long-term basis can fail [Thiyagarajan et al., 2018a]. Two-dimensional (2D) images for manual or automated analysis through closed-circuit television (CCTV) cameras are somewhat useful for monitoring corrosion conditions [Haurum and Moeslund, 2020; Wang et al., 2023], however, it does not provide information to infer the depth of corrosion. As a result, it is essential to transition from 2D image-based corrosion condition assessments to three-dimensional (3D) depth-based inspection, aided by recent improvements in reliable, low-power depth sensors. In the field of wastewater pipe geometry 3D reconstruction, laser profilometry is one of the most often utilized sensing methods [Bahnsen et al., 2021].

This sensing technique works well in dewatered pipes, but it has challenges to utilize in live wastewater pipelines with turbulent water flow. LIDAR-based sensing techniques [Kolvenbach et al., 2020] are also afflicted by problems of a similar kind. Therefore, RGB-D sensors, which can provide real-time pipe color information in addition to depth information, are very useful sensors.

Developing sensing and robotics for wastewater assets is not trivial due to the complexity of the environment. Most concrete gravity sewers are underground, and they are accessed through small manholes with diameters ranging from 400 mm to 650 mm. The depth can range from a couple of meters to hundreds of meters. The assets can be circular, rectangular, oval, or some other shape. There is a large range of cross-sectional sizes, from several centimeters to tens of meters. Sewer flow generally cannot be stopped (in some rare cases, this could be possible by diverting the flow), but it may be able to be regulated. There can be silt deposited on the pipe floor. There can be foreign objects, tree roots, snakes, and collapsed walls in sewers. The effluent can be very corrosive and dirty and consist of human hair, tissues, and other human shedding. Therefore, the development of a robotic system that can inspect all sewer environments may not be feasible. However, the development of a robotic system targeted at a particular asset type is a feasible step forward for taking humans out of the sewers.

The rest of this book chapter is structured as follows: Section 13.2 describes the nondestructive sensing system, and Section 13.3 presents the robotic tool design for field deployment. Section 13.4 describes the laboratory evaluation of the robotic sensing system. Section 13.5 elaborates on field deployments and evaluations, while Section 13.6 discusses lessons learned as well as future directions. Finally, Section 13.7 provides a summary of the work.

12.2 Nondestructive Sensing System for Condition Assessment of Sewer Walls

The Corrosion and Rebar Assessment Floating Tool (CRAFT) robot was developed in collaboration with Sydney Water to traverse and evaluate the interior condition of concrete wastewater pipes with a diameter of 900 mm–1,500 mm. The CRAFT robot shown in Fig. 12.1 was designed with three expandable arms for hosting sensing units and was compact enough for it to deploy through manholes with a minimum internal diameter of 600 mm and reach the walls of the wastewater pipe with an internal diameter of 900 mm–1,500 mm. Linear actuators are utilized to control the extension of the expandable arms, and encoders are used to track the current position of the expandable arms. A skid plate with compliance is attached to the end of each expanding arm, allowing the CRAFT to glide along the surface of the pipe when the arms are fully extended. The speed at which the robot descends down the wastewater pipe is regulated by a motorized cable reel attached to the

Figure 12.1 The CRAFT robot.

robot. In Sydney wastewater pipes, wastewater generally has a flow rate between $1\,\mathrm{ms}^{-1}$ and $2\,\mathrm{ms}^{-1}$, and it is designed that this flow rate will keep the cable tethering the CRAFT taut during its deployment. The cable reel is equipped with an encoder, which allows the CRAFT to track the distance that it has traveled down the wastewater pipe.

For assessing the state of corrosion conditions in concrete wastewater pipes, the CRAFT robot includes three main sensor systems. The sensors are (a) a ground penetrating radar (GPR), which is used to nondestructively detect any substantial depth of corrosion levels, (b) a pulsed Eddy current (PEC) sensor, which is used to estimate the distance from the exposed surface to the rebar, and (c) RGB-D sensors for 3D mapping of wastewater pipe internal surfaces, and this book chapter primarily focuses on the development of a 3D mapping sensing suite. The GPR and PEC sensors are housed at the tips of the expanded arms.

The CRAFT is equipped with a camera mount that holds three RGB-D sensors (Intel®RealSense™D435) as well as a CCTV camera with a built-in pan-tilt device on the front panel of the robot. In order to provide illumination for the sensors and cameras, three lights with adjustable intensity are included on the CRAFT. One of them is mounted on the CCTV camera itself, and the other two have been mounted on the red supports that are connected to the RGB-D sensors mount. The RGB-D sensor unit consists of two infrared cameras and one RGB camera. Each of the cameras is configured for a frame rate of 15 Hz. Using the two published infrared images and the camera's intrinsic parameters, a depth image is generated for each RGB-D sensor. In front of each RGB-D sensor, there is a 3 mm thick transparent acrylic layer, which serves as a component of the sensor housing. The RGB-D camera was chosen over other cameras as it provides pixel-to-pixel mapping of

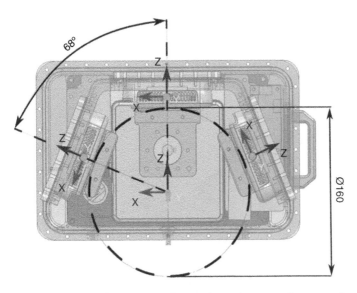

Figure 12.2 3D CAD model showing sensing unit camera distance and angle separation.

RGB data and depth information, which is useful for quantifying pipe defects with real-time color information. The sensor's depth accuracy and the depth range are suitable for the application.

Figure 12.2 depicts the 3D CAD drawing of the three RGB-D sensor arrangements within the housing, whereas Fig. 12.3 shows the physical prototype of the RGB-D sensing suite that can be attached to the CRAFT robot. The calibration of each RGB-D sensor is performed in this setup using Intel's Intel®RealSense D400 Series Dynamic Calibration Tool. The CRAFT is designed for inspecting wastewater pipes with diameters less than 1,500 mm, with an expected depth accuracy of approximately less than 5 mm.

Figures 12.4, 12.5, 12.6, and 12.7 show the images from the three cameras of the center RGB-D sensor (facing upward) as well as the generated depth image. When the arms of the CRAFT are expanded, the front edge of the skid plates attached to the end of the extendable arms comes into view in the infrared and depth images. Due to the smaller field of view of the RGB camera, the skid plate cannot be seen in the RGB camera view. This is shown more clearly in Figure 12.8, where the color image is overlaid on top of the depth image. The ability to visualize the position of the skid plates relative to the walls allows the CRAFT operators to know when the arms of the CRAFT are approaching the walls of the wastewater pipe if the pipe dimensions are not known prior to the deployment of the CRAFT robot.

Figures 12.4 and 12.5 show the left infrared image and right infrared image of the RGB-D sensor, respectively. On the left-hand side of the infrared image in Figure 12.5, a bright spot can be seen. This bright spot is caused by the internal

Figure 12.3 Physical prototype of the RGB-D sensing unit.

Figure 12.4 Left infrared image from the RGB-D sensor displays artifacts (white), pipe wall (dark), and a projected light pattern (white dots).

reflection of the infrared emitter on the acrylic plate above the RGB-D sensor. Although this is undesirable, it is inevitable since the acrylic plate is needed to guarantee that the camera mount stays waterproof throughout its deployment in a wastewater pipe without impeding the camera's field of view. The effect of this bright spot can be seen in the generated depth image in Figure 12.7, where a portion of the depth image is missing. The inner diameter of the pipe and the water level in the pipe have an effect on the amount of overlap in the field of view of the three RGB-D sensors. A portion of the missing information can be retrieved using the images from the other two RGB-D sensors.

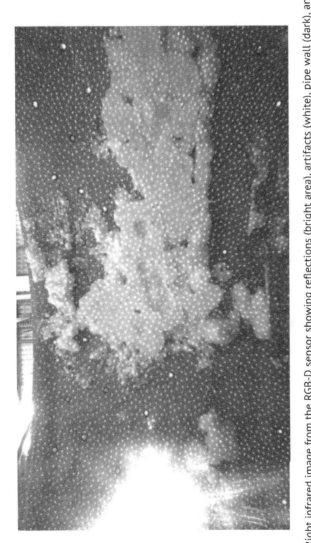

Figure 12.5 Right infrared image from the RGB-D sensor showing reflections (bright area), artifacts (white), pipe wall (dark), and a projected light pattern (white dots).

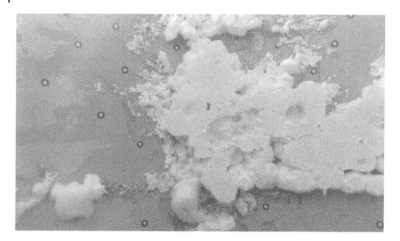

Figure 12.6 RGB image from the RGB-D sensor showing pipe walls (gray) and artifacts (white).

Figure 12.7 Depth image from the RGB-D sensor shows the pipe wall and defects.

Three-dimensional reconstruction of the pipe is achieved by converting the depth image using the following formula.

$$P_D = P_V/1000 \qquad\qquad (12.1)$$

where the value of P_V in the depth image is in the range $-32768 \leq P_V \leq 32767$. Only $P_V \geq 0$ is considered a negative value, implying that the points are behind the camera and are thus treated as noise. 1,000 is the maximum distance in mm between the pipe wall and sensor. The pixels are filtered using minimum and

Figure 12.8 RGB image from the RGB-D sensor (located in the center) of the camera mount is overlaid on the depth image generated from the infrared images of the same sensor.

maximum depth thresholds to minimize the computational requirements. The minimum depth is based on the extension of the arms seen in Fig. 12.1, as the surface of the pipe is expected to be further away from the camera than the skid plate attached to the end of the expandable arm. Applying the threshold also reduces the noise in the reconstructed image caused by the lights on the CRAFT reflecting off the splashing water and the wet walls of the wastewater pipe. The maximum distance for a point is set at 1,500 mm, which is the maximum diameter of the wastewater pipe that the CRAFT was designed for.

Once the distance to each pixel was calculated, the position of the point in 3D space was calculated. Using the field of view of the RGB-D sensors and the number of pixels in an image, a unit vector for each pixel in the depth image was calculated. Using the unit vector and the value of P_D, the position of the point relative to the RGB-D sensor was calculated using the equations below.

$$P_X = P_D * \hat{X}_{[X,X]} \tag{12.2}$$

$$P_Y = P_D \tag{12.3}$$

$$P_z = P_D * \hat{Z}_{[X,Z]} \tag{12.4}$$

where X and Z represent the column and row of the pixel in the depth image, respectively, and $\hat{X}_{X,Z}$ and $\hat{Z}_{X,X}$ represent the X and Z components of the unit vector for the pixel $[X, Z]$, respectively. Doing this for each pixel in the map will generate a point cloud in the camera's reference frame.

This calculation is repeated for each of the RGB-D sensors to obtain a point cloud for each camera of the sensor. After generating the point clouds for each of the cameras, the point clouds are then transformed from each camera's reference frame to the global reference frame to create the 3D reconstruction for the pipe. As the robot moves down the pipe, the 3D reconstruction is updated as new data becomes available.

12.3 Robotic Tool for Field Deployment

The flotation system on CRAFT is composed of two subsystems: one is the watertight aluminum hull, which houses the control electronics required for operating the individual components of the platform and provides positive buoyancy; the other is the extendable, profiled floaters on both sides of the hull, which are made of PVC pipe sections and a custom-profiled nose cone to increase hydrodynamic efficiency when cruising through the water. Figure 12.9 shows the CRAFT flotation system.

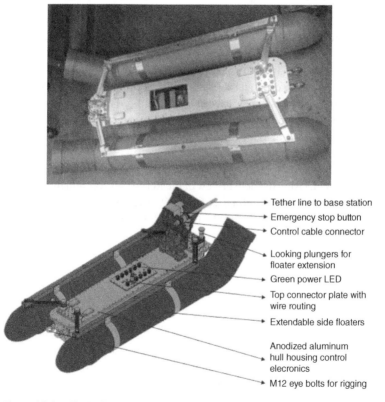

Figure 12.9 CRAFT floatation system.

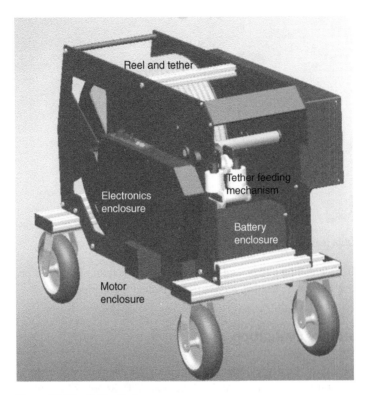

Figure 12.10 Cable drum.

For controlling the CRAFT robot from the base station located at ground level, a customized cable reel drum has been designed and rigorously tested for robustness during deployment. The cable drum is composed of a neutrally buoyant tether line, drive motor, battery system, power transmission, and electronics enclosure. Figure 12.10 shows the 3D CAD model of the cable drum, which possesses the 120 m cable reel. The robot was intended to navigate from manhole to manhole along a straight length of pipe with no or limited curvatures. A 120-meter cable tether covers only the distance between the manholes.

The cable feeding mechanism has been designed with ease of deployment and operation in mind without compromising the working strength under the extreme loads that the CRAFT and the tether line will be exposed to. The stand-alone feeding mechanism is fully collapsible, fits through the 600 mm diameter manhole, and weighs around 10 kg so that one person is capable of deploying it in the pipeline. Structurally, the strength of the mechanism comes from the three scissor arms combined with the bracing rubber pads at the ends. The slotted plates at the

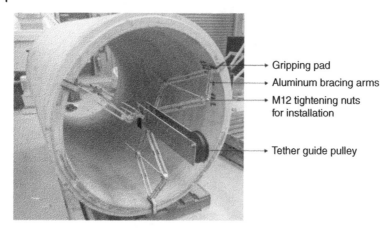

Gripping pad

Aluminum bracing arms

M12 tightening nuts
for installation

Tether guide pulley

Figure 12.11 Cable feeding mechanism.

center house, the free-rolling pulley, and the rollers guide the tether into the pipe. Figure 12.11 shows the cable feeding mechanism attached to the pipe.

12.4 Laboratory Evaluation

Given the complexity of the deployment environment, it is always appropriate to carry out prior laboratory evaluations. At the iPipes Lab, UTS Robotics Institute, a reinforced concrete pipe with an inner diameter of 1,200 mm was used to evaluate the CRAFT robot and the 3D reconstruction generated using the array of RGB-D sensors. Metal plates were attached to each end of the concrete pipe, allowing the pipe to be filled with water to 25% of its maximum capacity, as in Figure 1. Expanding foam was used to randomly form pipe defects on the crown of the pipe. The CRAFT robot was placed in the concrete pipe, and the arms of the CRAFT were expanded to touch the pipe surface. Then the CRAFT was pulled through the tethering mechanism to navigate through the pipe. The robot arm lengths can be adjusted to contact or be in close proximity to the pipe surface as the water level rises in the field settings. The robot is not deployed if the water levels are low since it relies on the water flow for navigation.

While the robot is navigating, it captures RGB-D data, which is then used to reconstruct the color-embedded 3D reconstruction of the inner walls (see Fig. 12.12).

With encoders and inertial measurement unit data, these point clouds are combined to achieve 3D reconstruction of longer sections. This is targeted at estimating defects and their dimensions rather than a comprehensive reconstruction of their assets; therefore, overall drift and bias corrections were not crucial requirements.

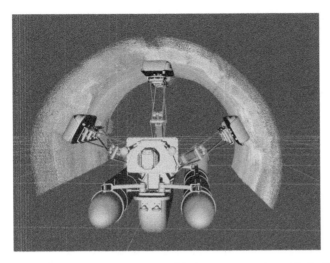

Figure 12.12 Image showing the 3D pipe reconstruction with the 3D model of the robot.

A comprehensive in-pipe robot localization method leveraging radio frequency identification (RFID) sensors was developed by the team as in Gunatilake et al. [2022b] and Gunatilake et al. [2022a] to address such issues in the pipeline environment. However, this method requires the installation of RFIDs. The CRAFT robot is equipped with GPR for corrosion detection. This sensing modality can also map the rebars present in the pipe subsurface. Our team presented a method for localizing the robot using GPR subsurface features [Wickramanayake et al., 2022]. Although this research is in its early phases, its relevance to wastewater pipe conditions requires more investigation.

Three-dimensional reconstruction of the lab concrete pipe is shown in Figure 12.13. In the next phase of the robotic prototype development, we plan on using Sonar to sense and map the areas of the pipe covered by water in order to detect the existence of slits and debris.

12.5 Field Deployment and Evaluation

Field deployment is a significant step toward a robust robotic inspection tool for sewers. With all the previously discussed challenges, sewer inspection is considered to be classified as a conventional 3D (dirty, dull, and dangerous) environment where robotics can play a significant role.

Planning and executing sewer deployments is a long and complex logistic operation. With comprehensive occupation, health, and safety processes in place, highly skilled field crew members are required to prepare the site with appropriate robots,

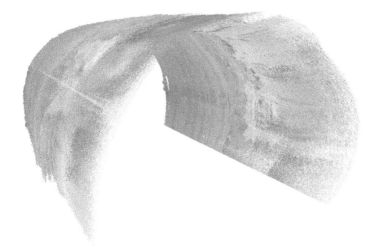

Figure 12.13 Image showing the 3D reconstruction of the concrete pipe.

Figure 12.14 Remote console.

tethers, and equipment, controlled air quality, manhole deployment mechanisms, and tether guiding mechanisms. A snapshot of such a deployment is shown in Figure 12.15 and Fig. 12.16. Once the robot was positioned on flowing effluent, floaters were expanded to achieve higher stability. The CRAFT movement was controlled by the tension of the tethering mechanism through a remote operating console. The remote console shown in Figure 12.14 has full access to all the sensory data, operational status, and actuators.

A three-dimensional reconstruction of a part of the pipe is shown in Figure 12.17. It can be clearly observed that the larger pipe section is transitioning into a smaller pipe section. The field evaluation demonstrates that the developed

Figure 12.15 The CRAFT robot is being lowered into a 600 mm diameter manhole.

Figure 12.16 The CRAFT robot enters the nontraversable section of the concrete sewer pipe.

Figure 12.17   3D reconstruction of a section of the nontraversable sewer pipe.

Figure 12.18   3D reconstruction of a 100 meter long section of the nontraversable sewer pipe.

robotic sensing suite is suitable for 3D mapping inside nontraversable concrete wastewater pipelines. Figure 12.18 shows the 3D reconstruction of a 100 meter long pipe section of the sewer pipe.

12.6   Lessons Learned and Future Directions

Research and development of complex systems like sewer inspection robots are iterative processes. It is nearly impossible to create a lab-based sewer system at this scale, and therefore, the first deployment is always a test in which challenges and surprises are expected. This section presents the major lessons learned during the field operation and possible future directions. This includes the following:

- The robotic sensing system for wastewater pipe inspection and condition assessment was deployed through a 600 mm diameter manhole. Although the majority of manholes are approximately 600 mm in diameter, there can be protruding step irons that can make the usable cross sections smaller than expected. The sizes of the step irons range from 10 to 15 cm. This constrains the diameter of the manhole pipe segment during robot deployment. Further, some other manholes can be as small as 400 mm. This issue may be addressed by replacing the robot's three detachable arms with a robotic manipulator and redesigning the robot with a smaller footprint.

- The GPR and PEC scanning were restricted to linear scanning along the robot's navigation. This is restrictive, even with multiple sensor units. A mechanism that can facilitate radial, as well as longitudinal scanning, is preferable. This may be realized through a robotic arm-mounted sensing head. We are additionally looking at customizing GPR antennas to improve resolution for our application.

- The tether mechanism serves several purposes, including data transmission, power transmission, robot control, and acting as an emergency retrieval mechanism. The tether creates a sag while in operation, making tether encoder-based localization prone to errors. This could be addressed by implementing conventional simultaneous localization and mapping (SLAM) approaches or nonconventional sensor-based localization methods such as [Wickramanayake et al., 2022].

- Pulling the CRAFT robot against the effluent flow for retrieval is nontrivial given that it allows more debris to be trapped on the platform, causing instabilities due to silt and other deposited items in the pipe. Therefore, manhole-to-manhole deployment may be preferable.

- The CRAFT was designed for a particular cohort of sewer pipes: circular pipes with diameters ranging from 900 mm to 1,500 mm. Sewer networks are less standardized and hence they are inherent with many other shapes and sizes of pipes. Therefore, a generalized modular design may be the future solution to keep humans away from sewers.

12.7 Concluding Remarks

In this work, the development of an in-pipe robotic 3D mapping sensing suite using multiple RGB-D sensors for nontraversable wastewater pipe infrastructure was presented. It was necessary to calibrate and combine each RGB-D sensor in order to develop a 3D mapping sensing suite, which was then incorporated into a CRAFT robotic platform. The system's capacity to create 3D maps was evaluated in a laboratory test pipe environment. After achieving acceptable results, the CRAFT robot platform was successfully deployed in a live wastewater pipe operated by Sydney Water. The results of the field tests show that the developed system is capable of carrying out efficient inspections in the field. Although this is a significant step toward keeping humans away from sewers, there are still many challenges to overcome.

Bibliography

Chris H. Bahnsen, Anders S. Johansen, Mark P. Philipsen, Jesper W. Henriksen, Kamal Nasrollahi, and Thomas B. Moeslund. 3D sensors for sewer inspection: A quantitative review and analysis. *Sensors*, 21(7):2553, April 2021. doi: 10.3390/s21072553.

Nicolas Giovanangeli, Lasitha Piyathilaka, Sarath Kodagoda, Karthick Thiyagarajan, Steve Barclay, and Dammika Vitanage. Design and development of drill-resistance sensor technology for accurately measuring microbiologically corroded concrete depths. In *ISARC*, pages 735–742. International Association for Automation and Robotics in Construction (IAARC), May 2019. doi: 10.22260/isarc2019/0099.

Amal Gunatilake, Sarath Kodagoda, and Karthick Thiyagarajan. A novel UHF-RFID dual antenna signals combined with Gaussian process and particle filter for in-pipe robot localization. *IEEE Robotics and Automation Letters*, 7(3):6005–6011, July 2022a. doi: 10.1109/lra.2022.3163769.

Amal Gunatilake, Sarath Kodagoda, and Karthick Thiyagarajan. Battery-free UHF-RFID sensors-based SLAM for in-pipe robot perception. *IEEE Sensors Journal*, 22(20):20019–20026, October 2022b. doi: 10.1109/jsen.2022.3204682.

Joakim Bruslund Haurum and Thomas B. Moeslund. A survey on image-based automation of CCTV and SSET sewer inspections. *Automation in Construction*, 111:103061, March 2020. doi: 10.1016/j.autcon.2019.103061.

Hendrik Kolvenbach, David Wisth, Russell Buchanan, Giorgio Valsecchi, Ruben Grandia, Maurice Fallon, and Marco Hutter. Towards autonomous inspection of concrete deterioration in sewers with legged robots. *Journal of Field Robotics*, 37(8):1314–1327, May 2020. doi: 10.1002/rob.21964.

Mahyar Mohaghegh Montazeri, Niels De Vries, Akpedze D. Afantchao, Allen O'Brien, Paul Kadota, and Mina Hoorfar. Development of a sensing platform for nuisance sewer gas monitoring: Hydrogen sulfide detection in aqueous versus gaseous samples. *IEEE Sensors Journal*, 18(19):7772–7778, October 2018. doi: 10.1109/jsen.2018.2866305.

Bharathi Raju, R. Kumar, M. Senthilkumar, Riza Sulaiman, Nazri Kama, and Samiappan Dhanalakshmi. Humidity sensor based on fibre bragg grating for predicting microbial induced corrosion. *Sustainable Energy Technologies and Assessments*, 52:102306, August 2022. doi: 10.1016/j.seta.2022.102306.

Bruno Rente, Matthias Fabian, Miodrag Vidakovic, Louisa Vorreiter, Heriberto Bustamante, Tong Sun, and Kenneth T.V. Grattan. Extended study of fiber optic-based humidity sensing system performance for sewer network condition monitoring. *IEEE Sensors Journal*, 21(6):7665–7671, March 2021. doi: 10.1109/jsen.2021.3050341.

Anna Romanova, Mojtaba Mahmoodian, and Morteza A. Alani. Influence and interaction of temperature, H2S and pH on concrete sewer pipe corrosion.

International Journal of Civil, Architectural, Structural and Construction Engineering, 8:592–595, 2014.

Shima Taheri, Martin Ams, Heriberto Bustamante, Louisa Vorreiter, Michael Withford, and Simon Martin Clark. A practical methodology to assess corrosion in concrete sewer pipes. *MATEC Web of Conferences*, 199(06010):4, 2018. doi: 10.1051/matecconf/201819906010.

Karthick Thiyagarajan, Sarath Kodagoda, Linh Van Nguyen, and Ravindra Ranasinghe. Sensor failure detection and faulty data accommodation approach for instrumented wastewater infrastructures. *IEEE Access*, 6:56562–56574, 2018a. doi: 10.1109/access.2018.2872506.

Karthick Thiyagarajan, Sarath Kodagoda, Ravindra Ranasinghe, Dammika Vitanage, and Gino Iori. Robust sensing suite for measuring temporal dynamics of surface temperature in sewers. *Scientific Reports*, 8(1):16020, October 2018b. doi: 10.1038/s41598-018-34121-3.

Karthick Thiyagarajan, Sarath Kodagoda, Ravindra Ranasinghe, Dammika Vitanage, and Gino Iori. Robust sensor suite combined with predictive analytics enabled anomaly detection model for smart monitoring of concrete sewer pipe surface moisture conditions. *IEEE Sensors Journal*, 20(16020):8232–8243, August 2020. doi: 10.1109/jsen.2020.2982173.

Xing Wang, Karthick Thiyagarajan, Sarath Kodagoda, and Miao Zhang. PIPE-CovNet: Automatic in-pipe wastewater infrastructure surface abnormality detection using convolutional neural network. *IEEE Sensors Letters*, 7(4):1–4, April 2023. doi: 10.1109/lsens.2023.3258543.

Sathira Wickramanayake, Karthick Thiyagarajan, and Sarath Kodagoda. Deep learned ground penetrating radar subsurface features for robot localization. In *2022 IEEE Sensors*, pages 1–4. IEEE, October 2022. doi: 10.1109/sensors52175.2022.9967350.

Min Wu, Tian Wang, Kai Wu, and Lili Kan. Microbiologically induced corrosion of concrete in sewer structures: A review of the mechanisms and phenomena. *Construction and Building Materials*, 239:117813, April 2020. doi: 10.1016/j.conbuildmat.2019.117813.

13

A Climbing Robot for Maintenance Operations in Confined Spaces

Gibson Hu*, Dinh Dang Khoa Le, and Dikai Liu

Robotics Institute, University of Technology Sydney, Sydney, Australia

13.1 Introduction

Inspection, cleaning, and painting inside confined spaces in bridges, ship hulls, and pipelines are high-risk tasks. As an example, steel bridge maintenance requires humans to work at height or in confined and dusty spaces and use harmful materials and hazardous tools [Burlet-Vienney et al., 2015]. The serious health and safety risks associated include oxygen deficiency, the dangers of fires or explosions, and physical injuries caused by working in uncomfortable poses. There has been a growing trend towards utilizing robots to support people doing high-risk tasks [Botti et al., 2017].

Figure 13.1 shows an example of confined tunnel spaces. This space is about 800 mm wide and many hundred meters long. There are many partition plates with manholes inside the space. Rivets can be found in many areas inside the space as well. The current health and safety regulations do not allow humans to enter such spaces.

Many researchers have conducted research on climbing robots for work activities in confined spaces. Botti et al. [2017] presented a survey on automated solutions for maintenance tasks in confined spaces. Suggestions on design using emerging technologies are provided. Several researchers [Xu and Ma, 2002; Pham et al., 2022; Ward and Liu, 2012] developed wall climbing robots equiped with magnetic feet or tracks for work activities in confined spaces with steel surfaces. Unique approaches were developed to enable and disable magnetic

*Email: gibson.hu@uts.edu.au

Infrastructure Robotics: Methodologies, Robotic Systems and Applications, First Edition.
Edited by Dikai Liu, Carlos Balaguer, Gamini Dissanayake, and Mirko Kovac.

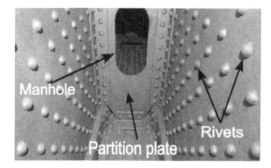

Figure 13.1 An example of a confined tunnel space with partition plates, manholes and rivets.

adhesion, allowing robot locomotion along the steel wall surface. A disadvantage of magnetic adhesion systems is the unreliability in adhesion due to imperfections of steel surfaces, e.g., rivets, debris, thick paint, and unevenness.

Buchanan et al. [2021] presents a multilegged robot designed to operate in confined spaces. A hierarchical planner that samples a random trajectory of body positions is used for robot navigation in the environment. The inchworm-type robot presented by Kotay and Rus [2000] is a design for climbing and maneuvering through tight and confined spaces due to its unique locomotion capability of bidirectional movement. According to the survey by Chu et al. [2010], many researchers have designed this type of robots over the years. Yang et al. [2016] focused on motion planning of this type of robots.

Maintenance tasks (e.g. inspection, cleaning, vacuuming, and painting) in a confined space require a robot to carry a significant payload, such as a painting or cleaning tool. However, a compact robot form is critical for easy maneuvering through confined spaces. Therefore, the robot design requires a balance between the size and capability.

Researchers from the Robotics Institute at the University of Technology Sydney have developed a Wall-climbing Autonomous Maintenance roBOT (WAuMBot) to perform inspection, cleaning, and vacuuming operations autonomously inside the confined space shown in Figure 13.1. The robot was designed to specifically traverse through artifacts from historical steel bridge designs, such as manholes and rivets.

This chapter is organized as follows. Section 13.2 presents the design of the robot. Section 13.3.1 discusses the methodologies for robot perception of the environments including confined tunnel spaces, manholes, and rivets. Section 13.3.2 presents the methods for robot motion planning and control, including avoiding rivets and traversing through manholes and areas with rivets. Section 13.4 details the experiments conducted in a lab test rig and a steel bridge to verify the design and methodologies. Finally, in Section 13.5, the challenges, lessons learnt, and future works are discussed.

13.2 Robot Design

When designing a robot for maintenance tasks in a confined space, several factors need to be taken into consideration, including (1) knowledge and understanding of the confined space, (2) the maintenance activities and their requirements, (3) the tools and methods that are available for the maintenance activities, (4) operation procedure and requirements, and (5) safety measures.

In collaboration with the Roads and Maritime Services of NSW, an intelligent climbing robot, i.e. WAuMBot (Figure 13.2), was developed and tested. This robot consists of two scissor lift units (front and rear units, each with a pair of scissor lifts to provide pushing force for adhesion), a 3 degree of freedom (DOF) body, a 6-DOF arm, four foot-pads with toes, and many sensors.

The unique aspects of this robot design include its method of climbing using force generated by a scissor-lift mechanism, its 3-DOF body, and a 6-DOF robot arm with a 3 kg payload. The robot has the required intelligence for perception, navigation, localization, mapping, planning, and control, enabling it to operate autonomously in the confined space, as shown in Figure 13.1. The robot achieves adhesion by exerting a pushing force. The advantage of this adhesion method is that it provides high stability for the robot during maintenance operations.

Wall-Pushing Mechanism

There are many options when designing a mechanism that can achieve linear movement to pull or push the footpads against a wall. A scissor lift mechanism

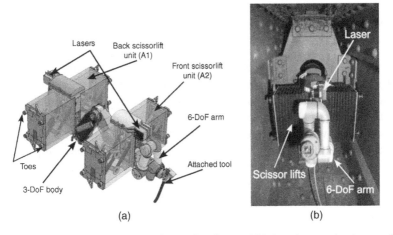

Figure 13.2 (a) The overview of the WAuMBot, and (b) the robot moving in a confined space.

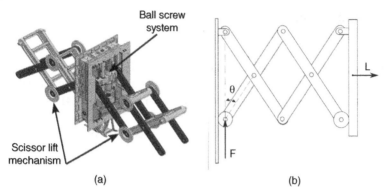

Figure 13.3 (a) A scissor lift unit design, and (b) free-body diagram of a scissor lift mechanism.

was chosen due to its compact design and large pushing force (Figure 13.3a). This is important for the robot to be able to maneuver through tight spaces. A pair of scissor lifts in each scissor lift unit are controlled by two separate motors which allow each side of the scissor lift to move individually. When the robot is moving forward or backward, only the **rear** or the **front** scissor lift unit applies forces to push the footpads to the wall surface. This means that each scissor lift must be strong enough to carry half of the weight of the robot at full extension at any given time. The force that each scissor lift unit needs to generate can be calculated from Eqs. 13.1 and 13.2.

$$F = \frac{L}{\tan\theta} \tag{13.1}$$

F: The force applied by the actuator
L: The load (or pushing force)
θ: The angle

The adhesion method uses the friction force between the toes of the robot and the wall surface. Eq. (13.2) calculates the friction force f_n, where μ_s is friction coefficient, $K = 8$ is the number of toes on the rear or front scissor lift unit, and N_n is a normal force that is applied to a toe by scissor lift mechanism. This friction force supports the robot weight and the reaction force from the maintenance tools. To achieve a large friction force, the friction coefficient needs to be as large as possible. After testing a variety of materials, polyurethane was selected due to its high static friction coefficient (0.8) against the steel surface.

$$P = \sum_{n=1}^{K} f_n = \sum_{n=1}^{K} \mu_s \cdot N_n \tag{13.2}$$

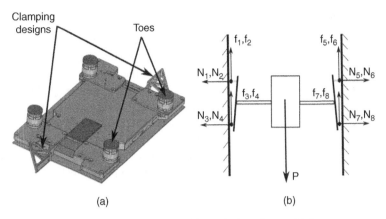

(a) (b)

Figure 13.4 (a) Foot-pad design, and (b) toe designed to avoid rivets.

The design of the footpad is shown in Figure 13.4(a). Each footpad has four toes with polyurethane material at the contact points. As shown in Figure 13.4(b), the length of the two top toes is slightly longer than the two bottom toes in each footpad. The reason for this design is that the pushing force of the scissor lift is not centered, and a chamfer is required to obtain the highest friction force.

An important requirement in toe design is that the robot needs to be able to step onto the space between rivets. The shape of the rivets is semi-spherical with a 25 mm radius. The space between rivets varies from 102 mm to 152 mm. The patterns of the rivets vary along the wall surface. Therefore, the footpad design must ensure that the toes always engage with the space areas among rivets (Figure 13.5a). Based on the data collected from the confined space, the distance between the toes was calculated for the footpad design (Figure 13.5b).

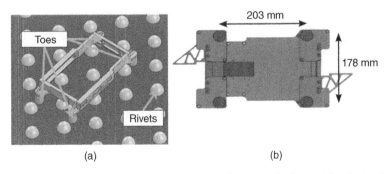

(a) (b)

Figure 13.5 (a) Design of toes, and (b) toe positions on the footpad for rivet patterns.

Design of the 3-DOF Robot Body

The robot is required to move in both horizontal and vertical directions. There-fore, a 3-DOF body was designed (Figure 13.6a). The 3-DOF mechanism consists of three joints: Joints 1, 2, and 3. During the motion of the robot, when the **rear** scissor lift unit is attached and the **front** unit is detached, the torque applied to Joint 1 is the highest torque, shown in Figure 13.6(b). Eq. 13.3 calculates the torque T applied to Joint 1, where P_1 and P_2 are the gravitational forces of two scissor lift units, and P_3 and P_4 are the weights of the two links of the 3-DOF body, respec-tively. P_a is the weight of the 6-DOF robot arm, and R is the sum of the weight of the maintenance tool and the reaction force. Based on the torques calculated, a suitable motor and gearbox can be selected.

$$T = \sum_{n=2}^{4} P_n \cdot d_n + P_a \cdot d_a + R \cdot d_r \tag{13.3}$$

When the 6-DOF robot arm (11 kg) and the maintenance tool (about 3 kg) (Figure 13.7a) are attached to the front scissor lift unit (Figure 13.2), the torque on Joint 1 is significantly increased. Using a larger motor can increase the capacity of Joint 1, but it also increases the weight and size of robot, which may make it more challenging (or impossible) for the robot to go through manholes. To address the torque capacity issue of Joint 1, a spring is added between Joint 1 and Joint 2 to provide passive assistance (Figure 13.6a). When the 3-DOF body mechanism is in a compact position, the spring rests and does not apply any force. When the 3-DOF body is extended, the spring is pulled out and provides up to 50 percent more torque to Joint 1.

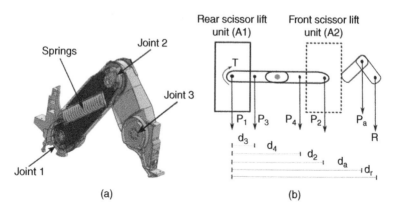

Figure 13.6 (a) 3-DOF body design, and (b) free-body diagram of the 3-DOF mechanism.

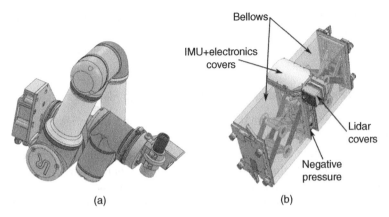

Figure 13.7 (a) A 6-DOF arm carrying maintenance tools, and (b) protective elements of robot.

Protection of the Robot

When conducting maintenance operations, dust and debris are often prevalent in the environment. Therefore, the robot needs to be protected to minimize potential damage to the electronics, motors, and sensors, as shown in Figure 13.7(b).

- **Bellows and Covers** are used to protect the mechanical and electrical components from dust and debris in the confined space.
- **Negative Pressure** is applied to ensure that dust does not leak into the bellows.
- **Protection** is applied to shield the sensors.

Sensors and Their Placement on the Robot

Many sensors are mounted to the robot to enable its perception, navigation, localization, mapping, safety, and control. The list below describes the sensors and how they are used:

- **LIDAR**: Two 2D LIDAR sensors are installed on the robot for building 3D maps of the environment. Each LIDAR sensor is mounted to a mechanism that allows for rotation controlled by a motor. When the sensor is rotated, a high-density 3D point cloud is produced.
- **IMU**: One IMU is housed on each scissor lift unit to measure the pitch and roll of the robot. This measurement is critical since the robot has mechanical sag during motion, and the sag can be corrected by using the IMU measurements.
- **Force sensors**: Force sensors are attached to the clamping mechanism (Figure 13.4a). These sensors are used to detect whether the clamping onto the manhole plate is successful.

- **Cameras**: Cameras are attached to both ends of the robot to capture information on the environment and relay real-time data to the operators. They are also used for inspecting the confined space, both before and after maintenance operations.

Design of a Clamping Mechanism for Traversal Through Manholes

In order for the robot to navigate through the manhole shown in Figure 13.1, it is necessary for the fully stretched 3-DOF body to have a length greater than 1.5 meter. This requirement arises from the presence of a nontraversable area near the manhole, where the footpad of the robot cannot make contact. However, from Eq. (13.3), the longer the robot, the greater the torque is applied to the Joint 1. Designing a 3-DOF body for the robot that exceeds 0.8 m in length has proven challenging due to restrictions on size, weight, and torque capacity. To overcome this challenge, a clamping mechanism has been introduced to enable the robot to successfully navigate through the manhole. Figure 13.4(a) shows the clamping mechanism, which allows the robot to attach itself onto the manhole plate, and then go through the manhole.

Software Architecture

The software architecture has four layers, see Figure 13.8. In this design, hierarchy and communication between different layers are defined. The system core delegates user input and interprets high-level commands to lower level software modules. Raw sensor data is handled by a perception module, and motor controls are handled by the hardware interface. This modular design is very common in robotic systems. It allows each module to be developed individually without affecting the system as a whole.

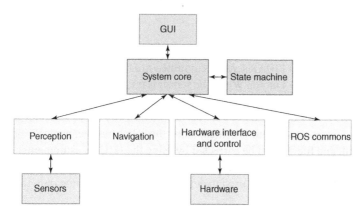

Figure 13.8 Software architecture.

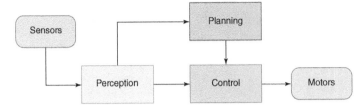

Figure 13.9 Process diagram.

As shown in Figure 13.9, perception algorithms translate the sensor data into usable descriptive data, e.g., generating a map, extracting features, or template matching. The control algorithms process this data and execute specific actions. Planning algorithms utilize perception data to plan optimal robot motions.

13.3 Methodologies

This section discusses the methodologies implemented for robot perception, planning, and control.

13.3.1 Perception

The LIDAR sensors mounted on the robot are Hokuyo UTM-30 which is capable of achieving 10 mm accuracy with a maximum detection range of 30 meters. Each LIDAR is attached to a motor that can rotate the LIDAR to conduct 3D scans of the confined space environment. This sensor is calibrated using the method proposed by Alismail and Browning [2015].

Method for Map Building of the Confined Space
To map the confined space environment, the scans from both the front and rear LIDARs are combined to create a pointcloud. The major planes (Top, Bottom, Left, Right, Front, and Back) of the confined space are extracted by a random sample consensus (RANSAC) plane fitting algorithm [Cantzler, 1981]. The dimensions including height (Top, Bottom), length (Front, Back), and width (Left, Right) are calculated. An orthogonal origin transform can also then be calculated from the major planes: Top, Left, and Front. Due to sensor occlusion and lack of accuracy, the nontraversable areas are determined using prior knowledge of the confine space which can be obtained through CAD drawings and prior exploration of the environment. Figure 13.10 shows the detected regions obtained from a single scan.

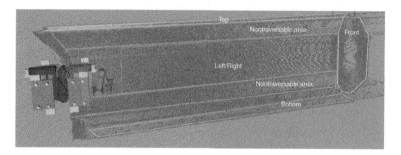

Figure 13.10 Plane detection.

The robot can plan its motion based on major plane dimensions and nontraversable areas. When the robot moves to a new location, it relocalizes itself within the map by scanning the environment again.

Method for Manhole Detection

For the robot to traverse through the manhole, detecting its dimensions is critical for planning and control of robot motion. When the robot is positioned in the right place the entire manhole can be detected (Figure 13.11a, b). However, when the robot is in transition (Figure 13.11d), only partial detection is feasible. This

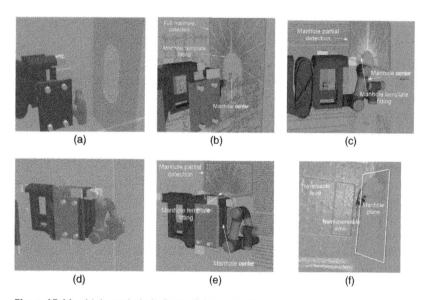

Figure 13.11 (a) A manhole in front of the robot, (b) a manhole detected by the robot, (c) a manhole detected partially, (d) a manhole observed from behind, (e) a manhole partially detected from behind, and (f) a nontraversable area in-front of the manhole.

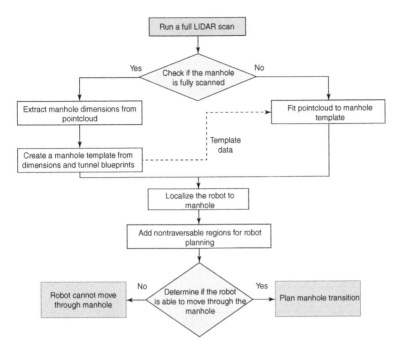

Figure 13.12 Manhole detection method.

is caused by occlusion from the robot itself, the limitation of the field of view, or the angular resolution of the LIDAR scans. This issue can be addressed by using a template (or prior knowledge) of the manhole. This template is generated by combining knowledge of the tunnel blueprints and a full scan of the manhole (when the robot is able to observe the whole manhole) (Figure 13.11b). When the robot can only partially observe the manhole (Figure 13.11c, e), the pointcloud can then be fitted based on the template model.

Often in the tunnels there are nontraversable areas next to the manhole (Figure 13.11f). The width of this area varies. As the area can be known from the blueprints, a rough detection of the plane's width can help to confirm the nontraversable area. Figure 13.12 details the process of manhole detection.

Method for Rivet Detection

Detection of rivets in the confined space is a perception challenge for the WAuM-Bot. The rivets can only be partially detected/observed due to the viewing angle and the limited angular resolution of the LIDAR, not to mention sensor noise. Occlusion also occurs when the rivets are under the robot's footpads.

Similar to the manhole detection method, templates of rivet patterns are can be applied for rivet detection. One major assumption is that the real rivet patterns

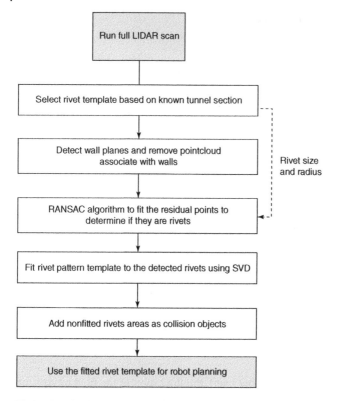

Figure 13.13 Rivet detection method.

are consistent and well engineered when building the structure. For the rivet patterns that do not match the templates, they need to be identified by exploring the confined space before deploying the robot into the environment.

The steps to detect rivets is explained in the diagram shown in Figure 13.13. This rivet detection method relies on a RANSAC fitting algorithms that extract features from the pointcloud and fits circles to the segmented points.

Figure 13.14 shows how the data is fitted and how the template is aligned to the data using single-value decomposition (SVD) [Klema and Laub, 1980]. Once the template is fitted, the robot can plan its motion.

13.3.2 Control

To successfully traverse through the confined tunnel environment, three control problems need to be addressed: (1) accurate positioning and motion, (2) precise robot control for traversing through manholes, and (3) precise robot control for traversing through dense rivet areas.

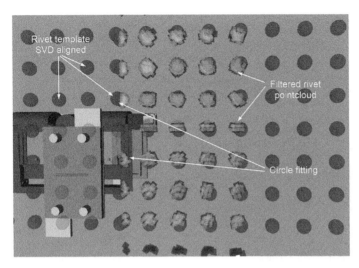

Figure 13.14 Template-based rivet detection.

Accurate Robot Positioning and Motion Control

A real-time control strategy with sensor feedback is needed to achieve precise position and motion.

Horizontal motion of the robot often leads to significant mechanical sag due to backlash in the joints and deformation of the links (Figure 13.15). This sag results in that the footpad placement (e.g. the front scissor lift unite $A2$) to be different from the estimated location calculated via the forward kinematic model. The largest sag occurs when the distance between the two scissor lift units, d, is at maximum in the horizontal direction.

A two-stage control approach was developed for addressing the mechanical sag problem.

Stage 1: active correction of mechanical sag when the robot is moving forward or backward. Using the onboard IMU, it is possible to measure the actual orientation $\theta_{ori}^{(t)}$ at time t at any position along the trajectory until time n. With this sensor input correcting the robot angle, solving the inverse kinematic problem to get the desired joint angles is straight forward. Equation (13.4) derives the angular error θ_e. Given the current robot transform $\hat{T}^{(t)}$, the desired transform $\hat{T}$ is given by Eq. (13.5). Since the IMU can only measure θ_e, $[x_e, y_e]$ can be treated as 0.

$$\theta_e = \hat{\theta}^{(t)} - \theta_{ori}^{(t)} \tag{13.4}$$

$$\hat{T} = \hat{T}^{(t)} + [x_e, y_e, \theta_e] \tag{13.5}$$

$$Q_e^{(t)} = Q^{(t)} - IK(\hat{T}) \tag{13.6}$$

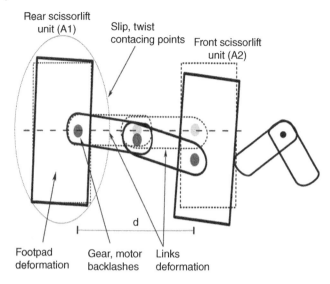

Figure 13.15 Mechanical sag of the robot (solid line: with mechanical sag; dotted line: without sag).

Applying inverse kinematics (*IK*), joint error $Q_e^{(t)}$ can be calculated from the expected joint angles (Eq. 13.6). Joint error $Q_e^{(t)}$ is continuously updated in real time, and the error is reduced by a modified joint position controller.

Stage 2: Correction of the final footpad position. In Stage 2, $[x_e, y_e]$ in Eq. (13.5) is determined by Eq. (13.7). Sensor data is collected before starting Stage 1, $S^{(t=0)}$, and after completing Stage 1, $S^{(t=n)}$. The second measurement, i.e. after completing Stage 1, is transformed into the same frame of reference as the first measurement (i.e. before starting Stage 1) through T_{goal}. The function (**g**) represents the data processing technique (e.g. feature matching, RANSAC fitting) to find the true transformation of the robot.

$$[x_e, y_e, \theta_{env}] = \mathbf{g}(S^{(t=0)}, T_{goal}^{-1} S^{(t=n)}) \tag{13.7}$$

It is possible to compare θ_{ori} from the IMU sensor and $\hat{\theta}_{env}$ returned from **g**. if $\hat{\theta}_{ori} - \hat{\theta}_{env} > threshold$, then $[x_e, y_e, \theta_{env}]$ needs to be recalculated.

The two-stage process is illustrated in Figure 13.16.

Robot Control for Going Through Manholes

Given the WAuMBot design, for the robot to go through the manhole shown in Figure 13.1, the feasible solution is to attach itself to the manhole (or partition)

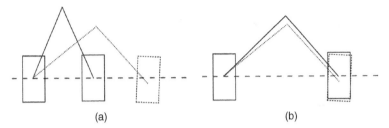

Figure 13.16 Two-stage correction process. Solid lines: start pose, dotted lines: end pose. (a) **Stage 1**, active correction of mechanical sag; (b) **Stage 2**: correction of the final footpad position.

plate using its clamping design (Section 13.2). The approach to navigating through the manhole involves the following steps:

- **Step 1: approaching a manhole**: When the robot gets close to a manhole (Figure 13.17a), it first scans the environment using the front LIDAR and localizes the manhole. Then the robot aligns itself to the manhole center by moving its front footpads to the position that is closest to the nontraversable area in front of the manhole (Figure 13.17b).
- **Step 2: clamping on the manhole plate with the front footpads**: The robot takes a scan to get the accurate location of the manhole. Then, it moves the front footpads to the manhole center and clamps the manhole plate. The touch

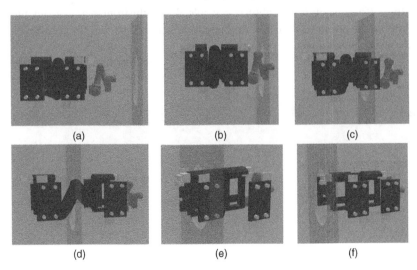

Figure 13.17 (a) Initial robot position, (b) Step 1, (c) Step 2, (d) Step 3, (e) Step 4, and (f) Step 5.

sensors on the clamps confirm whether it is securely clamped or not. The rear footpads can then move forward (Figure 13.17c).

- **Step 3: shuffle step**: The robot detaches its front footpads from the manhole plate and moves forward to the other side of the manhole plate. The rear footpads will then move to the closest feasible position to the manhole (Figure 13.17d).
- **Step 4: clamping on the manhole plate with the rear footpads**: The robot moves its rear footpads onto the manhole center. Then it scans the manhole plate using the rear LIDAR (Figure 13.17e) to get a more accurate position of the manhole. The rear footpads are then clamped onto the manhole plate. Once it is safely clamped, the robot moves the front footpads forward.
- **Step 5: final step**: The rear footpads are detached from the manhole plate and moved forward, resulting in the robot successfully traversing through the manhole (Figure 13.17f).

These steps are repeated when the robot moves forward or backward through a manhole.

Motion Control in Areas with Dense Rivets

Based on the design of the footpads and toes (Section 13.2) and the detection of rivets (Section 13.3.1), a control strategy is designed to adjust robot motion. It consists of the following steps (Figure 13.18):

- **Step 0**: Determine the direction of motion of the robot, i.e. forward or backward.
- **Step 1**: Scan using the front/rear LIDAR to detect rivets and search for rivet collision-free positions near the expected location.
- **Step 2**: If a collision-free location is available, detach the front/rear footpads and move to the position while performing sag correction such that the accuracy of the step is maintained.

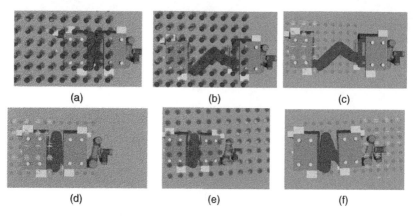

Figure 13.18 Motion control in areas with rivets: (a) Step 1, (b) Step 2, (c) Step 3, (d) Step 4, (e) Step 5, and (f) Step 6.

- **Step 3**: Attach the front/rear footpads and detach the other footpads.
- **Step 4**: Move the detached footpads to a compact robot position.
- **Step 5**: While detached, scan with the alternate LIDAR and run rivet detection again. Search for rivet collision-free positions that are close to the location of the detached footpads.
- **Step 6**: If a collision-free location is available, move the detached footpads to the location.

Repeat s 1–6 until the desired location is reached.

13.3.3 Planning of Robot Body Motion

To perform a maintenance task, the robot must first plan a sequence of motions to reach a location where the task needs to occur.

The robot body is a 3-joint co-planar mechanism and therefore has closed-form inverse kinematic solutions. From the current robot location to a goal location, a series of intermediate locations are selected using an A-star algorithm [Yang et al., 2016].

13.4 Experiments and Results

13.4.1 Experiment Setup

To evaluate the robot a laboratory test rig was designed based upon a real confined tunnel section inside a steel bridge (Figure 13.19). Four specific experiments were conducted in the lab test rig to evaluate the robot for:

- Accuracy in motion and localization: evaluated by assessing rivet avoidance and manhole stepping capabilities.

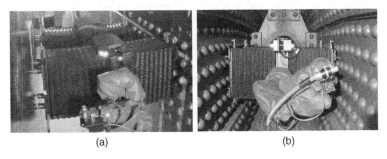

(a)   (b)

Figure 13.19   (a) WAuMBot in the lab test rig, (b) WAuMBot inside the still tunnel of the Sydney Harbor Bridge.

- Efficiency: measured by the time required for the robot to move and perform various tasks.
- Robustness: measured by the number of times the robot has an error state.

After the robot was extensively tested in the test rig, field trials were conducted in a steel bridge. In field trials, WAuMBot was tested at different locations of a confined tunnel section (Figure 13.1).

13.4.2 Lab Test Results

Accuracy in Robot Motion and Positioning

The control section discussed a two-stage control approach for addressing the mechanical sag and improving the accuracy of motion and footpad positioning during robot motion. To evaluate this approach, the robot was programmed to move in a straight line in the lab test rig. Displacement and angular errors relative to the horizontal line of the test rig were measured.

From Figure 13.20, it can be seen that with the sag control (or correction) approach, both the displacement and angular errors are significantly reduced. A less than 10 mm error of displacement was achieved, which is accurate enough for robot motion in this environment.

Testing for Going Through the Manhole

To go through the manhole, the robot needed to take the steps discussed in Section 13.3.2. The success rate of the robot's attempts to navigate through a manhole was evaluated through 20 tests, presented in Table. 13.1. A high success rate (95%) was achieved when the robot went through the manhole in forward and backward motions. When there were rivets on the other side of the manhole, the success rate was slightly slower because of the effect of the rivet detection accuracy. In some instances, even if the rivets could be detected accurately, no physical footpad location could be found, which resulted in a fail.

Performance of the Rivet Detection Method

An experiment was conducted to evaluate the performance of the rivet detection method by measuring the number of false positives (i.e. the method "detected" a rivet but in fact it is just sensor noise) and false negatives (i.e. a rivet is not detected by the method). The ground truth was obtained from manual inspection of rivets. Table 13.2 shows the results. The method was able to detect all rivets with zero false negatives. Without the use of templates, the accuracy was less because some rivets were not detected due to sensing and noisy data.

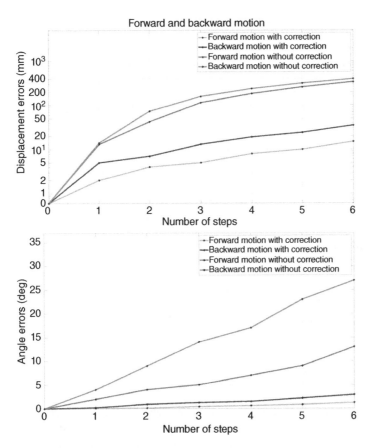

Figure 13.20 Displacement and angular errors over 10 runs.

Table 13.1 Success rate of going through a manhole over 20 trials.

Manhole execution type	Successful executions(%)
Forward through manhole without rivets	95
Backward through manhole without rivets	95
Forward through manhole with rivets	85
Backward through manhole with rivets	80

Table 13.2 Percentage of false positives and false negatives over 100 trials of rivet detection.

Direction	Using template	False positives (%)	False negatives (%)	Accuracy (%)
Forward	No	5	20	80
Backward	No	5	25	75
Forward	Yes	5	0	100
Backward	Yes	5	0	100

Table 13.3 Robot movement efficiency.

Robot movements	Time taken (minutes)
Forward movement	1.5
Backward movement	1.5
Forward movement through rivets	4
Backward movement through rivets	4
Forward manhole traversal	15
Backward manhole traversal	15
Typical arm movement	3

Robot Movement Efficiency

Table 13.3 shows the time that the robot takes for each movement, including forward movement, backward movement, forward movement in areas with rivets, backward movement in areas with rivets, moving forward through a manhole moving backward through a manhole and typical arm movement. Increasing the speed of motions can reduce the movement time, but it may reduce the reliability.

13.4.3 Field Trials in a Steel Bridge

The robot underwent four field tests on a steel bridge, attempting to traverse through four different confined spaces during each test. These trials included:

- Traversing an 8-meter confined tunnel section
- Traversing through three confined tunnel sections divided by manhole plates.
- Traversing through rivet-dense areas.
- Traversing between areas with and without rivets
- Sandblasting and painting tests in a confined space.

These trials helped the researchers to improve the robot reliability and robustness. As a result, the robot is ready for deployment in the field.

13.5   Discussion

Regarding the mechanical sag of the robot, it has been observed that there is a direct correlation between the width of the confined tunnel and the extent of sag experienced. If the mechanical sag is too significant, the robot's maximum stretch will be less than what was designed, which may cause issues when going through manholes and in avoiding collision with rivets.

The robot was designed for a specific confined tunnel space discussed in this chapter. If the confined space is different, or the size and/or geometric shape of the manhole is different, modifications to the robot design are needed.

An assumption made for rivets is that the rivets are well engineered and the tolerances meet the requirements of the original design. In the future, using better sensing technologies could enhance the robustness of rivet detection. It is critical that no false negatives are detected, otherwise colliding with a rivet could be catastrophic for the robot.

13.6   Conclusion

This chapter presented a climbing robot that can navigate inside a confined space for maintenance tasks including inspection, cleaning, vacuuming, and painting. An emphasis was placed on navigating in rivet-dense areas, going through manholes and robot control. The methodologies and the robot design have been extensively tested in a lab test rig and in a steel bridge, confirming the robotic system's reliability and robustness.

Acknowledgments

This work was supported in part by the Transport for New South Wales and the University of Technology Sydney, Australia. The authors would also like to acknowledge Mattew Abbon, Buddhi Wijerathna, Craig Burrows, Chia-han Yang, Ravi Ranasinghe, and Gavin Paul for their contributions to this project.

Bibliography

Hatem Alismail and Brett Browning. Automatic calibration of spinning actuated LiDAR internal parameters. *Journal of Field Robotics*, 32:723–747, August 2015. doi: 10.1002/ROB.21543.

Lucia Botti, Emilio Ferrari, and Cristina Mora. Automated entry technologies for confined space work activities: A survey. *Journal of Occupational and Environmental Hygiene*, 14:271–284, April 2017. doi: 10.1080/15459624.2016. 1250003.

Russell Buchanan, Lorenz Wellhausen, Marko Bjelonic, Tirthankar Bandyopadhyay, Navinda Kottege, and Marco Hutter. Perceptive whole-body planning for multilegged robots in confined spaces. *Journal of Field Robotics*, 38:68–84, January 2021. doi: 10.1002/ROB.21974.

Damien Burlet-Vienney, Yuvin Chinniah, Ali Bahloul, and Brigitte Roberge. Design and application of a 5 step risk assessment tool for confined space entries. *Safety Science*, 80:144–155, December 2015. doi: 10.1016/J.SSCI.2015.07.022.

H Cantzler. Random sample consensus (RANSAC). *Institute for Perception, Action and Behaviour, Division of Informatics, University of Edinburgh*, March 1981.

Baeksuk Chu, Kyungmo Jung, Chang Soo Han, and Daehie Hong. A survey of climbing robots: Locomotion and adhesion. *International Journal of Precision Engineering and Manufacturing*, 11(4):633–647, August 2010. doi: 10.1007/ S12541-010-0075-3.

V. Klema and A. Laub. The singular value decomposition: Its computation and some applications. *IEEE Transactions on Automatic Control*, 25(2):164–176, 1980. doi: 10.1109/TAC.1980.1102314.

Keith Kotay and Daniela Rus. The inchworm robot: A multi-functional system. *Autonomous Robots*, 8:53–69, 2000. doi: 10.1023/A:1008940918825.

Anh Q. Pham, Cadence Motley, Son T. Nguyen, and Hung M. La. A robust and reliable climbing robot for steel structure inspection. In *2022 IEEE/SICE International Symposium on System Integration (SII)*, pages 336–343, 2022. doi: 10.1109/SII52469.2022.9708747.

Peter Ward and Dikai Liu. Design of a high capacity electro permanent magnetic adhesion for climbing robots. In *2012 IEEE International Conference on Robotics and Biomimetics (ROBIO)*, pages 217–222, 2012. doi: 10.1109/ROBIO.2012.6490969.

Zeliang Xu and Peisun Ma. A wall-climbing robot for labelling scale of oil tank's volume. *Robotica*, 20:209–212, 2002. doi: 10.1017/S0263574701003964.

C.-H.J. Yang, G. Paul, P. Ward, and D. Liu. A path planning approach via task-objective pose selection with application to an inchworm-inspired climbing robot. In *IEEE/ASME International Conference on Advanced Intelligent Mechatronics, AIM*, September 2016. doi: 10.1109/AIM.2016.7576800.

14

Multi-UAV Systems for Inspection of Industrial and Public Infrastructures

Alvaro Caballero*, Julio L. Paneque, Jose R. Martinez-de-Dios, Ivan Maza, and Anibal Ollero

GRVC Robotics Lab, University of Seville, Camino de los Descubrimientos, s/n, Seville, Seville, Spain, 41092

14.1 Introduction

Aerial robots offer the possibility to inspect hard to access locations without putting personnel at risk. Today, teams of specialized crews climb using ropes, scaffolding, or elevated platforms, to perform asset inspections which at times require shutdowns of the facility with extremely high costs. This approach is not only time consuming but also very costly. Instead, detailed images and data can be collected by aerial robots to better assess and forecast the asset's maintenance needs, eliminating dangerous human work, reducing the risk of breakdowns, and improving competitiveness by minimizing production downtime. This impacts safety, environmental, and financial benefits.

Inspection and maintenance of industrial and public infrastructures include several activities such as general global inspection, close or very close inspection, contact inspection, and manipulation for structure construction [Augugliaro et al., 2013] or maintenance such as cleaning, installation of devices, and others. Thus, for example, in the H2020 AEROARMS project [Ollero et al., 2018], aerial robots have been tested in refineries performing contact inspection of pipes and tanks to determine the wall thickness at height as an application of the aerial robotic manipulation [Suarez et al., 2018] developed in this and in the FP7 ARCAS project. Additionally, these technologies have been applied to public infrastructure and particularly for bridge inspection in the H2020 AEROBI project. An analysis of the application of aerial robotic manipulation [Ollero et al., 2022] to the inspection and maintenance is in Part VI of the book in reference Ollero and Siciliano [2019]. This chapter is mainly devoted to inspection without contact.

*Corresponding Author: Alvaro Caballero; alvarocaballero@us.es

Infrastructure Robotics: Methodologies, Robotic Systems and Applications, First Edition.
Edited by Dikai Liu, Carlos Balaguer, Gamini Dissanayake, and Mirko Kovac.

The inspection of the electrical power system involves lines of many thousands of kilometers in Europe requiring accurate tracking of the lines and localization of dangerous situations, such as thermographic fault detection, insulator fault detection, and track clearance inspection including building of 3D models and computation of distances to buildings and vegetation.

Utility lines, such as electricity transmission lines and pipelines, are currently inspected once in every 3–18 months. Currently, the inspection of electrical power lines is performed with manned helicopters, which is costly and risky, or involves on-the-ground teams with ground vehicles, which are highly constrained due to the terrain. The application of unmanned aerial vehicles (UAVs) offers significant advantages. Some companies have applied multirotor systems to perform local inspections within the visual line of sight (VLOS) as shown, e.g. in https://youtu.be/Bj5ByZKVNao. In these inspections, different cameras (infrared, visual, and ultraviolet) are used. However, applications like beyond visual line of sight (BVLOS) for long-range inspection have been constrained by the flight endurance, safety, and regulations.

As it has been demonstrated in the H2020 AERIAL-CORE project, aerial robots could drastically reduce the inspection costs while increasing the monitoring frequency and quality. Particularly, the detection of isolation losses can be performed by means of infrared detection and mapping applications running onboard these aerial robots.

Today, multirotors, with small range and endurance, and usually with remote pilots in the visual line of sight, are applied for local inspections of transmission towers [He et al., 2019; Baik and Valenzuela, 2019] or grid segments [Iversen et al., 2021a]. However, this is not enough for long-endurance inspection of the electrical grid composed of medium- and high-voltage lines. Motivated by this, new research lines are arising to mitigate the limitations of these robots [Iversen et al., 2021b]. There are also long range and endurance UAVs, usually large and heavy, flying over 300 meters, which are used in military application, surveillance, maritime applications, environmental surveillance, and others. In addition, small fixed-wing systems, usually flying below 150 meter, are also applied for surveillance and mapping of relatively large areas. However, they have constraints related to the accuracy due to the limitations of the onboard sensing systems. In the last years, new systems with vertical take-off and landing and flight as fixed wing have been developed. Different types of systems exist, including those with different propellers to fly as fixed wing and vertical take-off and landing (VTOL), as well as systems that can change the orientation of the propellers after take-off to fly as fixed wing and later changing again to the initial configuration for landing. However, these platforms are currently not efficient to perform the hovering required for detailed very close view and aerial manipulation for maintenance operation.

The inspection of large infrastructures such as electrical grid systems can be performed more efficiently with multiple UAVs [Deng et al., 2014]. The advantages when comparing with the use of a single UAV are: decreasing the total time for the whole inspection; minimizing delays for event detection; application of teaming techniques involving multiple specialized platforms, for example, to obtain closed view in local inspection and have long-range inspection; and improved reliability by avoiding dependencies of a single UAV. Finally, it should be noted that BVLOS flights require to perform specific operations risk assessment (SORA) to obtain the flight permissions. This has already been obtained by several companies to perform regular inspections.

This chapter presents the use of a multi-UAV system composed of heterogeneous robots for inspection of large electrical power-line infrastructures. The chapter proposes two use cases and describes the architecture, a multi-UAV planning method, and a vegetation mapping method and shows experimental results in real power-line inspection experiments. The main contributions of the work are

- the selection of a suitable architecture for the multi-UAV system under study,
- a planning method that has been developed to optimize power-line inspection missions for team of UAVs with heterogeneous capabilities in terms of flight speed (VTOL fixed-wing UAVs and multirotors), considering also the existence of recharging stations along the operation area,
- a semantic mapping method integrating the LIDAR-based object classification method presented in Valseca et al. [2022] within a LIDAR-based geometrical mapping method and adapting it for online onboard execution,
- the implementation of the described multi-UAV system with VTOL fixed-wing platforms and multicopters,
- the validation of architecture, methods, and system in the ATLAS Flight Test Center in Villacarrillo (Spain).

In Section 14.2, the use cases in the inspection of the electrical lines and the architecture of the inspection system are described. In Section 14.3, the multi-UAV planning system is formulated. Section 14.4 presents the vegetation mapping. Finally, Section 14.5 closes the chapter with the conclusions.

14.2 Multi-UAV Inspection of Electrical Power Systems

14.2.1 Use Cases

The goal is to develop an autonomous multi-UAV system for power lines inspection and maintenance, which is aimed at drastically reducing both operational costs and human risks. Two use cases have been identified (see Figure 14.1):

1. Periodic detailed inspection and accurate 3D modeling of the power lines. This includes unattended continuous operation of the multi-UAV system. It improves current inspection systems, such as the use of manned conventional

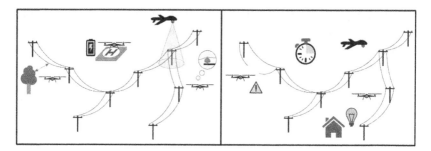

Figure 14.1 Uses cases: Periodic detailed inspection and accurate 3D modeling of the power lines (left) and localization of faults in the grid that cause power outages in a given area (right).

helicopters, providing, in addition to the information obtained with these systems, new detailed views, i.e. images from the grounding system in the base of the transmission towers not registered nowadays (the manned helicopters cannot fly close enough to the towers near the ground). These detailed views are used to generate maps that are used to predict future potential risks to the infrastructure and to detect high-priority maintenance scenarios (i.e. pruning a tree before it reaches the line).

2. Localization of faults in the grid that cause power outages in a given area. The multi-UAV system should take off and autonomously find the fault in minimum time. This requires searching for particular faults (short-circuits, fires, etc.) and also comparing the coherence of the grid with the last available model (to detect structural faults).

The considered multi-UAV system integrates heterogeneous drones and cognitive planning and sensing algorithms that have been designed in the H2020 AERIAL-CORE project to tackle both use cases. The aerial robots are heterogeneous since the system includes multirotors and the fixed-wing VTOLs. This configuration of the team covers long-range operation along the power lines, but also a detailed and close inspection of the transmission towers, including their grounding system in the base. However, as the flight time of the aerial robots is nowadays limited, compared to the large size of the power lines, the system also includes battery recharging stations where the UAVs autonomously land and take-off to extend their range and perform continuous operation.

14.2.2 Architecture

In general, the group architecture [Cao et al., 1997] of a cooperative robotic system provides the infrastructure upon which collective behaviors are implemented and determines the capabilities and limitations of the system. In particular, for multi-UAV systems, a classification of the different architectures can be found in Maza et al. [2015], and the scheme that can fit better for the application addressed in this chapter is an architecture for intentional cooperation. In the intentional cooperation approaches, each individual executes a set of tasks (subgoals that are necessary for achieving the overall goal of the system and that can be achieved independently of other subgoals) explicitly allocated to perform a given mission in an optimal manner according to planning strategies. In this case, problems such as multi-UAV task allocation, high-level planning, plan decomposition, and conflict resolution [Alejo et al., 2009] should be solved, taking into account the global mission to be executed and the different UAVs involved.

An example of this type of architecture applied to a multi-UAV team in field tests can be found in Maza et al. [2011], and we have adopted that approach. The architecture is endowed with different modules that solve the usual problems that arise during the execution of multipurpose missions, such as task allocation, conflict resolution, task decomposition, and sensor data fusion. The approach had to satisfy two main requirements: robustness for operation in power line inspection scenarios and easy integration of different autonomous vehicles. The former specification led to a distributed design, and the latter was tackled by imposing several requirements on the execution capabilities of the vehicles to be integrated in the platform.

Next section is focused on the planning techniques for the multi-UAV team to achieve the inspection application previously presented.

14.3 Inspection Planning

The envisioned use cases require the coordinated operation of a team of UAVs in order to provide an effective service. This is motivated by the long distances that need to be covered over power lines that usually include many branches. At the same time, the limited battery capacity offered by UAVs forces them to frequently recharge their batteries. Moreover, these robots could have heterogeneous capabilities. All the aspects above do not transform the team coordination in a straightforward task. Consequently, the use of planning methods should be considered as a suitable way to generate efficient inspection sequences that optimize certain metrics such as the energy consumption or the operation time. In this manner, the coverage over the power line without changing batteries can be maximized or the time needed to find a fault minimized.

14.3.1 Vehicle Routing Problem

As it is indicated in Cacace et al. [2021], power lines are linear infrastructures that can be modeled as graphs. Also, these power lines are frequently located in open areas without high densities of obstacles, usually trees, but with a relevant influence of the wind, which affects the energy consumption of aerial robots. Considering the previous features, the planning problem can be led properly as a vehicle routing problem (VRP) without the consideration of the obstacles [Nekovář et al., 2021]. Then, the obstacle avoidance can be solved locally by applying any state-of-the-art reactive technique, while the plan is executed without affecting significantly its optimality.

This section addresses the VRP by presenting a novel planning method that has been developed to optimize inspection missions for team of UAVs with heterogeneous capabilities in terms of flight speed. Thus, the operation for a combination of several VTOL fixed-wing UAVs and multirotors can be efficiently exploited. Moreover, due to the limited battery capacity of these aerial robots, the existence of recharging stations along the operation area is also considered. The visit to these stations puts the UAVs back into operation, recharging their batteries without mission interruptions. The expected results are optimal and feasible tours that allow the team of aerial robots to fulfil the inspection task in the minimum time.

The fundamentals of the planning method are presented in the following subsections. After that, the validation results are showcased.

Capacitated Min–Max Multidepot VRP

The planning method consists of a capacitated min–max multidepot VRP that is formulated over an abstraction of the inspection problem under analysis. For that, a graph-based representation of the problem is addressed first. Then, a mixed-integer linear programming (MILP) formulation is created from the graph representation, and subsequently solved to find the optimal assignment and sequence of power-line segments to be inspected by each UAV.

Graph-Based Representation of the Problem

The graph that models the inspection problem, represented in Figure 14.2 (left), is a directed weighted multigraph given by the tuple $G = (\mathcal{V}, \mathcal{E}, \mathcal{W}, \mathcal{D})$, where $\mathcal{V}$ is the set of graph vertices, $\mathcal{E}$ is the set of edges connecting the vertices, $\mathcal{W}$ is the set of weights corresponding to such edges, and $\mathcal{D}$ is the set of δ UAVs used to cover the graph. At the same time, $\mathcal{V} = \{\mathcal{O}, \mathcal{R}, \mathcal{S}\}$, with $\mathcal{O}$ embedding into a single vertex (vertex 0), the set of depots where each robot is located at the beginning of the mission, $\mathcal{R}$ is the set of recharging stations r_j, and $\mathcal{S}$ is the set of power-line segments s_i to be inspected. Since these segments can be inspected following two directions, the associated set of graph vertices is doubled in such a way that $\mathcal{S} = \{\mathcal{S}^a, \mathcal{S}^b\}$, where $\mathcal{S}^a$ encodes one particular direction of inspection and $\mathcal{S}^b$ the opposite. Thus, each pair of vertices $s_i^a \in \mathcal{S}^a$ and $s_i^b \in \mathcal{S}^b$ belongs to the same power-line segment s_i. For the sake of clarity, Figure 14.2 (right) depicts an example for the graph on the left where vertex $s_1^a \in \mathcal{S}^a$ represents the segment s_1

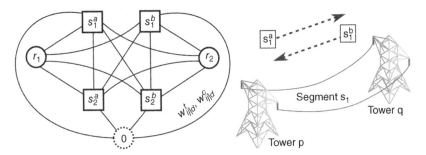

Figure 14.2 Graph-based representation of the inspection problem (left) and encoding for each power-line segment (right).

from transmission tower p to transmission tower q, and vertex $s_1^b \in S^b$ represents the segment s_1 from transmission tower q to transmission tower p.

Concerning the graph connectivity, all the vertices in S are connected to each other and to every vertex in $\{\mathcal{O}, \mathcal{R}\}$ by δ edges each pair, with the only exception of each pair of vertices $s_i^a \in S^a$ and $s_i^b \in S^b$ since they belong to the same power-line segment s_i. In this context, each edge represents not only the trip from vertex i to vertex j by UAV k but also the operation to be performed in vertex j; this is, the inspection of a power-line segment if $j \in S$ or the recharging of batteries if $j \in \mathcal{R}$.

Associated with the edges, the set of weights $\mathcal{W}$ can be decomposed as $\mathcal{W} = \{\mathcal{W}^t, \mathcal{W}^c\}$, with $\mathcal{W}^t$ the set of times required to cover the edges, and $\mathcal{W}^c$ the corresponding energy consumption in terms of battery percentages. In this manner, the weight from vertex $i \in V$ to vertex $j \in V$, with $i \neq j$, using the UAV $d \in D$ can be expressed as $w_{ij|d}^t \in \mathcal{W}^t$ for the time and as $w_{ij|d}^c \in \mathcal{W}^c$ for the battery consumption. For the computation of $w_{ij|d}^t$, the time can be estimated according to the Euclidean distance between vertices i and j (considering the reference flight speed v_d^{ref} for each UAV d), and adding the time required to complete the operation in vertex j. In contrast, $w_{ij|d}^c$ can be computed making use of the models of energy consumption presented in Bauersfeld and Scaramuzza [2022].

MILP Formulation

The MILP problem consists of finding the optimal assignment and sequence of power-line segments over the graph G to minimize the total mission time. For that, the time spent by the multi-UAV team should not only be minimized but also distributed evenly between the robots as the mission finishes with the last UAV completing its sequence. Under this optimization objective, the use of all the available UAVs is considered.

Associated with the edges in $\mathcal{E}$, $\mathcal{Z}$ can be defined as a set of binary variables $z_{ij|d}$, where i and j ($i \neq j$) represent vertices in V, and d corresponds to UAVs in D. These variables $z_{ij|d}$ encode if their associated edges are selected in the MILP solution ($z_{ij|d} = 1$) or not ($z_{ij|d} = 0$). Additionally, variables σ_{rp}, with $\{r, p\} \in D$ and $r \neq p$, quantify the difference between the total times spent by UAVs r and p. These variables are defined to be real and positive in the interval $[0, \sigma_{max}]$, being $\sigma_{max} \geq 0$ their maximum acceptable value.

With all the previous definitions, the MILP problem can be formulated as follows:

$$\underset{z_{ij|d}, \sigma_{rp}, y_{j|d}}{\text{Minimize}} \sum_{\{i,j\}\in\mathcal{V}, i\neq j, d\in D} w^t_{ij|d} z_{ij|d} + \sum_{\{r,p\}\in D, r\neq p} \sigma_{rp} \tag{14.1a}$$

$$\text{s.t.} \sum_{i\in\mathcal{V}, i\neq j, d\in D} \left(z_{ir|d} + z_{ip|d}\right) = 1, \ \forall j : \{r = s^a_j \in S^a, p = s^b_j \in S^b\} \tag{14.1b}$$

$$\sum_{i\in\mathcal{V}, i\neq j, d\in D} \left(z_{ri|d} + z_{pi|d}\right) = 1, \ \forall j : \{r = s^a_j \in S^a, p = s^b_j \in S^b\} \tag{14.1c}$$

$$\sum_{i\in\mathcal{V}, i\neq j} \left(z_{ij|d} + z_{ji|d}\right) = 2y_{j|d}, \ \forall j \in S, \ \forall d \in D \tag{14.1d}$$

$$\sum_{i\in S} z_{0i|d} = 1, \ \forall d \in D \tag{14.1e}$$

$$\sum_{i\in\mathcal{V}, i\neq 0} z_{i0|d} = 1, \ \forall d \in D \tag{14.1f}$$

$$\sum_{i\in\mathcal{T}, j\neq\mathcal{T}, d\in D} \left(z_{ij|d} + z_{ji|d}\right) \geq 2h(\mathcal{T}), \forall \mathcal{T} \subset S \tag{14.1g}$$

$$\sum_{\{i,j\}\in\mathcal{V}, i\neq j} \left(w^t_{ij|r} z_{ij|r} - w^t_{ij|p} z_{ij|p}\right) - \sigma_{rp} \leq 0, \ \forall \{r, p\} \in D (r \neq p) \tag{14.1h}$$

$$\sum_{\{i,j\}\in\mathcal{V}, i\neq j} \left(w^t_{ij|p} z_{ij|p} - w^t_{ij|r} z_{ij|r}\right) - \sigma_{rp} \leq 0, \ \forall \{r, p\} \in D (r \neq p) \tag{14.1i}$$

where Eq. (14.1a) is the objective function, consisting of two terms that encode the total time spent by the multi-UAV team and a penalty for uneven time distribution between robots. Constraints Eqs. (14.1b) and (14.1c) force that each power-line segment j is inspected exactly once and only in one direction. This means that every pair of vertices $\{s^a_j, s^b_j\}$ associated with segment j should be visited, which is set by (14.1b), and left, which is set by Eq. (14.1c). The latter requirement is complemented by constraints Eq. (14.1d), being $y_{j|d} \in \{0,1\}$ auxiliary integer variables guaranteeing that, if UAV $d \in D$ reaches vertex $j \in S$, the same UAV must leave it. Constraints (14.1e) and (14.1f) ensure, respectively, that each UAV starts and finishes the mission in its depot. Constraints (14.1g) prevent the formation of tours exceeding the battery capacities of the UAVs or with no connection with the depot vertex. For that, the function $h(\mathcal{T})$ should impose the minimum number of times that the robots must enter and leave every subset $\mathcal{T} \subset \mathcal{S}$. However, since the number of this kind of constraints can be excessive, they can be omitted initially and added dynamically as they are broken (see [Laporte, 2007]). In addition, to speed up the convergence, the following capacity constraints can also be added to the formulation

$$\sum_{i \notin R, j \in R} z_{ij|d} \geq \sum_{\{i,j\} \in \mathcal{V}, i \neq j} w^c_{ij|d} z_{ij|d}, \ \forall d \in \mathcal{D} , \tag{14.2}$$

which put a lower bound on the number of tours for each UAV. Finally, constraints Eqs. (14.1h) and (14.1i) balance the time spent by each UAV as much as possible.

Once the MILP problem is formulated, it can be solved by conventional solvers. Then, the selected edges in the graph G are known (they fulfil $z_{ij|d} = 1$) and with them, the optimal assignment and sequence of depots, power-line segments and recharging stations can be deduced.

Validation of the Planning Method

This section is devoted to the validation of the planning method presented above. In that context, the method generates a plan to autonomously inspect a power grid with a fleet of heterogeneous UAVs in the minimum time. The selected power lines for real validation are located in the surroundings of Flight Test Center ATLAS in Villacarrillo (Spain). These power lines, considered as input for the planning method, have a length of 12 km, distributed in several branches, and connects 80 transmission towers, whose positions are georeferenced. Figure 14.3 shows the associated map. Concerning the team of UAVs, it consists of the two fixed-wing VTOL UAVs and the two multirotors presented in Figure 14.4. This heterogeneous team includes a DeltaQuad Pro (left), denoted as FW1, a FuVeX Marvin (center), denoted as FW2, and two DJI Matrice 210 (right), denoted as RW1 and RW2,

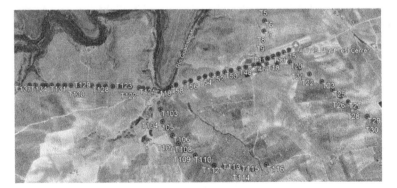

Figure 14.3 Power line located in the surroundings of flight test center ATLAS in Villacarrillo (Spain). Each transmission tower T_i is represented by a marker and its identification number i.

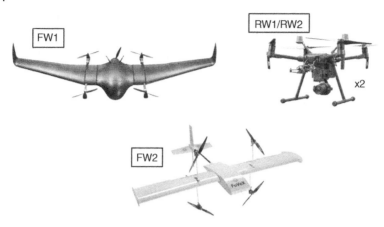

Figure 14.4 Fleet of heterogeneous UAVs selected for power-line inspection: one DeltaQuad Pro (left), one FuVeX Marvin (center), and two DJI Matrice 210 (right).

respectively. All of them start the mission from different known positions along the runway of ATLAS. Additionally, a recharging station has been installed inside the enclosure of the facilities of ATLAS.

The plan was generated by the approach presented above. For the computation of the set of weights $\mathcal{W}$, the reference flight speeds $v_{FW1}^{ref} = v_{FW2}^{ref} = 15$ m/s have been selected for the VTOL fixed-wing UAVs, while those corresponding to the multirotors have been fixed in $v_{RW1}^{ref} = v_{RW2}^{ref} = 5$ m/s. Once the graph G has been constructed together with the MILP formulation, the planning problem has been solved using MATLAB. For this, the *intlinprog* solver, which is specially oriented to the efficient computation of MILP problems, has been applied.

Figure 14.5 depicts the resulting plan, while Table 14.1 summarizes both the time required by each UAV to complete the mission and its flight distance. As can be seen, the planning method computes efficient assignments and sequences of

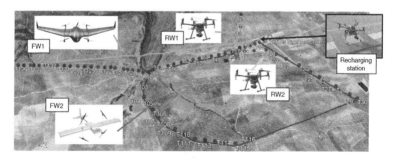

Figure 14.5 Plan computed for the inspection of a power line located in ATLAS flight test center using a fleet of heterogeneous UAVs.

Table 14.1 Planning results associated with Figure 14.5.

UAV	Time [min]	Distance [km]	Flight speed [m/s]
FW 1	9.22	8.30	15
FW 2	9.19	8.27	15
RW 1	9.73	2.92	5
RW 2	8.80	2.64	5

power-line segments for the heterogeneous team of UAVs. Thus, the total inspection time is minimized, allowing the inspection of 12 km of power line in less than 10 min. For that, the different flight speeds are exploited in such a way that the fastest UAVs (VTOL fixed-wing FW1 and FW2) are focused on the farthest power-line segments, while the multirotors RW1 and RW2, with a lower flight speed, are steered to the nearest ones. At the same time, the existing power-line branches are properly distributed according to their lengths and proximity to the UAV depots. The result is a plan with balanced flight times between UAVs.

In addition to the efficiency, the plan leads the muti-UAV team to complete the inspection mission successfully. In order to illustrate the latter, Figure 14.6 shows a snapshot of the video transmitted in flight to the Ground Control Station and recorded on-board by one of the UAVs. The video streaming allows the detection of an unexpected element (piece of plastic sheeting) hanging from a transmission tower that might produce a disruption in the power-line service.

Figure 14.6 View of a transmission tower when the UAVs follow the computed plan. An unexpected element (piece of plastic sheeting) is detected and georeferenced.

Moreover, since the video is georeferenced, an operator can go directly to the point of interest, which helps to provide a fast response in emergency situations. Finally, it is worth mentioning that, due to the large-scale nature of the inspection problem, the multi-UAV team should fly in BVLOS conditions. Consequently, the computed plan contributes substantially to ease the operation in these conditions by endowing the multi-UAV system with a higher level of autonomy.

14.4 Onboard Online Semantic Mapping

For the first use case previously presented in Section 14.2.1, an automatic mapping system should generate maps that can be used to extract quantitative measurements that enable predicting and avoiding potential risks to the power grid. These risks come mainly from the surroundings of the power line, in the form of vegetation of any kind coming close to the line. The mapping system should classify all the elements in the power line and its surroundings and map them accurately using the LiDAR sensors onboard each of the UAVs of the fleet. The map should enable taking distance measurements between the vegetation and the elements of the power grid such as electric lines and electric towers. The geometrical consistency between maps should be granted via fusing global GNSS measurements in the creation of these maps. This section is structured in three parts. The first one covers the selection of a mapping method and its adaptation to include GNSS measurements. The second presents the semantic classification performed to each LiDAR scan. The validation of the mapping system is presented in the third part. For the sake of space, the developments in the first and second parts are mainly covered in previous works [Valseca et al., 2022; Paneque et al., 2022] and cited accordingly. However, in this chapter, we address their combination to perform fully onboard, drift robust, and multi-UAV semantic mapping in power lines. During validation, we show how the GNSS sensor fusion allows combining results from different flights in a seamless way, since the individual maps are globally consistent. Using the resulting map, the evaluation of distances between power lines, vegetation, and other elements becomes straightforward by computing minimum distances between clusters of points.

14.4.1 GNSS-Endowed Mapping System

The required mapping system must be accurate, robust to different types of scenarios and vegetation, able to be executed in real time on computers on board the UAVs, and have low memory footprint for allowing long-range flights. Different state-of-the-art mapping software were analyzed to select the most suitable for the proposed application, namely *LOAM-Livox* [Lin and Zhang, 2020], *Livox-Mapping*

[Livox SDK, 2020], *LiLi-OM* [Li et al., 2021], and *FAST-LIO2* [Xu et al., 2022]. By performing an experimental comparison in the test scenarios [Paneque et al., 2022], *FAST-LIO2* was found to provide in general the best accuracy, computational cost, and memory consumption between the analyzed libraries. Thus, it was chosen as the basis of our mapping system.

FAST-LIO2 leverages an error-state manifold-based iterative Kalman filter (denoted as *ikFoM*) state estimation scheme. We refer the reader to Xu et al. [2022] for further details about on-manifold state estimation. To integrate GNSS measurements into *FAST-LIO2*, we can treat them as generic vector measurements $\in \mathbb{R}^3$. The measurement model for this sensor is obtained by estimating the position of the GNSS sensor at the current instant given the state of the robot:

$$\mathbf{h}\left(\mathbf{x}, \mathbf{v}\right) = \mathbf{T}_G^W \mathbf{T}_W^I \left(\mathbf{p}_I + \mathbf{v}_I\right), \tag{14.3}$$

where $\mathbf{x}$ is the state of the robot that includes $\mathbf{T}_G^W = \{\mathbf{t}_G^W, \mathbf{R}_G^W\}$ and $\mathbf{T}_W^I = \{\mathbf{t}_W^I, \mathbf{R}_W^I\}$, which, respectively, denote the transformations between the GNSS frame G and the global map frame of all UAVs W and between W and the inertial frame of the specific robot I. The position of the GNSS sensor in the inertial frame of the robot is $\mathbf{p}_I$ and is always fixed and calibrated with high precision. The random vector $\mathbf{v}_I \sim \mathcal{N}\left(\mathbf{0}, \mathcal{Q}_{\text{GNSS}}\right)$ models the uncertainty of the measurement.

The integration of the GNSS inside the *ikFoM* requires deriving two Jacobians. Since only the transformations in (14.3) are used to compute these Jacobians, we assume the full state vector is $\mathbf{x} = \{\mathbf{t}_W^I, \mathbf{R}_W^I, \mathbf{t}_G^W, \mathbf{R}_G^W\}$ for brevity. The Jacobians are computed as

$$\mathbf{H} = \left.\frac{\partial \mathbf{h}\left(\mathbf{x} \boxplus \delta\mathbf{x}, 0\right)}{\partial \delta\mathbf{x}}\right|_{\delta\mathbf{x}=0} = \left[\mathbf{R}_G^W \; -\mathbf{R}_G^W \mathbf{R}_W^I \lfloor \mathbf{p}_I \rfloor \; \mathbf{I}_3 \; -\mathbf{R}_G^W \lfloor \mathbf{T}_W^I \mathbf{p}_I \rfloor\right] \tag{14.4}$$

$$\mathbf{D} = \left.\frac{\partial \mathbf{h}\left(\mathbf{x}, \mathbf{v}\right)}{\partial \delta\mathbf{v}}\right|_{\delta\mathbf{v}=0} = \mathbf{R}_G^W \mathbf{R}_W^I \tag{14.5}$$

As shown in the validation in Section 14.4.3, map association between different UAVs is seamlessly performed by adding the GNSS measurements in the mapping system.

14.4.2   Reflectivity and Geometry-Based Semantic Classification

Elements surrounding a power line can be classified in four relevant categories: *Powerlines*, *Towers*, *Vegetation*, and *Soil*. Our method, see scheme in Figure 14.7, classifies LiDAR points using a combination of the LIDAR reflectivity and the spatial distribution of the acquired LiDAR points, see Valseca et al. [2022].

In short, the reflectivity is a measure of the ability of a surface to reflect radiation. Reflectivity depends on the material of the object that caused the reflection, the surface roughness and disposition, and also the incidence angle. Hence, reflectivity serves as a good property to distinguish between types of objects.

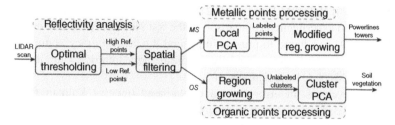

Figure 14.7 Scheme of the proposed LiDAR-based real-time segmentation method.

In the power-line inspection case, elements belonging to the power grid (electric lines and towers) tend to have low reflectivity since they are thin metal structures and braided cables, reflecting only a fraction of the diffused LiDAR ray, while vegetation and soil have a more consistent surface shape and provide higher reflectivity.

Knowing all this, the method works as follows. The first stage is to obtain a histogram of the current points' reflectivity and apply optimal thresholding to separate metallic and organic points into two sets, improving this classification with a local spatial filter to avoid outliers. Next, the points in the metallic set MS are classified using a local principal component analysis (PCA) knowing that power lines are distributed in one direction while the towers are not, and then applying a Mahalanobis distance-based region growing to generate clusters of points of type *Powerlines* and *Tower*. The final stage is to clusterize the points in the organic set OS by performing region growing and then classifying these clusters using PCA, obtaining *vegetation* and *soil* clusters.

14.4.3 Validation

The two presented methods for GNSS-endowed and semantic mapping in power-line inspection and maintenance have been validated both, separately and combined, in different experiments to evaluate their performance. The LiDAR sensor used in all experiments was a Livox Horizon, which is a high-performance solid-state 3D LiDAR with nonrepetitive horizontal scanning patterns. It features a field of view of $81.7° \times 25.1°$ and a detection range of 260 meter, giving up to 240,000 point measurements per second.

First, to validate the effect of integrating GNSS in the mapping software, we performed two different inspection flights and generate a joint map with and without integrating this sensor. Figure 14.8 shows how the resulting map is affected by some drift between both estimations, which can be canceled out with ease by integrating GNSS in both platforms.

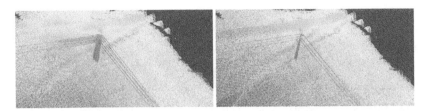

Figure 14.8 (Left) Original map obtained from two inspection experiments using *FAST-LIO2* without modifications. (Right) Resulting map when including GNSS.

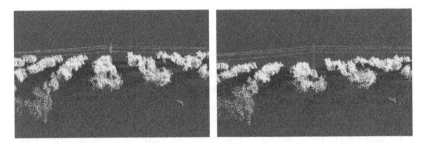

Figure 14.9 (Left) Resulting segmentation of a flight map classified by *PointNet++*. (Right) Segmentation obtained by the presented method.

Second, we validated the semantic classification method against *PointNet++*, a state-of-the-art classifier for point clouds using deep learning. Quantitative results in Valseca et al. [2022] show an overall improvement in classification performance, specially regarding the *Powerlines* (10% improvement) and *Tower* (89% improvement) classes. Figure 14.9 shows how our method had better performance even at intersections of different categories, which is the most difficult case in these scenarios.

Finally, we combined both systems and performed several (>30) mapping experiments in different power-line scenarios to assess the quality of the combined solution. The resulting scheme is able to produce high-quality, globally consistent, and semantically labeled maps for the studied scenarios. We first validated the accuracy of the maps in controlled scenarios where a ground-truth map was obtained with a Leica Total Station (4 mm scanning accuracy), obtaining an *RMSE* error of 3.8 cm against it. Then, we validated the mapping results in the different available power-line inspection scenarios. Figure 14.10 shows the resulting map obtained with two UAVs in an area where their flight plans partially overlap around an electric tower. It can be seem that no drift between the map appears and that all the elements are correctly classified.

Figure 14.10 Segmented map using the proposed method in an area where two UAV flight plans partially overlap.

14.5 Conclusion

This chapter has analyzed the application of systems with multiple UAVs to the inspection of large infrastructures. Particularly, the application to the inspection of a grid of electrical power lines has been considered. These applications involve long-range flights (many kilometers) and also detailed close views to achieve high accuracy in the inspection of some elements of the line. Conventional multirotor systems can obtain these detailed views hovering near the elements but do not have the required range and time of flight. On the other hand, light fixed-wing UAVs cannot have the payload required to carry sensor systems for accurate inspection. The new fixed-wing VTOL systems cannot be used yet to perform the above-mentioned hovering and large UAVs with significant payloads for accurate inspection have high cost and difficulties to be applied under the current regulations. Then, a current efficient solution, proposed in the H2020 AERIAL-CORE project and presented in this chapter, is to combine different platforms in a system with heterogeneous UAVs. Moreover, taking into account range and time of flight constraints, the system includes stations to recharge automatically the batteries of the UAVs by applying autonomous take-off and landing.

This chapter has pointed out the interest of aerial robotic systems with planning and perception capabilities. The planning method can compute the optimal trajectories for the heterogeneous UAVs taking into account their constraints and the location of the recharging stations. Moreover, the chapter has presented the results of a semantic mapping method that can be applied onboard the UAVs to compute online the vegetation map near the electrical line, which is very relevant in the inspection of the electrical lines. Both the planning and the mapping techniques have been integrated in an inspection system and successfully validated in the ATLAS UAV experimentation center of FADA-CATEC in Villacarrillo (Jaen, Spain) in the framework of the above-mentioned H2020 AERIAL-CORE project.

The validation of the integrated system in real environments and BVLOS conditions posed significant challenges and led to interesting lessons learnt. First, the

BVLOS nature of the experiments required extra levels of safety and robustness. Although ATLAS has a segregated space of 30 × 35 km, BVLOS missions constrain the operation of the safety pilots that are typically involved in these complex missions. To cope with that, additional onboard software components were added to monitor the fulfillment of the mission and abort the mission in case of deviation over the plans. In addition, the difficulties of integrating heterogeneous autonomous aerial robots are higher in case of BVLOS missions, in which the aerial robots are distant from each other and communications are more prone to failures. In fact, the communications between the aerial robots were carefully designed including the communication technology and hardware components, the selection of the data to be interchanged (to reduce bandwidth), and the use of communication buffers to cope with temporary communication interruptions. In the experimentation, we also found difficulties due to errors in the available power-line maps. Although regulation imposes that power lines should be accurately mapped, we found many power lines whose mapping errors were tens of meters and also some power lines that were not included in the maps, which can involve high risks. We noticed these power-line mapping errors when performing the first preliminary experiments, and this motivated the redesign of the robots to endow them with basic reactive obstacle detection and avoidance capabilities, not described in this chapter for the sake of brevity.

The presented experiments have shown the interest of the proposed techniques for the industrial application. These techniques have been integrated in an AERIAL-CORE inspection and maintenance system with other technologies dealing with aerial manipulation and coworking with human operators. This integrated system has also been validated in the ATLAS Center during the second half of 2023. Regarding future work, two main lines can be highlighted in the search of increasing the TRL (technology readiness level) of the proposed solution. First, the integration of long-range communication capabilities into the inspection system. In this sense, 5G or satellite communications can help to mitigate current limitations in the communications between the GCS (Ground Control Station) and the multi-UAV team when the latter operates at long distances. This will allow the extensive inspection of power lines, covering hundreds of kilometers of these infrastructures in an effective way. Second, the integration of perception methods for power-line tracking. As it was advanced above, power-line mapping errors, which are common, can lead to important deviations from the real power lines when UAVs follow computed plans. In the worst case, this can even force to abort the operation, making the executed mission useless. In order to address this problem, current techniques consist of correcting the position of those transmission towers, which are georeferenced badly. However, this is a time-consuming task that can become impractical easily as the number of errors increases. Alternatively, preliminary results based on visual servoing show a promising solution to detect the power line and provide the proper control references to the UAVs while inspecting it. Thus, robust power-line tracking could be ensured in spite of uncertainties in available maps.

Bibliography

D. Alejo, R. Conde, J.A. Cobano, and A. Ollero. Multi-UAV collision avoidance with separation assurance under uncertainties. In *2009 IEEE International Conference on Mechatronics*, pages 1–6, 2009. doi: 10.1109/ICMECH.2009.4957235.

F. Augugliaro, A. Mirjan, F. Gramazio, M. Kohler, and R. D'Andrea. Building tensile structures with flying machines. In *2013 IEEE/RSJ International Conference on Intelligent Robots and Systems*, pages 3487–3492. IEEE, 2013.

H. Baik and J. Valenzuela. Unmanned aircraft system path planning for visually inspecting electric transmission towers. *Journal of Intelligent & Robotic Systems*, 95(3):1097–1111, 2019.

L. Bauersfeld and D. Scaramuzza. Range, endurance, and optimal speed estimates for multicopters. *IEEE Robotics and Automation Letters*, 7(2):2953–2960, 2022. doi: 10.1109/LRA.2022.3145063.

J. Cacace, S.M. Orozco-Soto, A. Suarez, A. Caballero, M. Orsag, S. Bogdan, G. Vasiljevic, E. Ebeid, J.A. Acosta Rodriguez, and A. Ollero. Safe local aerial manipulation for the installation of devices on power lines: AERIAL-CORE first year results and designs. *Applied Sciences*, 11(13):6220, 2021.

Y.U. Cao, A.S. Fukunaga, and A. Kahng. Cooperative mobile robotics: Antecedents and directions. *Autonomous Robots*, 4(1):7–27, 1997. doi: 10.1023/A: 1008855018923.

C. Deng, S. Wang, Z. Huang, Z. Tan, and J. Liu. Unmanned aerial vehicles for power line inspection: A cooperative way in platforms and communications. *Journal of Communications*, 9(9):687–692, 2014.

T. He, Y. Zeng, and Z. Hu. Research of multi-rotor UAVs detailed autonomous inspection technology of transmission lines based on route planning. *IEEE Access*, 7:114955–114965, 2019.

N. Iversen, A. Kramberger, O.B. Schofield, and E. Ebeid. Pneumatic-mechanical systems in UAVs: Autonomous power line sensor unit deployment. In *2021 IEEE International Conference on Robotics and Automation (ICRA)*, pages 548–554. IEEE, 2021a.

N. Iversen, O.B. Schofield, L. Cousin, N. Ayoub, G. Vom Bögel, and E. Ebeid. Design, integration and implementation of an intelligent and self-recharging drone system for autonomous power line inspection. In *2021 IEEE/RSJ International Conference on Intelligent Robots and Systems (IROS)*, pages 4168–4175. IEEE, 2021b.

G. Laporte. What you should know about the vehicle routing problem. *Naval Research Logistics (NRL)*, 54(8):811–819, 2007.

K. Li, M. Li, and U.D. Hanebeck. Towards high-performance solid-state-LiDAR-inertial odometry and mapping. *IEEE Robotics and Automation Letters*, 6(3): 5167–5174, 2021. doi: 10.1109/LRA.2021.3070251.

J. Lin and F. Zhang. Loam livox: A fast, robust, high-precision LiDAR odometry and mapping package for LiDARs of small FoV. In *2020 IEEE International Conference on Robotics and Automation (ICRA)*, pages 3126–3131. IEEE, 2020.

Livox SDK. Livox-Mapping. URL https://github.com/Livox-SDK/livox_mapping, 2020.

I. Maza, F. Caballero, J. Capitan, J.R. Martinez-de Dios, and A. Ollero. A distributed architecture for a robotic platform with aerial sensor transportation and self-deployment capabilities. *Journal of Field Robotics*, 28(3):303–328, 2011. ISSN 1556-4959. doi: 10.1002/rob.20383.

I. Maza, A. Ollero, E. Casado, and D. Scarlatti. *Classification of Multi-UAV architectures*, pages 953–975. Springer, Netherlands, 2015. ISBN 978-90-481-9706-4. doi: 10.1007/978-90-481-9707-1_119.

F. Nekovář, J. Faigl, and M. Saska. Multi-tour set traveling salesman problem in planning power transmission line inspection. *IEEE Robotics and Automation Letters*, 6(4):6196–6203, 2021. doi: 10.1109/LRA.2021.3091695.

A. Ollero and B. Siciliano. *Aerial robotic manipulation research, development and applications*. Springer Tracts in Advanced Robotics, 2019. ISBN 978-3-030-12944-6.

A. Ollero, G. Heredia, A. Franchi, G. Antonelli, K. Kondak, A. Sanfeliu, A. Viguria, J.R. Martinez-de Dios, F. Pierri, J. Cortes et al. The AEROARMS project: Aerial robots with advanced manipulation capabilities for inspection and maintenance. *IEEE Robotics & Automation Magazine*, 25(4):12–23, 2018. doi: 10.1109/MRA.2018.2852789.

A. Ollero, M. Tognon, A. Suarez, D. Lee, and A. Franchi. Past, present, and future of aerial robotic manipulators. *IEEE Transactions on Robotics*, 38(1):626–645, 2022.

J. Paneque, V. Valseca, J.R. Martínez-de Dios, and A. Ollero. Autonomous reactive LiDAR-based mapping for powerline inspection. In *2022 International Conference on Unmanned Aircraft Systems (ICUAS)*, pages 962–971. IEEE, 2022.

A. Suarez, A.E. Jimenez-Cano, V.M. Vega, G. Heredia, A. Rodriguez-Castao, and A. Ollero. Design of a lightweight dual arm system for aerial manipulation. *Mechatronics*, 50:30–44, 2018. ISSN 0957-4158. doi: https://doi.org/10.1016/j.mechatronics.2018.01.005. URL https://www.sciencedirect.com/science/article/pii/S0957415818300011.

V. Valseca, J. Paneque, J.R. Martínez-de Dios, and A. Ollero. Real-time LiDAR-based semantic classification for powerline inspection. In *2022 International Conference on Unmanned Aircraft Systems (ICUAS)*, pages 478–486. IEEE, 2022.

W. Xu, Y. Cai, D. He, J. Lin, and F. Zhang. FAST-LIO2: Fast direct LiDAR-inertial odometry. *IEEE Transactions on Robotics*, 38(4):2053–2073, 2022. doi: 10.1109/TRO.2022.3141876.

15

Robotic Platforms for Inspection of Oil Refineries

*Mauricio Calva**

Chevron Technical Center, Chevron, USA

15.1 Refining Oil for Fuels and Petrochemical Basics

Large oil and gas producers recognize that the world of energy is transforming, and it cannot be any longer focused on fuels obtained from crude oil and natural gas. As the world moves to other forms of energy, the corporations are already changing and directing their knowledge and expertise to support a new generation of energy sources that will provide the basis for growth and development for all in the near and far future, responsibly.

During the transition to new energies, oil-derived fuels will be needed for twenty to forty more years, likely in decreasing quantities, but they will not disappear overnight. Maintaining the facilities so they can operate safely, efficiently, and with ever lower carbon footprint is of the highest priority.

Oil and natural gas are mined from reservoirs in the ground both using offshore and onshore facilities, once extracted they are preprocessed to separate water, liquid, and gaseous components and then pumped or transported to facilities for further processing. These facilities are known as oil refineries and transform crude oil and natural gas into different products including fuels and chemicals that are used by others to manufacture a myriad of products (Figure 15.1).

The process of oil in a refinery follows a sequence of operations intended to increase the value of the products obtained. It is not a linear process, but it normally starts by removing the remaining water and salts to be heated and distilled to separate components of the mixture that naturally forms the crude oil. The different streams obtained in this process continue to be transformed in different areas of the refineries known as process units or plants. Each unit will produce secondary streams of different liquids and gasses that feed additional processes or

*Email: mcalva@chevron.com

Infrastructure Robotics: Methodologies, Robotic Systems and Applications, First Edition.
Edited by Dikai Liu, Carlos Balaguer, Gamini Dissanayake, and Mirko Kovac.
© 2024 The Institute of Electrical and Electronics Engineers, Inc. Published 2024 by John Wiley & Sons, Inc.

Figure 15.1 Offshore floating oil production, storage, and offloading. Source: Chevron Picture.

get mixed as components of final products. Gasoline or petrol, liquefied petroleum gas for heating, diesel fuel, aircraft fuels, and lubricants are the main products, but other basics like alcohols, solvents, ethylene, and many other chemicals are also produced and processed by others to produce medicines, fertilizers, plastics, fabrics, tires, etc. It is hard to find now a product that in one way or another has not been influenced by oil and natural gas products.

The processing of the oil and natural gas in a refinery requires keeping it contained within the boundaries of the equipment to avoid releases and contamination of air or soil and to operate safely and efficiently. The oil received from production areas is stored in bulk liquid and placed in large storage tanks. The movement of liquids and gas is done through piping that conducts the different streams to process equipment of an incredible variety of sizes and geometries. Some of them are known as reactors, distillation columns, separators, desalters, heat exchangers, and storage vessels, among others. Mechanical components like pumps, compressors, and fans are also an important part, as well as complementary systems like cooling water towers for process, fire water service, sewer, steam, compressed air, electric power, controls, and wired and wireless communications. The amount of pieces of equipment is incredibly large, and the total distance of piping is easily calculated in the tenths of thousands of kilometers for each facility. Many of these components operate at high temperatures and are covered by insulation to preserve thermal energy. We normally classify the process equipment into two general categories: fixed and rotary equipment, this last one includes most pumps, compressors, and other machinery, the rest is considered fixed equipment.

The soundness and fitness for service of all components of a refinery are essential for its reliable and safe operation.

The number of tanks, vessels, pumps, motors, structural steel, piping, sensors, and other critical components is incredibly large. The sizes, configurations, heights of some of these components, and the shear extension on the surface of these facilities make them one of the largest and most highly complex industrial facilities. The technologies used for production and inspection are the product of no less than a 100 years of continuous improvement.

15.2 The Inspection Process

An inspection process that provides information about the condition of the equipment is essential to execute effective preventive maintenance, repairs, replacement, and modifications to preserve reliable and safe operations of oil refineries. In human terms, the inspection process is equivalent to a health plan that includes routine visits, exams, laboratory analysis, and consulting with specialists. The preferred solutions are noninvasive like using radiography or ultrasound, and other imaging techniques and only in those cases where not enough information can be obtained, or not possible, exploratory surgery and other invasive tools are used. All to maintain a happy and productive life. In most oil and gas facilities, several groups of engineers are involved in the processes necessary to maintain the health of the components. Maintaining the process equipment in conditions that allow the proper operation with a clear understanding of the damages or degradation that it may have suffered is essential. The risk associated with failure in process equipment cannot be underestimated. A failure can have an operational impact that reduces the output or affects the efficiency of the process.

The rupture of process equipment that would allow the release of materials may have serious consequences. As most of these components are flammable or explosive, the risk of fire and catastrophic failure are always present. Similarly, the release of gas or vapors into the atmosphere or liquids into the soil is not acceptable and needs to be avoided. Multiple events in the history of oil refineries have shown the terrible impact on operators, contractors, and their neighbors. It is one of the tenets of the company to maintain safe operations. The companies are aware of the consequences and know they need to invest in the process of maintenance to avoid or reduce the risk. Most inspection plans are based on international codes like the American Petroleum Institute API and similar organizations around the world and are subject to regulatory controls by the government or their agencies.

Oil refinery owners' and operators' work is based on a commitment to maintain safe operation and have to be ready to demonstrate compliance. In most organizations, the maintenance process includes multiple activities and processes that go from time-driven actions to predictive and advanced risk-based inspection among others. Each component is analyzed and studied to understand the susceptibility of damage, the operational reliability, the damage mechanisms, and the specific maintenance process needed. One of the tools of maintenance is the inspection process.

The inspection of equipment means collecting information directly from it, measuring characteristics and properties that will give us an indication of their health. Techniques like radiography and ultrasound are also used in the industry, together with electromagnetic and physical chemical methods, these four groups are the most important methods of inspection. Dozens of techniques and specific applications for each of these methods are available and continually evolving with new sensors, data collection and analysis, and software. Considering the large population of vessels and piping in a facility, making decisions of where to inspect is done with care by specialists that can relate the history of the vessel in question, or others similar. We want to go look at the locations that are more vulnerable or likely to fail, and those that carry the worst consequence if a failure or release happens. A combination of these factors, likelihood, and consequences is normally a good definition of risk. These locations may be where the vessel or pipe circuit has the highest temperature, at an injection point, or the flow velocity is the lowest, at elbows of injection points, at liquid–vapor interfaces, etc.

Each component is analyzed, and a specific inspection plan is prepared. For piping, sections are grouped in circuits that are exposed to the same fluid and operational conditions. Representative points of the circuit with the highest chance of being degraded are selected. These points are known as corrosion monitoring locations (CMLs) (Figure 15.2).

Once the inspection and testing plan have been completed, a second group takes over to select the proper nondestructive examination (NDE) methods that could be used to detect damage or degradation. The selection of the proper technique strikes a balance between the type of damage, sensitivity for detection, sizing accuracy, cost, and time available for the inspection process. Other variables like the configuration of the vessel, type of materials, thickness, presence of insulation, elevation, accessibility, and surface condition are also considered for the selection of the method and technique used for the inspection.

Out of all the inspection methods, the visual evaluation of components by an experimented and knowledgeable inspector has incredible value. In most cases, the inspection process starts by just looking at locations for visual detection of mechanical damage, coating damage, discoloration, and corrosion are some of the items they can discover. We dedicate a good part of the inspection process to providing inspectors with the means to access those areas so they can see and make a direct assessment of the materials and equipment. The inspection activities that can be performed during operations are preferred as there is no impact on

Figure 15.2 Partial view of an oil refinery. Note the number of vessels and piping. Source: Chevron Picture.

production and avoids lengthy turnarounds or shutdowns by having a good understanding of the condition of the vessel. Only in those cases where it is not possible, maybe because the reach from the outside does not provide all the information needed or is just not accessible the inspection process is left to the maintenance period and executed with equipment out of service. Multiple standards and codes are available like the American Society for Mechanical Engineers ASME and the American Society for Nondestructive Testing ASNT.

The data, images, and information collected during the inspection process are stored and used by specialists to define the damage detected needs to be further evaluated and additional information collected. Loops of inspection and data analysis are not uncommon, all those involved needs to be convinced that there is enough information to make a decision. If the damage or degradation is deemed of interest based on previous experience, a third step in the process starts. Specialists in materials, corrosion engineering, process, operations, and maintenance get involved, as well as fracture mechanics, critical assessment engineers, fitness for service, and others as needed to dictate the action to follow. If the damage is acknowledged but considered not critical to the point that the vessel can safely operate another cycle to the next shutdown for inspection, it will be recorded and used as a reference for future inspections. If the damage is such that needs repair, the action is started immediately to reduce the impact on the shutdown timing. The worst case is when damage or degradation is discovered, and the equipment is deemed not repairable and needs to be replaced. Nobody likes surprises, a new vessel or piping section will take at least a few weeks or months if the vessels are complex to manufacture and replace. The inspection process is intended to avoid these scenarios where unexpected repairs or replacement is needed.

The traditional inspection process is manual, both the visual inspection and NDE techniques have been designed for a person to use a camera or hold a probe in his hands to collect the data. Providing access to personnel to reach the locations where inspection is needed is time consuming and expensive. Many of these locations are at elevated structures, hard to reach requiring scaffold, cranes, or rope access. Also, these locations may be at high temperatures or places where dangerous materials may be present. One of the main drivers for the utilization of robotics is to avoid personnel to work in dangerous conditions. It has also been well documented the impact that working in dangerous or uncomfortable conditions has on humans and the effect on concentration, focus on tasks, performance, and overall quality of the inspection activity. Removing personnel from dangerous locations for data collection in inspection processes is the main goal of the utilization of the robotic platform.

Other factors like the large amount of equipment to inspect, demand actions from the inspectors that are dull, repetitive, and prone to errors. Even in the best conditions, humans are likely to make mistakes that will require rework and repetition of tasks. We want to use the talent of the inspectors to plan the inspection process, to analyze, to evaluate, to interpret, and to provide high-value opinions. The action of collecting data in dangerous and repetitive tasks can be done by machines. We have been exploring the use of automated, remotely operated, and autonomous platforms for inspection for about 10 years now.

15.3 Inspection and Mechanical Integrity of Oil Refinery Components

The process and storage equipment can be classified into four large groups: liquid storage tanks, process and storage vessels, heaters and heat exchangers, and process piping. This classification serves the purpose as some basic operational, maintenance, and degradation are similar. Let us look in more detail into the inspection activities for three of these groups, storage tanks for liquids, pressure vessels, and piping. The techniques and methods described here would in general be used for other components not directly grouped in these categories.

15.3.1 Liquid Storage Tank Inspection

Storage tanks are large vertical cylinders that contain liquids like crude oil, kerosene, diesel, gasoline or petrol, jet fuel, heating oil, and others. The storage tanks may have a fixed roof or the roof may float on the liquid. Some floating roof tanks may have an external dome or external roof. The diameters can be as small as 10 feet (3 meter) up to 330 feet (100 meter) and heights of several dozen feet. They are mostly constructed from carbon steel or low alloys. One of the main characteristics of these tanks is that access to the tank floor is not possible during operation, and access to floating roofs and internal floating roofs is also not normally possible. Internal components like support columns, roof structures, piping, and sumps are also not accessible from the outside. American standards like the API 650 require the tanks to be designed to operate free of damage and avoid leakage for at least 10 years. Certain designs are approved to operate for 20 and up to 30 years when the tanks are new. It is the floor of the tank the one that carries the highest risk, in most cases, the tanks lay on soil or gravel and any leakage of liquid may go undetected for a long time sipping straight into the ground. In most cases, the condition of the floor plates is the main driver to removing tanks of service and opening them to gain access to the floor and perform inspections. The roof structure is another component that carries risk, if the roof collapse, it is likely to create a major spillage or catch fire. All other components are also important, and the health of the tank requires maintenance and inspection of all of them (Figure 15.3).

When a storage tank is removed from service for inspection a sequence of activities starts, they may be different depending on the tank design and the product contained. Once the liquid is pumped out, the remaining vapors need to be oxidized or removed through a process called degassing. Once the tank is open, sediments or sludge can be removed, it normally requires cutting a large door on the tank shell to get front loaders or other machinery to push it out. Personnel equipped with high-pressure washers and manual tools get into the tank to complete the cleaning process. In most cases, the personnel has air supply and respirators as the residues and vapors need to be avoided.

After the floor is clean, the inspection crews will go in and perform visual inspections, for evidence of mechanical damage, coating degradation, and areas

Figure 15.3 Liquid storage tanks. Source: Chevron Picture.

of corrosion. The NDE crews will follow using magnetic flux leakage (MFL) on a manual scanner to detect areas of corrosion on the floor, those areas where the floor thickness is smaller are followed up with an ultrasonic UT technique to confirm the presence of corrosion, measure the remaining floor thickness and the extent of the areas of corrosion. The floor plates are either replaced or patches are welded over when the remaining thickness is such that it could leak before the next inspection period. A new coating is normally applied on the floor to extend the life. Piping, columns, and structure are also inspected in most cases requiring the construction of a scaffold for the inspectors and examiners to reach these components. The data collected is analyzed and repairs or modifications as needed are executed.

The complete evolution may take a few weeks for clean product tanks, and up to several months for large tanks with crude oil or other heavy sludge-forming products. Other than MFL and ultrasonic follow-up for the floor, visual inspection and ultrasonic thickness are the most common, magnetic particles when cracking is suspected supplements the suit of NDEs. When the tank is in service their access to the tank shell, external pipe sections, nozzles, and depending on the tank configuration to the tank roof. We also perform inspections to check for settlement and tank deformation. Preferential corrosion at the shell welds and corrosion at the liquid level is also checked using external ultrasonic techniques like corrosion scanners, also known as automated ultrasonics (AUT) or C-Scan are used.

15.3.2 Pressurized Vessels Inspection

In this category, we include cylindrical or spherical vessels that are used for the processing of liquids, gases, or mixtures, they can be in the horizontal or vertical position and operate at high pressure, which means the thickness is normally large. The dimensions vary from less than 3 feet (1 meter) in diameter up to 40 feet (15 meters) on cylindrical and up to 75 feet (25 meters) for spherical vessels. Most of them are constructed of Carbon Steel and in some cases are either coated with corrosion protective materials or have a corrosion-resistant alloy on the inside applied as a clad or welded overlay. Many of these vessels are insulated and have a jacket made of aluminum or steel to protect the insulation from water ingress.

In traditional inspection, an evaluation is made to define the likely damage mechanisms and their location, if possible on-stream or in-service inspection is preferred, as it can be done with the equipment in service, avoiding operational disruption and reducing the length of inspection during shutdowns or turnarounds. These devices are equipped with nozzles that connect to piping and valves meant to carry the product in and removed the processed fluids. Most process vessels have internal components from simple baffles, turbulence breakers, impact plates, and separation plates, to others with complex geometry and multiple components like demisters, distillation trays, etc. Special vessels like fluid catalytic converters have other vessels inside like cyclones. When vessels come out of service during maintenance periods, the remaining liquids are emptied and the vapors are oxidized or vented, crews go inside to clean and remove residues, sediments, and sludge as needed. For this operation, a scaffold may be needed to provide access to the crews. The inspectors will then go in to perform a visual inspection searching for mechanical damage, areas of corrosion, coating damage, cracking, and other signs of degradation (Figure 15.4).

NDE crews will come in to measure corrosion, weld inspections, evaluate the coating condition, and detect tight cracking-like stress corrosion or environmentally induced cracking. Techniques like ultrasonic methods based on phased arrays, time of flight diffraction, and angular or shear wave beams are used. Other methods like magnetic particles, eddy currents on single probes or arrays, and coating spark testing among others are used. On-stream or in-service inspection requires in many cases removing the insulation, after that, visual inspection, and ultrasonic techniques for corrosion detection using automated scanners known as AUT or "C" Scans are used. The detection of cracking at nozzle welds is also done with ultrasonic techniques like phased arrays, time of flight diffraction, and angular or shear wave angular beams. Moist or water trapped in insulation will degrade the coatings and increase the chance of corrosion. Technologies like pulsed eddy currents that can detect corrosion under insulation without the need of removing insulation are finding their place in the inspection processes. If the vessels are known to have accelerated corrosion in specific areas, or the access to certain areas is difficult permanent monitoring devices measuring thickness are

Figure 15.4 Pressurized process vessel. Source: Chevron Picture.

installed and data is collected either via radio or using radio-frequency induced device equipped sensors.

Access to this equipment requires scaffold construction either on the outside or the inside of the vessels, sometimes rope-access, man lifts or crane baskets are used. The inspection techniques are manual and require a person to place and manipulate the probes to collect the data. Inspectors getting inside vessels are exposed to higher risks as they are in dangerous confined spaces.

15.3.3 Process Pipping

The amount of piping in oil refineries is incredibly large, with diameters going from a half inch (12 mm) for instrumentation and through all other intermediate diameters for process piping and going up to 60 inches (one and a half meters). The distribution is normally linear where the small diameter lines are more common than the larger ones. Different specifications of piping depending on pressure ranking are used, most pipes are carbon or low alloy steels and externally coated. About half the pipes are insulated with a protective jacket made of aluminum or steel to protect the insulation to avoid water ingress. It is estimated that in a refinery that process about 200,000 bbls/day, there is about 100,000 km of piping. The challenge is then where to look, as attempting to inspect every inch of this piping would be impossible.

Techniques like risk-based inspection (RBI), analysis of circuits to identify sections that have a higher susceptibility to damage, and evaluation of the consequences of a leak or release are used to define the location of the inspection areas and also the frequency of inspection. For noninsulated piping, the detection and characterization of external damage and corrosion are done with visual inspection looking for areas where the coating has degraded, this includes supports and places where water mist or water dripping is present. For insulated piping, the external corrosion requires in most cases removing the insulation and performing a visual inspection for areas of corrosion. As mentioned above for pressure vessels, a metallic jacket protects the insulation from rain and water, and any perforation or damage on the protective jacket will allow water to moisten the insulation and create conditions for corrosion under insulation (CUI) (Figure 15.5).

Techniques like guided waves or long range ultrasonics as well as pulsed eddy currents can be used for the detection of CUI, but are highly dependent on the experience and knowledge of the technician, and have serious limitations on sensitivity and conditions. A well-written and careful execution of both techniques provides an assessment of CUI, and any detection of damage should trigger a follow-up removing the insulation at those locations to perform a direct visual and NDE inspection of the damage. Other alternatives like real-time radiography are capable of detecting corrosion product accumulation under the insulation, this method is subject to the technician's careful execution and patience, which in

Figure 15.5 Process piping in a refinery. Source: Chevron Picture.

some conditions becomes a serious obstacle as human factors are highly influential in this process. On smaller diameter piping, normally below eight inches, it is possible to use profile radiography for the detection of CUI and internal corrosion. Also, double-wall radiography is possible to detect internal corrosion on insulated piping.

For internal corrosion of the pipes, in addition to profile radiography and double-wall radiography, a well-designed program to measure the thickness using ultrasonics at specific points where the highest likelihood of internal corrosion exist is a common and well-established technique. The point thickness measurements are for general corrosion, and they will not find localized corrosion or other nonhomogeneous damage on the pipe. Thickness point measurements using permanently attached probes with radio communications and also those with radio frequency identification (RFID) tags. Other components that are included in this group are stacks, flares, and other slim vertical structures.

15.3.4 Heat Exchanger Bundles

Heat exchangers are process equipment that has two basic components an external shell and an internal tubing bundle, and two liquids at different temperatures circulate both inside the tubing and outside of it in contact with the shell surface. The largest surface of the tubing creates the most efficient heat transfer.

The shell itself can be inspected and falls within the requirements of most pressure vessels, except that some sections may have a complex geometry like the channel and the closure flange. The bundle tubing is open to one or both sides and is inspected by inserting different tools inside. The tubing is subject to corrosion and cracking because of the small diameter going from 1/2″ to 3/4″ and thickness of only a few thousand of an inch the inspection process is challenging. The most common technique used is eddy currents in different configurations depending on the type of metal. Normally for nonferromagnetic tubing, eddy current works very well, but for ferromagnetic tubing, specialized techniques like remote field and near-field eddy current are used. A rotating ultrasonics tool is also used on nonferromagnetic tubing, it provides a direct thickness measurement around the circumference using a rotating mirror, and as it is pushed and pulled inside it creates a full surface coverage. The eddy current techniques are a lot faster than the ultrasonic rotating head. The analysis and interpretation of the eddy current are harder and require highly trained and experienced personnel. One of the major issues with the inspection of the bundle is the cleaning process, most of the quality issues with the data can be related to the surface cleaning both inside and outside the bundle.

15.4 Plant Operations, Surveillance, Maintenance Activities, and Others

15.4.1 Surveillance, Operations, and Maintenance of Oil and Gas Refineries

In addition to the inspection process related to Mechanical Integrity, the Operations, Maintenance and Environmental groups in Oil Refineries perform tasks to verify that the system and process are operating correctly. These processes are intended to assure a reliable and efficient operation of each one of the components, avoid escapes on the operation that may reach conditions of operational upsets, damage due to operation control loss, and release of products through safety systems like pressure relieve valves or flares, and ultimately to reach conditions that the equipment and materials are not designed to support creating a risk of rupture and catastrophic events that would have a significant effect on the safety or health of the operators, contractors, our neighbors or the environment.

Maintaining control of the process within the equipment specification is a tenant of operations and must be observed. The activities related to these processes require both the presence of instrumentation to relay field operational conditions to the operators and computerized and automatic systems, variables like temperature, pressure, vibration, presence of hydrocarbons on the air, flow velocity, liquid levels, chemical and physical–chemical composition, and characteristics of feeds and products as they transit through the different segments of the refinery to be mixed and reprocessed is a complex process that requires high-tech components and most importantly expertise, experience, and knowledge of the operators and process engineers (Figure 15.6).

In addition to the information transferred by the instrumentation, the operators are required to walk the units and collect visual information and use sensors and

Figure 15.6 Petrochemical facility. General view. Source: Chevron Picture.

tools to verify that the values and plant conditions received in the control room match the measurements and observations in the field. Groups like maintenance will check that oil levels, lubrication, vibration, and other variables for rotary components comply. Environmental groups will check the abnormal presence of contaminants in the air, soil, and water. Others like design engineers, construction and repair crews, as well as turnaround planners will walk the refinery units to take measurements and prepare for changes, repairs, and replacement of equipment, or to execute the maintenance needed. When conditions in the plant become unsafe due to the presence of contaminants in the air, high explosive levels, or upset operational conditions, operators and first responders get dressed in protective gear and go to those locations to understand the conditions in the field and to address the issues. They become the ears, eyes, and hands of the control room operators to execute tasks in these incredibly dangerous conditions.

Keeping all the systems and components in good operational conditions is another high-priority activity, and failures and unscheduled shutdowns in equipment may affect the whole production unit and the refinery. A group of maintenance technicians and engineers create maintenance programs to maintain a reliable and safe operation. Using a combination of fixed instrumentation and human intervention to verify variables are used to maintain the plant components in check.

Visually confirming these conditions in the field by walking, measuring, observing, and executing routine tasks are part of this process. Some of them are simple like checking the oil level in a pump, and others are quite complex like replacing filters or washing equipment internals. Other activities related to checking the coating condition on process components, structural steel, and others are the responsibility of these groups. Roads and buildings' condition and repairs are also on their list of duties.

15.4.2 Safety and Security

Like any industrial facility, oil refineries keep the highest levels of industrial safety and physical security of the facilities.

All personnel, including contractors, are required to take safety training, have approved work permits, register at the control room before starting any job, have a radio for direct communication with the control room in case conditions change, and wear personal protective equipment adequate for the work and plant conditions. Work in elevated structures, confined spaces, or other dangerous conditions triggers additional requirements for conditions verification, protective measurements, and rescue personnel. Verifying that everyone executing physical work complies with these requirements is very important, refinery personnel in safety groups are an important part of this process, they walk the units when work is executed to verify safe conditions and execution per the approved procedures. The task becomes extremely complex during turnarounds, construction, and major repairs where hundreds of people both employees and contractors are needed.

The physical security of our premises is taken very seriously, avoiding the public to enter facilities without notice or intruders with intentions to steal or damage the facilities needs to be avoided. Patrolling the fences around the refineries and controlling access at the gates is an important activity that is executed by a combination of technology and personnel.

15.4.3 Utilities and Support Activities

The oil refineries are large industrial facilities that for the most part require utilities and services like a small city. Many of the services like water and power are either extracted and produced in-house or we have agreements with the local government and external suppliers to provide them. In most refineries, electric co-generation and substations are owned and operated internally, the distribution of power from domestic levels to very high voltage for large motors is part of this process, which requires transmission, transformation, and distribution of power across the facility.

We have electric switch rooms to operate this complex network. Service water for process and cooling water for heat transfer are also supplied across the refinery using piping and systems specifically for this application. We also have steam production and distribution across the refinery as needed for the different processes. The steam is supplied in a variety of pressures and moisture content. We have large warehouses that store parts, components, and materials that need to be transported to the work areas. Samples of intermediate and finished products are collected and transported to the analysis and chemical laboratories for quality control. Other services like food preparation and distribution, mail, and packages are also part of the operation.

15.5   Robotic Systems for Inspection

As it may have been noticed in the sections above, the inspection of oil refineries is an extensive and complex process that has been designed to be executed by people. The fact that data needs to be collected on the field from actual process equipment using visual and NDE methods, and that the data collected requires evaluation, analysis, and interpretation to make use of it, makes it in principle, and by design a human activity. As has also been noted above, the importance of the data collected in decisions that affect the safety of the operators, neighbors, and the environment and also to operate in a reliable, efficient, and safe way.

It is well documented that the quality of the data collected is very limited, the sensitivity, accuracy, and repeatability are affected by three main factors. The first one is the limits of the physics and technology behind the methods and techniques used. The second one is the human by itself, with biases produced by his knowledge, experience, and other factors related to boredom, tiredness, routine, activities performed in uncomfortable conditions, physically demanding, mentally exhausting, or at risky locations. The third one is the errors created by the process of inspection and documentation associated where misidentification of components and errors in orientation, flow direction, etc., as well as data mislabeling, file name errors, storage errors, and others are possible.

The use of automated systems in oil refineries is not new, we have a complete set of digital instrumentation and actuators that are connected to computers and interact with operators to process engineers to run the processes. The introduction of robotic platforms for inspection is also not new, since the 1990s mechanized devices and semi-automated scanners have been available in the market. Also in the early 2000s, robotics for tank floor inspection and other specialized devices became available.

The difference now is that we want to use the current technologies in robotic systems to eliminate or at least reduce the three sources of errors and associated limitations in the inspection process. We want to use all the abilities of the robotics platforms to increase the quality of the inspection to produce accurate, reliable, and free of errors. We will always start the process of new technology integration by evaluating and qualifying existing solutions, and we need to understand the true capabilities and limitations of these tools. Operating these systems within the boundaries where it is safe and efficient is of the most importance, once we understand where and how to use these tools we will look for use cases and applications, and a careful analysis of the economics and business case should lead to the intersection where robotics or other solutions can be explored. The fact that there is a new solution does not mean we will stop everything we are doing and replace it all. The integration and adoption process should follow a sequence that ensures we understand all the possible variables and specific conditions that make the proposed solution a success. Once the technology is identified, a series of tests both in the laboratory and plant environment are carried out, trying to present as many conditions and variables as possible, it is OK to fail and go back and explore other solutions, or to get back to the vendor to improve or modify certain aspects of the product or service, as long as the cost of the modification is justified by the business case. If for any reason a new integration or further development of the tool is needed, we are normally inclined to economically and technically support these suppliers to accelerate the outcome so we can solve our problems as quickly as possible. We will work with them to reach levels of development that at least solve some or partially solve a complex problem so that we can test the system and support our vendor. In those cases where no solutions are in the market and we see an opportunity to bring a solution, we go back to work with companies open to new ideas, research centers, and universities to explore new solutions. We have several systems that have been developed within the scope of this program, it has been very successful and we believe that the success of transforming our facilities is based on the high value our corporation gives to our partnerships.

15.5.1   Robotics for Storage Tanks

Let us explore the robotic systems available for tank inspection, we will look first at those for inspection of the floor with the tank in service, and then we will talk to those available for internal inspection when the tanks are removed from service and personnel is allowed inside. Finally, we will talk about robotic systems that can be used on the exterior of the tank, normally executed with the tank in service.

Several manufacturers of robotic systems for tank floor inspection with the tank in service, during operation, and with limited disruption of the tank use are offered in the market. These devices as indicated above are intended to detect the presence of corrosion on the bottom plate or floor of the tank, with the tank full of product, and it is done with minimal or no impact on the operation. All of these systems use ultrasonic sensors that measure the remaining thickness of the floor, and by plotting these values, a map of corrosion can be produced. There are differences in these devices, the main one is the level of certification to operate in explosive environments.

Most of them lack certifications that are required to operate by replacing the vapors of the tank with nitrogen to reduce the risk of fire or explosion. There are only two devices that are certified, both manufactured in the United States, one of them carries a Class 1 Division 2 Certification, or equivalent to ATEX Zone 1, while the other one has a Class 1 Division 1 Certification or equivalent to a Zone 0/Zone 1 per ATEX. The refinery owners have different tolerance to risk, and some of them will find a way to use not-certified systems while others may find that alternative out of the question. In addition to this, the surface coverage and number of measured points is likely another important one. While some systems will only capture about 2–3% of the surface of the floor, others can get measurements on a tight grid and are capable of carefully navigating the floor plates to capture a 100% of the surface accessible to these devices. These two factors, the explosive certification and the surface coverage, and sensitivity to minor damage are the basic criteria to select alternatives in the market. The technology for these devices has been evolving since the early 2000s when the first generation was offered. The latest versions of these systems are fully autonomous, floating in the liquid like mini-submarines and having the latest in ultrasonic packages with phased arrays. Several other manufacturers are coming to the market with other systems soon. We expect these devices to continue evolving for a long time.

My vision of these devices is that in the future they will permanently reside inside the tanks and scan the surfaces without human intervention, the data will be processed automatically, and the inspectors will be given the results in only those areas where follow-up actions are needed or the systems are incapable of making a decision.

The inspection process mentioned here is assuming that tanks are clean and free of residues, sediments, and sludge. This material that is naturally produced or induced by the process makes in some cases impossible to drive or navigate the robotic devices to reach the floor and take ultrasonic measurements or any other technique. There are currently no robotic systems capable of removing large amounts of sediments, residues, or sludge with the tanks in service and products on them. The robotic systems available can be used to fluidize, push, and vacuum them out of the tanks with the tank out of service and the side man-way open.

We also expect to see in the future systems capable of navigating through the sludge or residues collecting data, or capable of removing and vacuuming the tank floors as part of the same process.

Other than the floor, other internal components like roof structure and columns can now be partially inspected using Class 1 Division 1 video cameras. The development of scanners with magnetic wheels capable of entering the tank in service and making a proper visual inspection of the internal components is in our vision of future systems.

Additionally, we know of robotic devices currently in development, also with a Class 1 Division 1 certification, that will be used for the inspection of the pontoon tanks in the floating roof types. When the tanks are taken out of service for repairs or modifications, another group of robotic systems can be used. Aerial systems and magnetic wheeled platforms equipped with cameras and NDE probes allow inspectors to reach locations in these tanks that otherwise would require the use of scaffolds or ladders. The condition of roof structures on fixed roof tanks is of primary importance, discovering areas where corrosion, deformation, missing hardware, or coating damage is important as normally once inspected, repairs are scheduled to make sure the mechanical integrity and the structure of the roof are ready for the next 10–20 years of service.

Inspection of the interior of the cylindrical shell is focused on the bottom section near the weld to the floor, and also on the welds between plates and the general surface. The primary focus is on the detection of corrosion which is more likely where residues may accumulate and also the weld heat affected zones may become susceptible to corrosion, other areas to look at are the liquid-to-air interface if the tank level is maintained for a long time. The plates may also have corrosion areas due to construction imperfections and metallurgical characteristics. Visual inspection is the primary tool for corrosion detection, and for that, using drones capable of reaching close to the surface is preferred, using tangential lights or "shadowing" to see texture changes is important. Once the corrosion is detected, magnetic wheeled platforms can be used to take thickness measurements near the corroded areas and structured-light or laser-based cameras to capture the 3D image and measure the depth and extension of corrosion. We expect the future for drones to add probes for inspection. At this point, the first generation of confined space drones with ultrasonic thickness probes just arrived on the market, these drones will improve the efficiency of the inspection process, we also expect lidar cameras or other 3D cameras to become part of the inspection drones. We can see how these drones will be capable of doing both the visual inspection for discovery and the follow-up work characterizing the corrosion and thickness measurements in the walls, roof, and structures. The use of automated flights is common now on outdoor flights, but not in confined spaces, mainly because of the reliance on GPS to position the drones and execute the route. We are seeing now also the use of lidar in confined space drones for positioning and expect this will lead to autonomous flying soon.

When the tank is out of service, robotic systems that scan the floor surface using ultrasonics have been available for a few years, the most common of these tools evolved from robotic devices used for boiler inspections. They have been able to reduce the gap between transducers and offer a line of probes covering a length of 18 to 24 inches on a suspension that helps jumping overlap welds and repair plates of the floor. Currently, these systems are getting to the point where a full scan of the floor is possible with incredibly high speed and sensitivity. The software for placing the ultrasonic information as a map showing the location and depth of corrosion is very good. We anticipate these types of systems to become a normal way of executing inspection of tank floors and the use of MFL or similar will fade out. There is no need if the electromagnetic technique if the ultrasonic devices are becoming as fast as them in collecting data.

Other robotics for tank floors includes the blasting and removal of internal coatings and then the application of a fresh coating. We expect also these solutions to become more common soon. It should be noted that the benefits of robotics use for tank inspection include a reduction or elimination of personnel in confined spaces, scaffolds, and other dangerous locations. We can also see that by using robotic systems for tanks we can reduce the need for resources required during traditional inspections, so reducing the time and the cost associated with these operations. Inspection of the exterior of tanks while in service, also called a non-intrusive inspection or on-stream inspection, let the inspectors know most of the condition of the tank and use that information as part of a RBI program and to reduce the scope of inspection when the tanks are taken out for maintenance. Several robotic systems are used and many more coming to reduce the need to reach the tank roof, elevated areas of the shell, and internal floating roofs, among others. The most simple scanner with a single ultrasonic probe has been available for many years, several manufacturers have these types of devices and are used to take point thickness measurements normally to assess for general corrosion of the plates, and several inspection codes require these inspections. This type of inspection can be accomplished with mechanical poles for smaller tanks. Aerial systems capable of taking thickness measurements are an alternative but the relative cost is still an impediment. This device is only for point measurements, and it cannot be used if surface scanning that covers a wider area is needed.

Using automated corrosion mapping or also known as C-Scan ultrasonics or corrosion mapping is not new, these devices have been popular in the industry for 20 years or more now. These are needed for the detection of preferential corrosion at welds or liquid-to-air interfaces, and also corrosion on the plates, the wide scan allows for detect localized and pitting corrosion with the highest chance. The scanners are normally equipped with magnetic wheels that drive a bridge with a single ultrasonic transducer that moves left to right while the wheels move in one row, the movements are coordinated in a way that a grid of thickness measurements is collected, the spacing both on the horizontal and vertical direction can be programmed from fractions of an inch to several inches depending on the need. They move quite fast and can cover large surfaces in a few minutes for about two square feet per minute. Once the data is collected, it is displayed using a color palette that shows where the low thickness areas are. This method is capable of detecting plate laminations, dirty steel inclusions, corrosion, and certain specific damage like hydrogen-induced cracking, or low-temperature hydrogen damage. The scanner mentioned above for floor scanner using ultrasonics can also be used for the tank shell, it moves easily 10–20 faster than the traditional systems, the cost is higher but when time is part of the decision, it needs to be considered. A new generation of vacuum scanners for insulated tanks with a thick coating is in final development and should be on the market soon.

In the vision for the future, we anticipate that small, unattended scanners will roam the tank surface autonomously collecting thickness and visual information and returning to their cradle to recharge and feed the information back.

15.5.2 Robotics for Pressure Vessels

The inspection plan for pressure vessels, storage vessels, heat exchanger shells, and other process equipment like reactors, distillation columns, stacks, and flares can be grouped, but careful consideration depending on the geometry, environment, reach, and coverage, combined with the requirements in the inspection plan, should be considered.

We will explore the robotic inspection alternatives in two groups, those that require the equipment to be out of service and are open to introducing the tools inside and also those robotic devices that can be used during operation, with access only to the exterior of the equipment. It is preferred always to do as much inspection with the equipment in service so that the inspectors can assess the condition of the equipment and use RBI methods and also reduce the scope of inspection when the vessels are out for maintenance during shutdowns or turnarounds. It has been reported consistently that economic savings are obtained when areas or sections of the vessel are removed from the scope of internal inspections during turnarounds by doing external in-service inspections. Detailed analysis of the geometry, damage mechanisms, and configuration of the vessel is needed to fully replace the internal inspection with in-service inspection, the most obvious area that may require internal inspection are the nozzles weld, especially those with reinforcement pads, other internals like demisters, separators, distillation trays, vortex arresters, etc., can only be inspected when the vessel is open (Figure 15.7).

Inspection of the vessel interiors happens in two stages normally, the one that is done immediately after the vessel is open and before cleaning which is normally conducted by the operation and process engineers to check for major damage, type and location of residues, and other clues that can help them know how the vessel operated. After this initial check, the vessel is cooled down and cleaned to

Figure 15.7 Inspection camera accessing a restricted area. Source: Chevron Picture.

remove sediments and residues from the interior surfaces. The level of cleanliness required is linked to the type of damage mechanism and locations identified in the inspection plan. As described above, traditional inspection requires the erection of a scaffold or other staging for inspectors to reach the walls and other surfaces to check it up. Using robotic systems eliminates the need for personnel to enter these confined spaces with a direct impact on the safety of the crew while reducing or eliminating the need for scaffolds, avoiding the need for special permits, hole watch, and rescue teams normally required when personnel gets in confined spaces. The status of the technology currently is such that the large majority of vessels can be inspected with robotic systems avoiding people in confined spaces, but not all of them.

One of the main restrictions is the cleaning process, if a scaffold is needed for crews to clean and prepare the surface, the value of remote inspection is reduced. What is amazing is to know that remote cleaning of vessels and tanks using high-pressure nozzles operated from the outside has been around a lot longer than the inspection tools. Certain industries like pharmaceutical and food used these technologies but not oil and gas. These systems are used only sporadically and for special cases in oil refineries, we believe there is a case to explore robotic systems for cleaning that will supplement the inspection process. As with every dependency, we have to work in parallel to offer both the cleaning and the inspection solution. At this point in the technology evolution, it makes sense to continue working on the inspection solution and then present the cleaning solutions as a package. The process of adoption and integration is complex, but I believe it is more attractive as more value is presented. The next phase of the process is the repair and re-coating of the areas found with damage, this operation is currently done by hand, which also needs to be resolved to fully utilize the power of robotic systems. Cleaning, inspection, and repairs are the sequence needed in these and other operations around oil refineries that will bring the full potential of robotics.

There are three basic tools used for remote inspection of process equipment: inspection cameras mounted on poles, drones designed for confined spaces that are crash tolerant, and magnetic wheeled robotic platforms with special cameras and NDE sensors mounted on them. Let us explore each one of these systems both in their current development and also with a visionary mind looking into the future.

Inspection cameras are evolving exponentially, incredible resolution of the detectors with fast scanning shutters and global shutters, low light sensitivity, and increasing the size of the detector are making these sensors incredibly powerful, the form factor is much smaller and the cost is coming down. Of course, the camera sensor is only one of the elements, equally important are the optics that now offer 20X–30X in very compact arrangements. Additionally, the mechanics for camera tilt and pan-tilt are smaller and available in several configurations. These elements, the sensor, optics, and movement need integration to make a functional camera, as with many technologies, it seems that the speed of development of individual components is faster than the integrated inspection camera. One additional and very important part of the cameras nowadays is the software associated to operate, storing images, and post-process them to obtain the image quality and content that shows clearly the information that needs to be communicated. Inspection cameras are used both for video and still pictures. The debate on the benefits of a full video capture when considering the huge size of files, and the issues with storage, indexing, and search within several hours of video is ongoing. Traditionally as mentioned above, pictures are used to help understand the condition of equipment and likely will continue to be our normal way of communication. We expect that data management and tools for image identification in videos using machine learning and artificial intelligence will continue evolving and in the future assist in the search for specific pieces of equipment. Part of the issue for searching within video files for specific areas of interest lies in the tagging of that location in a three-dimensional global coordinate system. Geo-tagging using GPS is likely one of those tools, but also using digital twins and 3D map reconstruction using lidar cameras and advanced photogrammetry will be part of the identification process. The location of images as part of the metadata of the video is essential to take full advantage of it. Terabytes of data are captured every day in the form of video that likely will never be seen again. We need this data to become information with tools that allow us to search and find specific frames within hours of video to be useful. Inspection cameras can be used simply by placing them inside equipment through nozzles or manways. In its simpler way, using poles or ropes is a powerful tool, in many cases, the information needed can be reached from these locations and avoids the need for personnel to access those spaces. Placing inspection cameras through nozzles for visual inspection of confined spaces should be considered during the planning as one of the simplest, lower cost but powerful solutions to consider.

Aerial systems that are crash tolerant and capable of reliably operating in confined spaces with denied GPS and low light conditions have been in the market for a few years, dominated currently by a single provider but a few others coming behind them. These drones are protected with a geodesic cage that protects the propellers and the camera. This device is equipped with optical sensors using optical flow and other similar technologies to let the drone compensate for drifting and maintain very good stability in these conditions. The camera is 4K resolution and can be tilted, the rotation of the drone provides then an almost 360 global view of the environment, the camera lacks zoom, but that is compensated by the possibility of getting the camera as close as needed for a good image, lights on board are enough for navigation and also for good video and pictures. Lights are positioned and can be turned on in such a way that shadowing is possible on vertical surfaces. We have added after-market lights to the top and bottom to get the same effect on the floor and ceilings. The system is very well designed and has become an essential tool in the inspection of pressure vessels' interiors. Some navigation tools like fix-distance operation and dustproof lighting have greatly increased the value of this tool in real environments. We consider using a crash-tolerant drone as a complementary tool to inspection cameras on poles. If there are locations where the inspection cameras cannot reach or a closer view with shadowing is needed, using this drone will supplement that inspection.

Currently, a new generation of drones is coming to the market, this tool is equipped with lidar and a more powerful onboard computer that is opening the door for positioning, navigation, and autonomous flying in future releases. Having a 3D model of the interior of the vessel or the exterior, marking the points where the pictures or UT-Thickness needs to be gathered as a preprogram for automatic flying not only optimizes the flight but also ensures repeatability and consistency in the inspection strategies. We also expect the concept of drone-in-a-box to reach these applications in confined spaces and congested environments. Smaller drones, more stability, larger flight time, better cameras, additional NDE sensors, and capability to perform some minor repairs are part of the future of these tools.

The third tool, magnetic wheeled platforms are the only one capable of carrying cameras and NDE sensors close by the areas of interest. These powerful tools have a large payload capacity, and power is provided through a tether which makes them a powerful workhorse to carry sensors into confined spaces. Two basic configurations are currently available, the ones with magnetic wheels and the ones with a central magnet unit, they both work well and have their issues, likely the most concerning is the accumulation of ferromagnetic residues in the magnets which reduces their strength and exposes the device to fall off the surface. Other issues with these magnetic platforms include the difficulty of accessing inside equipment if avoiding breaking the confined space plane is required. Some of them are capable of going over 90° concave and convex corners, but the lack of flexibility in the suspension makes it tricky to align the cart for practical operation in those locations. Also, the speed of movement once inside the vessel is slow which makes the whole operation a process that takes much more time than the previous two alternatives. These limitations are greatly compensated when special cameras like structured light, lidar, or laser scanners are placed on top and also when shadowing lights and NDE sensors like UT-Thickness, UT-Shear wave, and UT-Phased Array are used, as well as surface eddy current, coating test, cleaning nozzles, rotary tools for grinding have been demonstrated. The magnetic platforms are a complement of the inspection cameras and flying systems, and they can be used to reach areas where damage was detected with either one of the two and obtain detailed information and characterization of the damage, and even take action for remediation and repairs. We expect these tools to continue evolving into more flexible devices capable of navigating less structured environments capable of jumping and going over beams, supports, and complex configurations. Advanced positioning, navigation, and autonomous behaviors will also continue improving to the point where it would be possible to similarly with drones we can plan a route and execute with minimal intervention which will result in improved reliability and repeatability.

We also expect smaller scanners capable to deploy over the inspection surface and go over selected areas or scan one-hundred percent of the area, returning to the base to recharge the battery and transmit data and continue the activity until completed. Multiple robots working together or swarm robotics promise collaboration that may improve the efficiency of these tasks. Unattended, resident, and minimal or no human intervention is a way to describe what we expect these devices to become in the future. Advancements in damaging detection and classification with Machine Learning and Artificial Intelligence are also an important part of these solutions and will evolve into fully integrated systems where the data captured can be analyzed and presented to the inspector in a friendly and succinct form extracted from all the data captured.

15.5.3 Robotics for Process Piping

As mentioned before the big challenge with piping in oil refineries and most process industries is just the enormous amount we have. The variety of sizes, temperatures, and damage mechanisms is a huge challenge when trying to optimize inspection processes. The fact that about 50 percent of the piping is insulated and a good percentage is underground, and also that the majority is in locations where access is difficult and requires scaffold, rope access, or other means for a person to reach complicates the inspection strategies and the techniques that can efficiently be deployed. Using robotic devices for difficult access is one of the original drivers to explore options, the cost of erecting scaffold or other alternatives is in most cases more expensive than the inspection itself, and carries a lot of risks to the workers and technicians when constructing and accessing scaffold.

There are several solutions that we have been exploring for inspection of piping, we can for simplicity split the population into groups that are not exclusive but may overlap like: high temperature/insulated piping, ambient temperature up to 180 °F (90 °C) without insulation, piping in elevated racks and vertical sections to process equipment, and underground piping or road crossings. In most cases, inspection is done with the piping in service, but there are some options to inspect them during a turnaround or plant shutdown. Flying drones at high altitude, of about 50 to 400 feet above ground providing a top view of piping has been available for many years, incredible cameras on these devices provide a close view of the pipes on the top layer of the racks, as well as flares, stacks, and piping around tall structures and process equipment. Using crash-tolerant drones capable of flying in congested environments provides images of the intermediate and lower layers of the pipe racks, the pipe supports, nozzles, flanges, and other components that are not seen from the top views.

Using crash-tolerant drones outside confined spaces is very recent as the lack of, or unreliable GPS in these locations requires advanced drone navigation using optical flow and other image and sensor navigation that has just matured in the last few years. It is possible to do the visual inspection at speeds that are at least ten times faster and one hundred of the cost than using conventional inspection methods. As these procedures evolve and our inspectors become more familiar to see things on a computer screen and not with their eyes we anticipate exponential growth in this area. Visual inspection process piping should and will be done with drones in the future. Requirements for operations in areas with explosive vapors and gas are always an issue and consideration. The evolution of drones to become certifiable will only be possible when the market is large enough that all the component manufacturers see the pull from the users to create these tools, for now, and until that happens we are using modified hot work permits to react when alarms are set off and equip drones with gas detectors that create one more point of detection, the lack of explosive certification should not become an obstacle for evolution, but it should be solved with solid engineering principles and consideration of the risks, nothing new in our industry.

The opportunity of automatic flights and increased autonomy makes it possible to predict that as with pressure vessels and tanks, it will be possible to mark on a 3D map the locations where pictures and videos are needed and capture the needed information. The opportunity of using machine learning and artificial intelligence tools to search in the multiple hours of video and thousands of feet of pipe to detect and rank areas with damage will come sooner than later. As with other applications, these aerial systems will also continue evolving to have better cameras and other NDE sensors including point contact ultrasonic thickness and other contact NDE. Placing RFID-based ultrasonic tags on elevated components is one of those initial technologies that is providing a solid perspective of the need for autonomous flights. Drones that float and take measurements in a steady condition will be supplemented by those that will land on the pipe and provide a platform for deployment of other NDE crawlers and arms to reach specific points, landing on pipes reduces the energy, and jumping from one point to the next instead of continuous flight offer other options like deployment of radiographic methods and other exposure type NDE. Other generations of drones will just drop sensors or prepare the surface for sensor installation, or even place wireless NDE scanners and systems that will be retrieved when the operation is completed, robotic collaboration, and complex systems with multiple elements are expected and needed. We expect also to see long-term operations with a resident, unattended and requiring minimal or no human intervention for long periods.

Other tools like magnetic wheeled systems are by themselves hard to use to reach piping if they are not placed by a drone or capable to move on vertical surfaces and sort the presence of flanges, valves, nozzles, and other accessories that are at or near the pipes to inspect. As we mentioned before, dropping these scanners from drones is an alternative, the other is to design highly flexible magnetic scanners that can climb on any geometry to reach the place for inspection, these flexible platforms may need to start on structural components and find their way to other pipes just to climb over to the pipe of interest. The magnetic wheeled platforms have the incredible advantage of the large payload and stability that is needed for many of the actions related to inspection and also for maintenance and repair. As the payloads and tools evolve, it is not hard to see how they can be mounted and quickly integrated into these platforms. Scanners for piping like the ones that are placed over horizontal lines insulated or not have been in the market for a few years, and those that hug the pipe and can move circumstantially are just now becoming available. The current systems are a great proof of concept of the benefit of having a robotic platform on the pipe and executing complex tasks like profile radiography or pulse eddy current, and the current state requires technicians to be near the tool and help navigate obstacles like flanges, supports, and valves. The scanners that wrap around the pipe can be used for nonmetallic or insulated lines with an aluminum jacket or similar and they also offer flexibility to rotate and avoid some support types and obstructions. We expect to see other forms of locomotion like multileg platforms and mixed locomotion as they transit over the complexity of the piping system we have.

For underground and road crossing piping sections, piping in heaters and boilers, the short pipe section that connects a pressure vessel to the first valve, and in general where access to the external surface is difficult, the pipes need to be taken out of service and perform inspection placing robotic devices in the interior. For several years, tools like in-line inspection tools have been available for long transmission pipelines, these systems are driven by fluid or gas and require the installation of a launcher that drops the system in and retrieves it at the other side. These technologies are difficult to use for short sections of pipe and those that have multiple elbows and bends. Especial tools, much shorter and developed specifically for process piping in refineries and other process facilities have been developed, there are two basic types: those that require air to push them down the pipe and have a tether to pull them back or free flow so they need to be caught at the end and those with electric motors that drive inside the pipe. Mechanical systems are complex pieces of machinery but have been successfully used in the industry for many years. The main issues are the ability to drive around bends and elbows while providing data in those locations and the friction the retrieval cable or the umbilical cord presents after a few turns. These tools are available, they have reached a level of development that has not evolved for some time, and it may require fresh ideas on the mechanics and design. Another issue is the cleanliness needed not only to travel but also to execute the NDE tasks normally limited to visual, MFL-based tools, ultrasonic thickness, and laser measurements. We are expecting the design of systems that can reside permanently inside these pipes with no operational disruption so allowing us to do inspections without the need to open every time needed.

15.5.4 Robotics Heat Exchanger Bundles

Heat exchangers are a very important piece of hardware in oil refineries and represent one of the most time consuming and leading efforts during a turnaround or plant shutdown. Most of these systems have small-diameter tubing arranged in bundles that transfer or receive heat from liquid around it. The use of small-diameter tubes is intended to increase the area of contact and heat transfer efficiency. Like any other component, these tubes are subject to degradation that requires inspection as presented above. The use of mechanized systems in large steam generators in nuclear plants to guide the eddy current probes to each of the tubes is not new. What is new is the use of these devices in refinery environments. Thousands of tubes are inspected, and the manipulation of the probe to be pushed in and then pulled out when done manually is subject to errors. As also discussed above, cleanliness and surface preparation are essential for a high-quality inspection. The most recent developments include both operations performed in sequence so that the cleaning nozzle is injected in one tube while in the next one the inspection probe is also pushed, and a computer programmed with the geometry and position of tubes moves the head in sequence to complete the inspection. The efficiency of this process is almost double or multiplied depending on the number of heads that are used to clean and inspect at the same time. As these devices become smaller, and easier to use and the cost of the process becomes comparable to the manual operation we anticipate that these tools will become a normal way of inspecting soon. Other opportunities include the insertion of the cleaning and inspection heads on the tubes using one of the side nozzles so that the head doesn't need to be taken out of service, this will drastically reduce the time and effort needed for the tubing inspection. Other similar tools to be inserted on the shell side for visual and surface inspection of the external surface of the bundle, the piping supports, and other mechanical parts are also needed if we want to reach a high level of efficiency. As with other systems discussed above, a permanently installed cleaning and inspection head that can perform while the exchanger is in service are also expected for this type of operation.

15.6 Robotics for Plant Operations, Surveillance, Maintenance, and Other Related Activities

15.6.1 Operations, Surveillance, and Maintenance of Oil and Gas Refineries with Robotic Systems

For a few years, several manufacturers have developed robotic platforms that can be used for operations, surveillance, and other similar activities in oil refineries. Some of these platforms have been specifically designed for this purpose and others come from a more general market offering where after some work they can be used for this purpose. The opportunity to have robotic platforms executing repetitive, simple, and dull activities is seen as a way to get back to the operators in the control room to improve the efficiency of the processes. In general, two types of platforms, the none certified and the Class 1 Division 2 or ATEX 1 certified platforms are now available and several others will soon come to the market. Like with other robotics using noncertified robotic platforms in classified locations requires a modified hot work permit that will use the reaction capabilities of these platforms to move them away from dangerous locations and/or shut down immediately after a dangerous environment is detected. The reason why we are exploring the noncertified platforms is the fact that they are less expensive and are evolving a lot faster as the process of certification delays and increases the cost to bring them to the market. Also, any new payload available like better cameras, sensors, and manipulators require the manufacturer to recertify these devices, while the others can move right to integrate and deploy to the market. There are specific use cases, like alarm responses and emergency responses where a fully certified system is the only alternative, for that reason we believe it is important to continue pushing the market to evolve both groups of solutions. Also, because of the extra protective armor needed, certified platforms tend to be larger and heavier than noncertified. The main difference between the platforms, other than the explosive certification, is the type of mobility (Figure 15.8).

We currently have systems on wheels, tracks, and quadrupeds, each solution has advantages and disadvantages both on the terrain they can traverse and the energy needed to move. As these are normally large platforms, they can carry a large amount of payload weight which increases the opportunities for integration of advanced inspection cameras, 3D mapping Lidar, and sensors as needed. Most of these platforms have an arm that can be used for simple manipulation in their current state. These platforms can be operated remotely or can be programmed to operate automatically with different levels of autonomy. We are currently exploring standard operational procedures where images, temperatures, sounds, smells, and vibration detection are needed.

Figure 15.8 Robotic platform for inspection. Source: Chevron Picture.

We will continue to consider those where one-hand manipulation can increase the value of the solution. The first generation or early second generation for some of these manufacturers has proven to be more difficult than expected. The lack of consistent Wi-Fi or phone data like 4G or LTE across the plant is making communications difficult, the images and video transfer, construction of routes, and execution of auto-walks with the precision needed are some of the issues we have. Weather conditions like temperature, wind, rain, and snow, different times of day, and a sunny or overcast day will alter the vision systems and need to be addressed. We know these will be resolved and expect that the next generation of systems will offer solutions to these problems. These types of ground robots are limited by design to the reach and access a person can have, with a single arm. Access to work platforms and reaching gauges, switches, and small valves are well within the expected capabilities. From the access side, most of our tall equipment requires access using a vertical ladder, which likely these platforms will not address. From the manipulation aspect, we know that the dexterity of the human hand is still far from what is possible, but we are seeing improvements in this area, so we expect to get into these platforms in the future. The need for a double arm for many operations is clear, the use of two platforms working in collaboration will also increase the value of these solutions. As we explore the use cases of these platforms we for sure will find areas of work that will benefit from these systems.

The progress of these systems in oil refineries depends on the value of the opportunity and the cost of the solutions. As many other manufacturers dedicated to oil and gas and others in the general industry come to the market, we will likely see hardware cost reduction that will accelerate the use of these devices. We are planning for facilities with a lot fewer people walking around and doing basic work, and we need our human colleagues to take their intelligence to help us solve advanced problems based on the information and data captured by these systems. We expect to see our operators having the information needed at their fingertips with minimal intervention in the data collection itself. As these platforms become more sophisticated and more capable, they will evolve from surveillance and data gathering to active operational and maintenance activities. They will receive orders to modify the process conditions based on instructions received from the operators, regardless of where these operators are located. Actions like collecting samples, replacing filters, changing the pump and compressors gaskets, lubrication, cleaning, and other maintenance activities are likely to be performed for the most part by robotic platforms with minimal or no human intervention. Robots collaborating with humans in complex operations as well as robotic swarm collaboration are part of the vision for these systems.

We see the future robotic platforms equipped with arms that are capable of adapting different tools as needed for the task to follow. It may be necessary to take a wrench or hydraulic tool from a rack to remove the bolts from a piece of equipment, pull from a drawer the scissors to open the packet where the new gaskets are located, or maybe replace the standard robotic arm with a smaller one of higher dexterity to replace a fuse in a panel. A multitool approach with a capable platform we believe is the logical solution, while we get there, we may need to have a fleet of multiple platforms to execute the tasks on hand.

15.6.2 Safety and Security Robotics

The activities related to security and safety controls in our facilities can be considered part of the operation and close to the surveillance our operators do while walking around the unit. The main difference is that the systems dedicated to the security of the premises require other types of robustness and redundancies that operational systems can avoid. If we have a robotic device doing the rounds checking on the physical fence is because it is capable of identifying a natural from an intentional act to breach that barrier. A deer trying to jump a fence should not trigger the same response as a person climbing it. Using flying systems on beyond line of sight and ground platforms for patrol is not new but requires integration with our current systems. The use of Machine Learning and Artificial Intelligence to process the information collected is likely to be an important part of this sequence.

Maintaining the safety of our operators and maintenance groups is of the highest importance, we need to make sure that the minimal requirements on personnel protective equipment, real-time location of personnel, and reaction to injuries and other incidents are executed with the highest level of competence. We expect both ground and aerial platforms to play a role in keeping an eye on the people and their actions. Simple activities like tying a ladder, using the lanyards on a high elevation harness, and wearing a gas monitor may be the difference between a good day or an injured worker or worst. As we try to make all our processes more efficient, the use of robotic systems in support of our health and safety groups is being considered and we expect to be part of our development soon.

15.6.3 Robotics for Utilities and Support Activities

Oil refineries are extensive areas of work with office buildings, shops, civil infrastructure, roads, railroads, power, communications, water, and sewer. We also provide other services like cooling water, process steam, and fire protection water networks. We provide food, drinking water, air conditioning, and basic office services to our employees and contractors. The work all these activities require is done by several hundreds of people that execute them daily to maintain, upgrade, and replace infrastructure. Robotic systems for inspection, maintenance, repair, and construction in each one of these areas will likely follow a parallel development as they are offered to towns and cities. Some inspection activities for electric systems like the inspection of power poles are already being done by drones that follow specific routines for inspection. Also in boiler houses, cooling towers, and chillers, robotics can be used to inspect the different components.

Inspection of structural steel, road bridges, buildings, and other infrastructure can be done with flying systems and ground robotics. We expect in the future some of these flying systems to discover coating failures in structures that can be immediately solved by cleaning and applying the new coating in a small area before extensive damage is produced and requires repairs. Food transportation from the internal cafeteria and external sources to lunchrooms using autonomous vehicles as well as personnel transportation using autonomous people movers are definitely in the future of our facilities. Building painting road maintenance by robotic platforms as well as the construction of a building using additive manufacturing are also part of this vision. The opportunities of robotic systems to assist in maintenance and other support activities are only limited by our imagination and the speed these solutions reach a market condition that makes them economically attractive and capable of solving our needs.

15.7 Conclusion

Robotic technology for inspection of oil refineries and other facilities is in early development, performing routine work using new technologies is challenging, but they will become accepted as the systems become more capable, the cost goes down, and the process of adoption of robotics becomes more common. Each oil refinery components: tanks, process vessels, piping, and heat exchangers has their own failure and degradation mechanisms that require specialized tools and processes to detect and characterize the damage, so that their condition can be assessed, and the integrity and safety of the process are maintained. The new solutions based on robotics have to be capable of performing at least as well as a person executing those tasks. Data analysis, process, and assistance from AI and machine learning will be part of this process. Using robotic platforms capable of transporting sensors and cameras to perform inspections not only remove people from exposure to dangerous conditions but also presents the opportunity to automate and continue the journey toward fully autonomous systems that will the future assist our inspection teams to collect data. Evolution of both, the robotic platforms and the sensors and cameras, is necessary. We are moving forward. We believe that the future requires of less resources, and robotics are tools that will assist on these goals. The future is bright, and there is ample room for growth, and hundreds of new companies are needed to support this development.

16

Drone-Based Solar Cell Inspection With Autonomous Deep Learning

Zhounan Wang[1], Peter Zheng[1], Basaran Bahadir Kocer[1,2]*, and Mirko Kovac[1,3]

[1]*Aerial Robotics Laboratory, Imperial College London, London, United Kingdom*
[2]*School of Civil, Aerospace and Design Engineering, University of Bristol, United Kingdom*
[3]*Laboratory of Sustainability Robotics, Swiss Federal Laboratories for Materials Science and Technology, Dübendorf 8600, Switzerland*

16.1 Introduction

16.1.1 Motivation

The use of renewable energy sources, particularly solar energy, has gained immense popularity in recent years due to the need for affordable and clean energy. With leading economies across the world showing favorable growth in their renewable energy sector, clean energy is the future. In the United Kingdom, electricity generated from renewable energy reached 36.9% in 2019. Solar installation is expected to increase significantly in the near future, making solar farms and panels a critical component of renewable energy – the estimated total capacity of solar energy is 15,674 MW by 2023, [Sustainable Development Goals, 2021; Regen Transforming Energy, 2019].

Solar farms, also known as solar parks, are large-scale installations that generate electricity by harnessing the power of the sun. They typically consist of numerous solar panels arranged in rows or arrays. Solar panels, on the other hand, are smaller units that convert sunlight into electricity. The panels are constructed using photovoltaic (PV) cells, which consist of semiconductor materials such as crystalline silicon. These materials absorb photons from the sun and convert them into electrical energy through the photoelectric effect.

Solar panels are crucial for the cost-effective production of electricity, and their performance is vital to the overall efficiency of solar farms. However, solar panels

*Email: b.kocer@bristol.ac.uk

Infrastructure Robotics: Methodologies, Robotic Systems and Applications, First Edition.
Edited by Dikai Liu, Carlos Balaguer, Gamini Dissanayake, and Mirko Kovac.
© 2024 The Institute of Electrical and Electronics Engineers, Inc. Published 2024 by John Wiley & Sons, Inc.

are subject to wear and tear and can be damaged during installation, transport, or operation. Environmental factors such as extreme heat, humidity, and weather events can also cause physical damage, leading to a reduction in solar panel efficiency. Therefore, regular inspection and maintenance are necessary to ensure the continued performance of solar panels.

Monitoring solar panels are traditionally conducted by manual inspection with visible fault identification [Golnas, 2012]. However, there are common invisible faults as illustrated in Figure 16.1 that can affect the performance of the module. To address these challenges, two main test strategies have emerged for solar panel inspection. The first involves current–voltage curve analysis at both module and string level. This method enables the detection of electrical faults that affect the performance of the panel. The second strategy involves ground and aerial imaging inspections using infrared (IR) thermography and luminescence imaging technologies. These techniques allow for the identification of nonvisual faults, such as cracks and shading, which can impact the panel's performance [Høiaas et al., 2022].

Solar panels are often installed in large clusters. For some settlements, solar panels are often installed on access restricted roof that is not easy to conduct operation. Therefore, it is difficult and costly to inspect with traditional methods, consuming excessive time and manpower [Hernández-Callejo et al., 2019]. With the development of aerial robots, the aerial inspection of solar panels has become popular due to its flexibility and efficiency [de Oliveira et al., 2020]. Equipped with

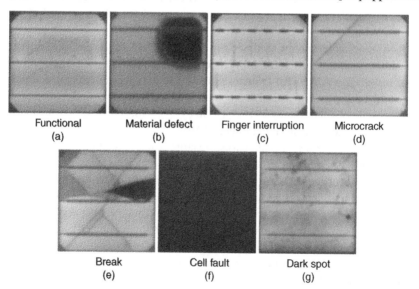

Functional
(a)

Material defect
(b)

Finger interruption
(c)

Microcrack
(d)

Break
(e)

Cell fault
(f)

Dark spot
(g)

Figure 16.1 Various defect types on the solar cell: seven types of solar cell defects are identified that will form the output of the learning approach.

specific types of cameras, an aerial robot can analyze the condition of the solar panel using various imaging methods including EL, PL, and IR imaging to inspect the solar panels [Berardone et al., 2018]. Luminescence is based on light emission and it can allow measuring the temperature on different IR radiation of objects with a different IR wavelength [Trupke, 2017].

Since EL imaging has an excellent performance in various defect detection especially in microcrack detection [Dunderdale et al., 2020; Zhao et al., 2021], in this study, EL images which provide resolution in semiconductor material level are used to detect the defects.

The current solar cells are mainly based on crystalline silicon and there are two main types: (i) monocrystalline and (ii) multicrystalline. While the former has a uniform background which became more prominent, the latter has a nonuniform background and more complex textures. There are several reasons why monocrystalline solar cells became more prominent in recent years. One reason is that monocrystalline solar cells are more efficient at converting sunlight into electricity than other types of solar cells. This means that they can generate more electricity from the same amount of sunlight, making them a more attractive option for people looking to power their homes or businesses with solar energy. Additionally, monocrystalline solar cells tend to be more durable and longer lasting than other types of solar cells, which makes them a good choice for use in solar panel systems that are exposed to the elements. To classify different defects, extracting features from defective solar cells is required. Deep learning is widely used for data processing. The CNN is a network architecture, commonly applied to image processing. Compared to traditional methods, CNN is more reliable and suitable for the regular inspection of solar panels.

Different maintenance strategies can be applied according to the defective condition. Most of the time, the detection could result in a more general output without providing details about the defect type. However, we would like to provide a more specific categorization for the defect types by leveraging the proposed framework with deploying aerial robots.

16.1.2 Related Works

Visual inspection of solar cells is an active research topic in recent years [Gallardo-Saavedra et al., 2018]. The damages of the solar cell are divided into three categories [Deitsch et al., 2019a]: shading/soiling, intrinsic defects, and extrinsic defects. Shading and soiling are usually caused by the surrounding environment, and they could significantly reduce the performance of the solar panel. The effects of dust accumulation in moderate dust conditions and the performance reductions of the solar system are approximately 15–30% [Sarver et al., 2013]. In desert areas with adequate lighting but dust accumulation, the reduction could be as high as 40% [Hussain et al., 2017]. Environment planning (e.g. trees, buildings, poles, or other objects) and regular maintenance are effective in avoiding shading and soiling. Intrinsic defects, caused by careless handling during production, transportation, and installation, may cause microcracks, finger interruptions, and material defects. While this may not degrade the initial performance of the panel, this may reduce its lifespan. Extrinsic defects are mechanical damages that are caused by external force or extreme weather such as falling rocks and strong wind. These result in broken solar cells which may not be functional anymore.

Several approaches have proposed the use of different technologies to detect the specific surface damages of the solar cell, which contains intrinsic and extrinsic defects. An image processing method is proposed in Fu et al. [2004] by enhancing the counter of the solar cells to detect cracks (edge cracks and inside cracks). The regions of interest approach is presented in Tseng et al. [2015] to detect the finger interruption in EL images of multicrystalline solar cells by extracting the features on both sides of the damage. A self-reference approach is used in Tsai et al. [2012] based on the Fourier image reconstruction to detect several extrinsic defects, such as small cracks, breaks, and finger interruptions in EL images of the solar cells. However, some defects with more complex shapes are difficult to be detected using this method. Buerhop et al. [2012] introduced IR-thermography imaging to investigate the actual performance of the solar cell by analyzing the different characteristics of the temperature profiles. IR imaging is reliable and fast to evaluate the quality of solar modules. However, it has limitations due to its low resolution, which makes it difficult to detect small defects, such as microcracks, as compared to the EL imaging [Deitsch et al., 2019a; Mantel et al., 2019].

Deep learning is a class of algorithms that can be used for data processing, such as audio recognition and image classification. In recent years, with the development of deep learning, it has gradually replaced traditional data processing methods and becomes one of the most preferred approaches to inspect surface defects. There are various types of learning with artificial neural networks, and CNNs are one of the most effective ones in image recognition [Cha et al., 2017]. The deep belief networks (DBN) is used in Ni et al. [2018] to train the model for detecting the cracks in the solar cells. A multispectral deep CNN is designed in Chen et al. [2020] by analyzing light spectrum features in color images. It can detect several defect types, in the visible light spectrum range, such as chipping, and broken gates, and the accuracy can reach 94.30%. Deitsch et al. [2019a] employed two automated approaches, the support vector machine (SVM) and the deep convolutional neural network (DCNN), to detect the possibilities of the defect in the solar cells. In Akram et al. [2019] and Tang et al. [2020], CNN-based frameworks to inspect the solar cell are considered with EL images. In Akram et al. [2019], various data augmentation methods are applied, and the CNN architecture (CNN layers, dropout layers, regularization layers, and batch normalization layers) is optimized. The CNN model can detect if the solar cell has defects, and the accuracy of the prediction achieves around 90%. Different defect types of solar cells can be detected in Tang et al. [2020]: microcrack, finger interruption and break. It compares different CNN parameters: CNN depth, stochastic pooling, and a number of kernels and finds the best CNN architecture which can achieve around 80% accuracy. It is also noted that this is an unsupervised-based approach where the computational aspects are left for future work.

16.1.3 Scope

The aim of this study is to present an overview of drone-based PV module inspection. We also present case studies for the field practitioners including defect detection based on a CNN. The trained CNN model can detect the condition of solar cells with seven types of defects. Different approaches to improve the accuracy of the identification are presented. Second, we also present a simple and practical framework that can autonomously find solar panels and detect solar panels. An evaluation of this framework is utilized in the NEST building model (https://www.empa.ch/web/nest) in the AirSim platform.

16.2 Aerial Robot and Detection Framework

Nowadays, with the recent development of aerial robots [Zheng et al., 2020; Xiao et al., 2021; Kocer et al., 2021b; Orr et al., 2021; Zhang et al., 2022], it is the desired approach in inspection and data collection missions [Kocer et al., 2019; Farrell et al., 2021; Ho et al., 2022; Stephens et al., 2022; Hauf et al., 2023]. In Bommes et al. [2021], an aerial robot is proposed for PV modules mapping and inspection operation considering thermal images. Therefore, for solar panels, the research is also moving toward inspecting using aerial robots. Quater et al. [2014] used a thermal camera and a HD photo camera installed on the drone to evaluate the performance of the solar panel. However, these two approaches cannot capture some internal damages, like microcracks. Pierdicca et al. [2018] presented the approach to detect whether the solar cell is damaged with DCNN, and the data is acquired from the thermal IR sensor on the drone. In summary, there is still a need for a systematic framework that overviews the autonomy of the flight, detection of the target module, and inspection and learning framework for the PV module with aerial imagery.

Figure 16.2 shows the flowchart of our case study. This approach allows the research and development stages to be conducted in a realistic simulation environment. The first block imports a 3D model into the physics-based simulation environment. Depending on the geometry and the mission, the trajectories can

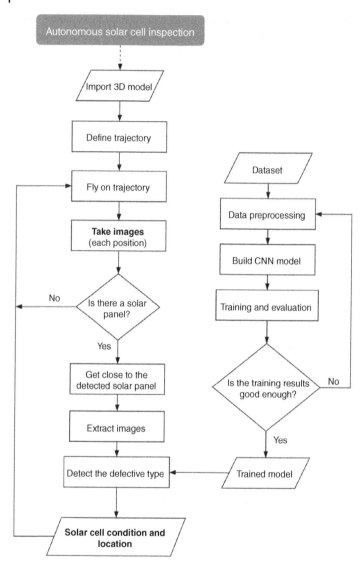

Figure 16.2 The flow chart of the proposed framework: the aerial robot first flies on the imported design for the structure and the solar cells to extract the images. This stage will be followed by solar panel detection. Once the solar panel is detected, the identification pipeline will be triggered that will identify the solar cell defects.

be fed into the system to extract solar panel images. While this is conducted, the learning-based framework takes the extracted images and provides the identification of the solar cell functionalities.

16.2.1 Simulation Environment

An aerial robot is simulated in the AirSim platform with a NEST building model. AirSim is an open-source platform to simulate realistic environments with robots, and it is built on Unreal Engine. The NEST building is one of the research hubs endowed with solar cells on the structure, which is shown in Figure 16.3. The solar panels are installed on top of the building. In the simulator, the solar panels' dimensions are set to be 1.6 m by 1 m according to the collected field data. Since the optimum angle of the solar panels depends on the location, season, and time, the roof angle of the solar panels is set to be 35 degrees, which is a common preference in the United Kingdom [Spirit Energy, 2019].

16.2.2 Solar Panel Detection

One solar panel consists of multiple solar cells. To conduct an inspection, the aerial robot needs to recognize the solar panel and get close to it for further detection, which is shown in Figure 16.2. The solar panel has the shape of a rectangle, which is an obvious feature and the location is fixed, in our case – on the roof of the building. Therefore, quadrilateral detection can be used to recognize the solar panel using OpenCV. Figure 16.4 shows the detection procedure.

Canny edge detection is a multistage algorithm to detect the edges in an image [Canny, 1986]. The first stage is to reduce the noise using the Gaussian filter and

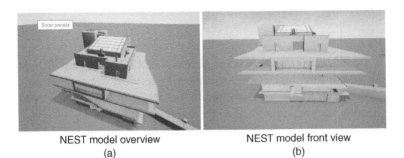

NEST model overview
(a)

NEST model front view
(b)

Figure 16.3 NEST model: a research and energy hub part of Swiss Federal Laboratories of Material Science and Technology (EMPA) which is built and represented in Airsim.

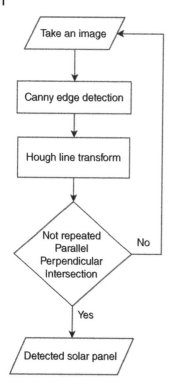

Figure 16.4 Flow chart of the solar panel recognition: because of the shape of the solar cells, quadrilateral detection using Canny edge and Hough line transform is implemented.

obtain a smoothened image. For each pixel, the intensity gradient and direction are calculated using Eqs. (16.1) and (16.2).

$$G = \sqrt{G_x^2 + G_y^2} \tag{16.1}$$

$$\theta = \tan^{-1}\left(\frac{G_y}{G_x}\right) \tag{16.2}$$

where G is the intensity gradient, θ is the direction, and G_x and G_y are the first derivative of the horizontal and vertical directions, respectively. Then, all the edge values in different directions are compared to find the most likely edges. After filtering according to the threshold value, the real edges are found.

Hough line transform is used to detect straight lines. In the polar coordinate system, lines that go through a point can be represented by

$$r_\theta = x_0 \cos\theta + y_0 \sin\theta \tag{16.3}$$

where (x_0, y_0) are points of state vector, and (r_θ, θ) are the vector parameters. It can be considered as a curve. For all the points in an image, the lines can be represented as Eq. (16.3). Therefore, if the curves of two points have an intersection, there will be a straight line between these two points in this image. The Hough line transform is susceptible to noise, so it performs better after edge detection.

The outputs of the Hough line detection are the coordinates of two ends of the lines and those coordinates will be given the decision box for the detection. Repeated lines mean the length and position of the lines are similar. To avoid repeated detection, only one in repeated lines will remain as the detected ones. To detect the pair of parallel lines, the slopes of the lines are calculated and compared. If the slopes of two lines are similar, these are referred to as a pair of parallel lines. Then the angle between each line in two different pairs of parallel lines will be calculated to find the two pairs of perpendicular lines. The last step is to detect if two adjacent lines in the two pairs have intersections. Passing through these filters, the remained lines are detected as the solar panel.

16.2.3 Aerial Robot Trajectory

To achieve autonomous inspection using this method of solar panel recognition, the trajectory of the robot is set considering the geometry of the NEST model.

The pipeline of the aerial robot trajectory includes (i) the free flight phase, (ii) identification of the solar panel position; (iii) the approaching phase; (iv) extracting images from AirSim; and (v) inspection with the trained model. The trajectory of the robot is divided into four stages: (i) getting close to the building, (ii) climbing the building, (iii) detecting the solar panel, and (iv) getting close to the solar panel. Each stage is marked in Figure 16.5. From the first to the last stages, the velocity is regulated and decreased in the last two stages to avoid a collision.

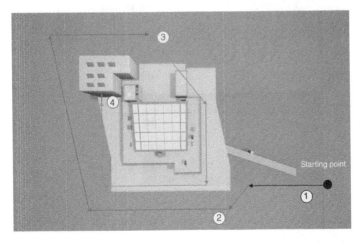

Figure 16.5 The trajectory of the aerial robot: the system starts from 1 and reaches 4 for the solar panel detection.

After detecting the solar panel, according to the coordinate of the solar panel, our approach computes the distance between the center of the solar panel and the center of the camera frame. Afterward, the robot flies to the calculated position in the horizontal plane and reduces the flight height by 2 m to conduct the inspection task. After the solar cell is marked with the defect type, the robot will fly to the next position, to detect another solar panel and conduct a solar cell inspection. The trajectory is set to detect solar panels on the roof considering battery life.

The detection of the field images is more challenging as compared to the data in the simulation. We have tested our approach on field images, and one of the examples is illustrated in Figure 16.6 together with the filtering stage.

16.2.4 Sensory Instrumentation for Aerial Robot

The aerial robot needs onboard sensors to conduct the studies in daylight and night conditions. For the imaging techniques, it is possible to leverage the daylight and night instrumentation as detailed in dos Reis Benatto et al. [2020] and Ang et al. [2018]. With the use of RTK-GPS, inertial measurement unit (IMU), and onboard vision sensors, the accuracy in centimeter ranges can be achieved which is sufficient for this application.

Onboard localization would be also achieved with the combination of visual-inertial [Chen et al., 2019] or lidar inertial odometry [Chow et al., 2019]. Visual odometry can run onboard with the assumption of having a static

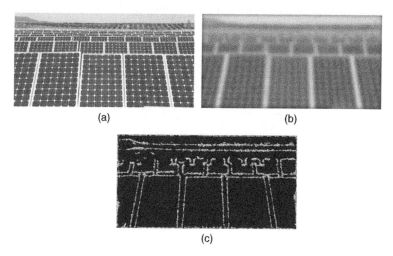

(a)

(b)

(c)

Figure 16.6 The solar panel detection with field data: (a) original image; (b) smoothing stage; and (c) detected solar panels.

environment with sufficient illumination. As an alternative to the visual inertial odometry (VIO), thermal inertial odometry can also enable flight in the night [Delaune et al., 2019; Khattak et al., 2020]. This could be further advantageous since the thermal camera can help to detect solar panels [Vega Díaz et al., 2020] and classify defect types [Henry et al., 2020; Bommes et al., 2021]. However, it is also found that more specific environmental conditions are required for thermal imaging, and it is more challenging to detect microcracks and identify the fault causing the hotspot [Rahaman et al., 2020]. A review has summarized the progress until 2016 in Tsanakas et al. [2016] focusing on common fault types and electrical & thermal responses of PV modules. After a year, available camera technologies are summarized in Gallardo-Saavedra et al. [2017] for aerial data capture.

Considering its advantages, EL-based images are reported in dos Reis Benatto et al. [2020] with 120 frame per second on a drone in daylight conditions. However, it is more common to measure at night to reduce the background noise [Jahn et al., 2018]. In Koch et al. [2016], a comparison study is presented considering EL settings with a tripod, multicamera construction, and aerial robot-based, illustrated in Table 16.1. There are available tools for connecting solar modules with power supplies that can enable connection up to 100 module strings (500 kWp) [Koch et al., 2016]. Therefore, an aerial robot with a set of string boxes can enable solar module inspection of up to 1 MW per night. For daylight EL imaging, tackling the sunlight noise can be done with optical filters, lock-in signal, and image postprocessing [Høiaas et al., 2022]. For example, a set of EL images considering aerial robot movements are collected in dos Reis Benatto et al. [2018] in daylight conditions and followed the stages: (i) detection; (ii) motion compensation; (iii) background separation; (iv) denoizing; and (v) background subtraction and perspective correction. Finally, this work is extended in dos Reis

Table 16.1 Comparison of EL measurement systems.

	Tripod	Multicamera	Aerial robot based
Throughput in MW/night	0.1–0.2	0.3–0.8	0.5–1
Module/image	1–4	10–15	1–100
Operators needed	2–3	4–6	2–3
Advantages	No lead time	Identical pictures	No plant design limitation, fast

Table 16.2 Comparison of measurement techniques.

	Advantage	Disadvantage
RGB images	Detect obvious defects (e.g. break and soiling)	Cannot detect other defects
EL images	Detect break, crack, and other micro defects	Expensive
	Widely used	Additional module required
PL images	Detect break, crack, and other micro defects	Not common
	Can be applied to the early stage of the cells	
IR images	Detect hot spots on the solar cell	Difficult to distinguish
	Penetrate smoke and fog better	defect types

Benatto et al. [2020] where an aerial robot equipped with InGaAs camera for the daylight EL image collection for solar module inspection.

Photoluminescence (PL) is another inspection method that refers also to the radiative signal emitted from the PV material. While PL occurs with photoexcitation via photon absorption, while EL is based on the material's response to electricity by emitting light. Table 16.2 summarizes the differences between common PV inspection methods.

16.3 Learning Framework

There are available data-driven solar module inspection approaches including model-based difference measurement, real-time statistical anomalies, output signal analysis, and machine learning-based methods [Pillai and Rajasekar, 2018]. It is also proved that machine learning-based approaches are prominent in identifying defects occurring on solar modules. Most available applications with shallow and deep neural networks (SNN and DNN) are listed in Li et al. [2021]. Most of the SNN approaches are based on multilayer perceptron and consider the data in measured features (e.g. the ratio of voltage or current). CNN is mostly preferred for DNN applications with well-known architectures including Visual Geometry Group (VGG) [Simonyan and Zisserman, 2014], R-CNN [Girshick et al., 2014], ResNet [He et al., 2016], AlexNET [Krizhevsky et al., 2017] and YOLO [Redmon and Farhadi, 2018].

Among the recent PV module inspection applications, EL images are the most preferred as compared to others with cell-level images [Li et al., 2021]. The performance of the detection is mostly associated with the availability of sufficient

Table 16.3 Open datasets for PV module inspection.

Source	Data type	Data size	Faults
Basnet et al. [2020]	1D data	3000	Line-to-line fault, open-circuit fault
Buerhop-Lutz et al. [2018]			
Deitsch et al. [2018]			
Deitsch et al. [2019]	EL images	2624	Defective cell
Karimi et al. [2019]	EL images	1031	Cracks
Mehta et al. [2018]	RGB images	45754	Soiling

data. Current practice with data augmentation for PV module inspection includes rotation and flip of the images. Currently available open datasets are presented in Table 16.3 [Li et al., 2021]. The following subsections include a case study based on the selected images from Buerhop-Lutz et al. [2018] and Deitsch et al. [2018, 2019].

16.3.1 Dataset Preparation

The open datasets referenced in this study can be found at Buerhop-Lutz et al. [2018] and Deitsch et al. [2018, 2019]. The dataset consists of 2624 EL images with a resolution of 300 × 300 pixels, each labeled with a probability for damage. The damage probabilities are categorized as 0, 0.33, 0.67, and 1, with a probability of 0 indicating a functional solar cell and a probability of 1 indicating a damaged cell. To differentiate between defective and functional cells, the dataset is divided into two categories based on the probability values. Cells with probabilities of 0 and 0.33 are considered functional, while cells with probabilities of 0.67 and 1 are considered defective.

In order to identify and categorize the different types of defects, the defective cells are further subdivided into six categories, as illustrated in Figure 16.1. These six categories include intrinsic and extrinsic defects, as these types are the most prevalent in the dataset. While material defects and finger interruptions do not always impact the performance of the solar cells, microcracks, breaks, and cell faults can reduce panel efficiency. Dark spots, as seen in Figure 16.1 (e), can be caused by various factors, such as material defects, dust, or breaks, and cannot be distinguished by EL imaging alone. As a result, all solar cells with dark spots are labeled accordingly.

The dataset is divided into two sets for the purpose of training and evaluation: the training set (90%) and the evaluation set (10%). Within the training set, there are two subsets: the training dataset (80%) and the validation dataset (20%).

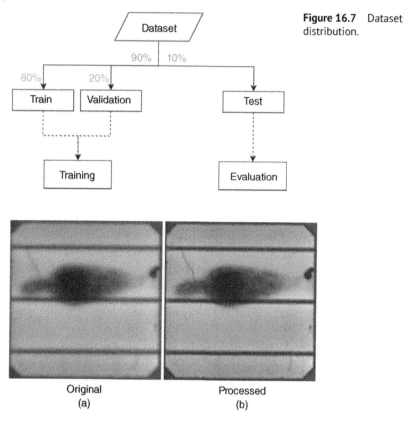

Figure 16.7 Dataset distribution.

Original
(a)

Processed
(b)

Figure 16.8 Dataset processing.

Image preprocessing techniques are applied to the dataset to reduce noise and enhance important features, including median blur and sharpening filters, as demonstrated in Figure 16.8.

While the current dataset includes 2,624 images, it may be prone to overfitting due to the small sample size and the presence of noisy images. To address this issue, various data augmentation techniques are applied to expand the dataset. These techniques include image rotation, random rotations within a range of $-2°$ to $+2°$, and vertical and horizontal flipping. The augmented dataset is balanced by applying augmentation operations to each defect type in different proportions. The effects of the data augmentation techniques on the dataset are analyzed and presented in Table 16.4. These techniques are used to improve the performance of the model and prevent overfitting during training by artificially inflating the size of the dataset. Further augmentations to avoid overfitting include data warping, which preserves the label of existing images through geometric and color

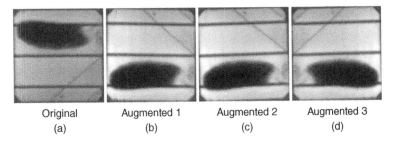

| Original | Augmented 1 | Augmented 2 | Augmented 3 |
| (a) | (b) | (c) | (d) |

Figure 16.9 Dataset augmentation: The original image and its augmented versions. Horizontal flip, vertical flip and rotation are applied to the image randomly.

Table 16.4 Data distrubution in dataset.

		Functional	Micro-crack	Finger interrup-tions	Material defect	Break	Cell fault	Dark spot	Total
Train	Original	701	226	575	35	71	15	34	1657
	augmented	1402	1356	1150	1400	1278	1410	1360	9356
Validation	Original	174	50	145	8	24	3	11	415
	augmented	348	300	290	352	384	348	352	2374
Test	Original	101	28	83	3	10	1	6	232
	augmented	202	168	166	198	200	202	192	1328

transformations, random erasing, adversarial training, and neural style transfer, as well as oversampling, which creates synthetic instances and adds them to the training set.

16.3.2 CNN Architecture

The CNN model, shown in Figure 16.10, is built with the basis of VGG16, an existing architecture built in TensorFlow 2.3.0. All the models are trained in the Google Collaboratory. The following subsections introduce the evaluation measures of the model and the approaches for improvement. Furthermore, the comparisons of the different methods are further shown in the following subsections. To enhance the generalization performance of the model, one could consider employing more intricate architectures and functional techniques, such as dropout regularization, batch normalization, transfer learning, and pretraining [Shorten and Khoshgoftaar, 2019].

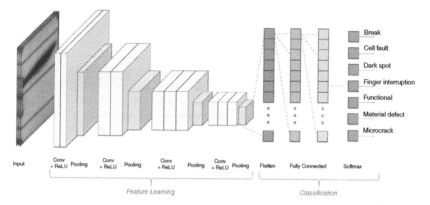

Figure 16.10 CNN architecture. There is an rectified linear unit (ReLU) layer between two fully connected layers. The output of the model is the array of possibilities for seven types.

16.3.3 Performance Evaluation Measures

When detecting one solar panel, the output of the CNN model is an array of numbers between 0 and 1, which represent the predicted possibilities for each defect category, respectively, for the input image. To measure the performance of the CNN model, we expect a high validation accuracy and low validation loss. The accuracy is the ratio of the correctly predicted number to the total predicted number [Sokolova et al., 2006], calculated with

$$A = \frac{t_n + t_p}{t_n + t_p + f_n + f_p} \tag{16.4}$$

where A is the accuracy, t_n is true-negative values, t_p is true-positive values, f_n is false-negative values, and f_p is false-positive values.

The loss function can show the difference between the predicted label and the true label. We used the cross-entropy loss function, which is commonly preferred in multiclass classification.

$$L = -\sum_{c=1}^{M} y_{o,c} \log(p_{o,c}) \tag{16.5}$$

where L is the loss, M is the number of the class, y is the binary indicator (0 or 1) if class label c is the correct classification for observation o, and p is the predicted probability observation o of class c.

Architecture Configuration

Overfitting often occurs in the early stage of the training. The model is overfitting when it has perfect results for the training dataset but performs poorly on the validation dataset. The overfitted model cannot generalize well for real-world data.

Therefore, avoiding overfitting is significant for improving the performance of the CNN model. We present the following approaches to prevent overfitting that are leveraged in our model.

To find an applicable CNN configuration with sufficient performance, models with different architectures and parameters are trained using the dataset. This compares the experimental results of the improved model against other baseline approaches. For this comparison, the models use different depths and convolutional layers: 7, 10, and 13. The results show the validation accuracy and validation loss for those three models indicating a better performance with 10 layers. The accuracies of the model with 10 and 13 convolutional layers are higher than that of seven convolutional layers and reach about 86%.

Data Augmentation The common practice is to use a larger and more complete dataset. A small dataset has fewer representative data, and its noise will influence the learning. Therefore, when the dataset is limited, data augmentation is an efficient way to increase the diversity of the data.

Model Size Reduction The size of the model depends on its capacity and is represented by the network size (number of layers and nodes). With a small capacity, the model cannot learn well, in contrast, with a bigger capacity, the model will learn all the inputs even including the noise, which will cause overfitting. There are various attempts for online architecture configurations [Kocer et al., 2021a] considering the variance and bias and batch simulations for hyperparameter optimization that are suggested for the future research.

Optimizers Optimizers play a critical role in changing the parameters of the CNN, such as the learning rate, to minimize the loss. Two of the most commonly used optimization algorithms are Adam and stochastic gradient descent (SGD), which are both based on gradient descent. Adam optimizer is widely popular due to its fast convergence. However, recent experiments conducted by Wilson et al. [2017] reveal that SGD optimizer generalizes better than other optimizers, including Adam.

To evaluate the performances of the two optimizers, the training model is run with both SGD and Adam optimizers. The results show that the model's accuracy is higher when using SGD optimizer, reaching the 80% range after 20 epochs. Similarly, the loss of the model with SGD optimizer reduces quickly, while the Adam optimizer does not converge within 200 epochs and has an increasing trend. In conclusion, while Adam optimizer may offer faster convergence, SGD optimizer can offer better generalization performance, making it a more practical choice for optimizing the neural network parameters.

Learning Rate Decay The learning rate is a critical parameter for the optimizer and one of the most important hyperparameters which control the tolerance of the loss [Bengio, 2012]. A model with a high learning rate converges fast but may have poor performance. A small learning rate has better performance but the convergence time could be very long. The learning rate decay is to train the model from a large learning rate and lower it after several epochs. Therefore, the training will first converge fast to avoid learning the noisy data and then learn complete features [You et al., 2019]. The inverse decay function can be used for the learning rate decay.

$$LR_d = \frac{LR}{1 + R_d \times \text{floor}\left(\frac{GS}{DS}\right)} \tag{16.6}$$

where LR_d is the decayed learning rate, LR is the initial learning rate, R_d is the decay ratio, GS is global steps and represents the total training epochs, and DS is the decay steps and determines the frequency to decay the learning rate.

There are three learning rates used in the model: 0.001, 0.005, and decay. For the decay, the inverse decay function is applied, and the learning rate reduces for every 50 epochs with an initial learning rate of 0.001 and a decay ratio is 0.5. The results show that the accuracies of the model training with different learning rates are similar and reach closer to 90% range. The accuracy of the training with a 0.005 learning rate is higher than the other two before 25 epochs since a higher learning rate helps the training for fast convergence. However, for the loss, the model with a learning rate of 0.005 has a much larger loss than the other two and has an increasing trend. The model with the decaying learning rate has a lower loss than that of a 0.005 learning rate after 160 epochs and convergence to the minimum range is achieved.

Dropout A dropout layer is to drop units and their connections in the neural network randomly with a defined probability in each batch during the training process [Srivastava et al., 2014], and these units will not be considered in the network either forward or backward. With a dropout layer, the data are trained in different models, and the output can be considered as the combination of the outputs from multiple neural networks.

Batch Normalization In the training, the input of each layer is calculated from the output of the last layers. Since the parameters of each layer will change during the learning, the distribution of layers' input will change too, that is called internal covariate shift. The layers need to adapt to this shift, and with the accumulation, the influence of the little change in the previous layers' parameters increases continuously. The batch normalization fixes the mean and variance to be close to

0 and 1, respectively, after each layer, and in the meantime, it keeps the learning features from previous layers. Batch normalization layers can reduce the shift without losing the learned feature from previous layers, and it is a good approach to avoid overfitting and reduce the training time. It occurs at the convolutional layer and activation function.

In our approach, three different models are used for comparison: with both batch normalization layers and dropout layers, without the batch normalization layers, and without the dropout layers. According to the results, the accuracy of the model without the batch normalization layers increases slowly and starts to converge until 160 epochs. The main reason for this is that the batch normalization layers can reduce the internal covariate shift and reduce the training time. For the loss, both the model with and without dropout layers converge within 50 epochs. Without the batch normalization layers, the loss of the model does not converge.

After the improvement, the detailed parameters of the improved CNN model are shown in Table 16.5. To reduce the training time and have sufficient performance, a batch normalization layer is added to each convolutional layer, and there is one dropout layer after each fully connected layer. The SGD optimizer is used with the learning rate decay for every 50 epochs and an initial learning rate of 0.001 and a decay ratio of 0.5. After applying approaches to avoid overfitting, the performance of the model is improved. The validation accuracy is around 85%, and the loss is left behind 0.5. The validation loss reaches 0.5 within 75 epochs. This validation performance of the model is acceptable for the multiclass classification.

To test the trained model, the test dataset and validation dataset are both used for the evaluation of the classification. Figure 16.11 shows six of the classifications – blue and red colors mean the correct and wrong predictions, respectively. It shows that types of finger interruption and microcrack are easy to be confused, and the wrong prediction is due to noises in the functional dataset. Figure 16.12 is the confusion matrix of the trained CNN model. Each column represents a true type that the image belongs to, while each row is the predicted label by the improved model. The highlighted cells in Figure 16.12 represent the correct prediction, and the accuracy rate of the evaluation dataset reaches 84.63%. The accuracies of the category of cell faults and material defects are much higher than others and reach 100% and 92.55%, respectively. This high accuracy is due to their obvious features. The accuracy of the type of microcrack is the lowest, which is 77.35%. For the wrong prediction of the solar cell with microcrack, it is mostly predicted to be the functional solar cell, broken solar cell, and the solar cell with dark spots. The main reason for the wrong prediction of the broken solar cells is the associated cracks. The noises in the dataset and similar features in multiple defect types lead also to degradation in prediction accuracies.

Table 16.5 Parameters of the CNN model for defective types classification.

CNN	Output shape	Setting
Input	$224 \times 224 \times 3$	
Convolutional 1	$224 \times 224 \times 8$	[ReLU]
Batch normalization 1	$224 \times 224 \times 8$	
Convolutional 2	$224 \times 224 \times 8$	[ReLU]
Batch normalization 2	$224 \times 224 \times 8$	
Max pooling	$112 \times 112 \times 8$	2×2 filter
Convolutional 3	$112 \times 112 \times 16$	[ReLU]
Batch normalization 3	$112 \times 112 \times 16$	
Convolutional 4	$112 \times 112 \times 16$	[ReLU]
Batch normalization 4	$112 \times 112 \times 16$	
Max pooling	$56 \times 56 \times 16$	2×2 filter
Convolutional 5	$56 \times 56 \times 64$	[ReLU]
Batch normalization 5	$56 \times 56 \times 64$	
Convolutional 6	$56 \times 56 \times 64$	[ReLU]
Batch normalization 6	$56 \times 56 \times 64$	
Convolutional 7	$56 \times 56 \times 64$	[ReLU]
Batch normalization 7	$56 \times 56 \times 64$	
Max pooling	$28 \times 28 \times 64$	2×2 filter
Convolutional 8	$28 \times 28 \times 128$	[ReLU]
Batch normalization 8	$28 \times 28 \times 128$	
Convolutional 9	$28 \times 28 \times 128$	[ReLU]
Batch normalization 9	$28 \times 28 \times 128$	
Convolutional 10	$28 \times 28 \times 128$	[ReLU]
Batch normalization 10	$28 \times 28 \times 128$	
Max pooling	$56 \times 56 \times 16$	2×2 filter
Flatten	25,088	
Fully connected 1	1024	[ReLU, 0.5 dropout]
Fully connected 2	1024	[0.5 dropout]
Output	7	[Softmax]

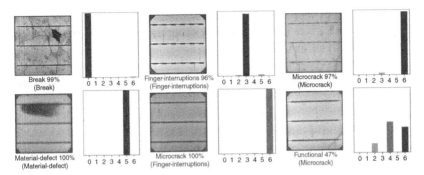

Figure 16.11 Classification results of the trained model. Blue means the correct prediction, and red means the wrong prediction. The defective type with a percentage behind is the prediction about the above image, and the percentage is the possibility of the prediction. Numbers of 0-6 represent the defective types (Break:0, Cell-faults:1, Dark-spot:2, Finger-interruptions:3, Functional:4, Material-defect:5, Microcrack:6).

		Expected						
		Break	Cell faults	Dark spot	Finger interruptions	Functional	Material defect	Microcrack
Predicted	Break	**488**	0	0	0	0	34	25
	Cell faults	1	**550**	0	0	0	6	0
	Dark spot	4	0	**430**	9	16	1	5
	Finger interruptions	2	0	3	**356**	58	0	28
	Functional	1	0	25	65	**437**	0	47
	Material defect	3	0	0	5	9	**509**	1
	Microcrack	85	0	86	21	29	0	**362**

Figure 16.12 Confusion matrix of the evaluation dataset.

16.4 Conclusion

This study presents an overview of drone-based PV module inspection and a case study focused on identifying defect types in solar modules. The framework aims to achieve autonomous inspection by detecting the solar panel and identifying the defective types of solar cells. For detection, shape detection techniques are used based on edge detection, Hough line transform, and coordinate calculations.

For identification, an optimized CNN model is presented, which can classify seven defect types for solar cells with an accuracy of approximately 85%. In this study, the model size, optimizers, learning rate, dropout layers, and batch normalization layers were parametrically studied for optimal performance. The proposed learning-based framework is simulated in the Unreal Engine platform using the NEST model, which is a building streaming energy and sensory data online.

To further enhance aerial systems' responsiveness and adaptability, teleoperation frameworks can be considered [Kocer et al., 2022; Stedman et al., 2023], which could facilitate the transition from manual methods to the use of aerial imaging technologies. Additionally, generating new and open datasets with multiple sensor modalities, including time-based data and thermal and luminescence-based images, could significantly improve overall inspection methodologies. Such datasets can provide a benchmark to evaluate current learning-based inspection approaches.

Acknowledgments

This work was partially supported by funding from EPSRC (award no. EP/N018494/1, EP/R026173/1, EP/R009953/1, EP/S031464/1, EP/W001136/1), NERC (award no. NE/R012229/1), and the EU H2020 AeroTwin project (grant ID 810321). Mirko Kovac is supported by the Royal Society Wolfson fellowship (RSWF/R1/18003).

Bibliography

Spirit Energy. Best Angle for Solar Panels in the UK and Beyond? 2019 URL https://blog.spiritenergy.co.uk/contractor/best-angle-solar-panels-uk. Accessed on 01/10/2023.

Regen Transforming Energy. A bright future: Opportunities for UK innovation in solar energy July 2019, 2019. URL https://www.regen.co.uk/wp-content/uploads/The-Solar-Commission-web.pdf. The Solar Commission.

Sustainable Development Goals, 2021. URL https://www.un.org/sustainable development. Accessed on 02/02/2021.

M. Waqar Akram, Guiqiang Li, Yi Jin, Xiao Chen, Changan Zhu, Xudong Zhao, Abdul Khaliq, M. Faheem, and Ashfaq Ahmad. CNN based automatic detection of photovoltaic cell defects in electroluminescence images. *Energy*, 189:116319, 2019. ISSN 03605442. doi: 10.1016/j.energy.2019.116319.

Kevin Z.Y. Ang, Xiangxu Dong, Wenqi Liu, Geng Qin, Shupeng Lai, Kangli Wang, Dong Wei, Songyuan Zhang, Swee King Phang, Xudong Chen, et al. High-precision multi-UAV teaming for the first outdoor night show in Singapore. *Unmanned Systems*, 6(01):39–65, 2018. doi: 10.1142/S2301385018500036.

Barun Basnet, Hyunjun Chun, and Junho Bang. An intelligent fault detection model for fault detection in photovoltaic systems. *Journal of Sensors*, 2020:1–11, 2020.

Yoshua Bengio. Practical Recommendations for Gradient-Based Training of Deep Architectures. In *Lecture Notes in Computer Science (including subseries Lecture Notes in Artificial Intelligence and Lecture Notes in Bioinformatics)*, 7700 LECTU, pages 437–478, 2012. ISSN 03029743. doi: 10.1007/978-3-642-35289-8-26.

I. Berardone, J. Lopez Garcia, and M. Paggi. Analysis of electroluminescence and infrared thermal images of monocrystalline silicon photovoltaic modules after 20 years of outdoor use in a solar vehicle. *Solar Energy*, 173:478–486, 2018. doi: 10.1016/j.solener.2018.07.055.

Lukas Bommes, Tobias Pickel, Claudia Buerhop-Lutz, Jens Hauch, Christoph Brabec, and Ian Marius Peters. Computer vision tool for detection, mapping, and fault classification of photovoltaics modules in aerial IR videos. *Progress in Photovoltaics: Research and Applications*, 2021. doi: 10.1002/pip.3448.

Cl Buerhop, D. Schlegel, M. Niess, C. Vodermayer, R. Weißmann, and C.J. Brabec. Reliability of IR-imaging of PV-plants under operating conditions. *Solar Energy Materials and Solar Cells*, 107:154–164, 2012. ISSN 0927-0248. doi: 10.1016/j.solmat.2012.07.011.

Claudia Buerhop-Lutz, Sergiu Deitsch, Andreas Maier, Florian Gallwitz, Stephan Berger, Bernd Doll, Jens Hauch, Christian Camus, and Christoph J. Brabec. A benchmark for visual identification of defective solar cells in electroluminescence imagery. In *European PV Solar Energy Conference and Exhibition (EU PVSEC)*, 2018. doi: 10.4229/35thEUPVSEC20182018-5CV.3.15.

John Canny. A computational approach to edge detection. *IEEE Transactions on Pattern Analysis and Machine Intelligence*, PAMI-8(6):679–698, 1986.

Young-Jin Cha, Wooram Choi, and Oral Büyüköztürk. Deep learning-based crack damage detection using convolutional neural networks. *Computer-Aided Civil and Infrastructure Engineering*, 32(5):361–378, 2017. doi: 10.1111/mice.12263.

Leong Jing Chen, John Henawy, Basaran Bahadir Kocer, and Gerald Gim Lee Seet. Aerial robots on the way to underground: An experimental evaluation of VINS-mono on visual-inertial odometry camera. In *2019 International Conference on Data Mining Workshops (ICDMW)*, pages 91–96. IEEE, 2019.

Haiyong Chen, Yue Pang, Qidi Hu, and Kun Liu. Solar cell surface defect inspection based on multispectral convolutional neural network. *Journal of Intelligent Manufacturing*, 31(2):453–468, 2020. ISSN 15728145. doi: 10.1007/s10845-018-1458-z.

Jiun Fatt Chow, Basaran Bahadir Kocer, John Henawy, Gerald Seet, Zhengguo Li, Wei Yun Yau, and Mahardhika Pratama. Toward underground localization: LiDAR inertial odometry enabled aerial robot navigation. *arXiv preprint arXiv:1910.13085*, 2019.

Aline Kirsten Vidal de Oliveira, Mohammadreza Aghaei, and Ricardo Rüther. Aerial infrared thermography for low-cost and fast fault detection in utility-scale PV power plants. *Solar Energy*, 211:712–724, 2020. doi: 10.1016/j.solener.2020.09.066.

Sergiu Deitsch, Claudia Buerhop-Lutz, Andreas K. Maier, Florian Gallwitz, and Christian Riess. Segmentation of photovoltaic module cells in electroluminescence images. Technical report, Springer, 2018. https://link.springer.com/article/10.1007/s00138-021-01191-9.

Sergiu Deitsch, Vincent Christlein, Stephan Berger, Claudia Buerhop-Lutz, Andreas Maier, Florian Gallwitz, and Christian Riess. Automatic classification of defective photovoltaic module cells in electroluminescence images. *Solar Energy*, 185:455–468, 2019a. ISSN 0038092X. doi: 10.1016/j.solener.2019.02.067.

Jeff Delaune, Robert Hewitt, Laura Lytle, Cristina Sorice, Rohan Thakker, and Larry Matthies. Thermal-inertial odometry for autonomous flight throughout the night. In *2019 IEEE/RSJ International Conference on Intelligent Robots and Systems (IROS)*, pages 1122–1128. IEEE, 2019.

Gisele Alves dos Reis Benatto, Claire Mantel, Adrian A. Santamaria Lancia, Michael Graversen, Nicholas Reidel, Sune Thorsteinsson, Peter B. Poulsen, Søren Forchhammer, Harsh Parikh, Sergiu Spataru, et al. Image processing for daylight electroluminescence PV imaging acquired in movement. In *Proceedings of the 35th European Photovoltaic Solar Energy Conference and Exhibition: EU PVSEC 2018*, pages 2005–2009. WIP Wirtschaft und Infrastruktur GmbH and Co Planungs KG, 2018.

Gisele Alves dos Reis Benatto, Claire Mantel, Sergiu Spataru, Adrian Alejo Santamaria Lancia, Nicholas Riedel, Sune Thorsteinsson, Peter Behrensdorff Poulsen, Harsh Parikh, Søren Forchhammer, and Dezso Sera. Drone-based daylight electroluminescence imaging of PV modules. *IEEE Journal of Photovoltaics*, 10(3):872–877, 2020. doi: 10.1109/JPHOTOV.2020.2978068.

Christopher Dunderdale, Warren Brettenny, Chantelle Clohessy, and E. Ernest van Dyk. Photovoltaic defect classification through thermal infrared imaging using a machine learning approach. *Progress in Photovoltaics: Research and Applications*, 28(3):177–188, 2020. doi: 10.1002/pip.3191.

Tucker Farrell, Kidus Guye, Rebecca Mitchell, and Guangdong Zhu. A non-intrusive optical approach to characterize heliostats in utility-scale power tower plants: Flight path generation/optimization of unmanned aerial systems. *Solar Energy*, 225:784–801, 2021. doi: 10.1016/j.solener.2021.07.070.

Zhuang Fu, Yanzheng Zhao, Yang Liu, Qixin Cao, Mingbo Chen, Jun Zhang, and Lee Jay. Solar cell crack inspection by image processing. In *Proceedings of 2004 International Conference on the Business of Electronic Product Reliability and Liability*, volume 200030, pages 77–80, 2004. doi: 10.1109/beprl.2004.1308153.

Sara Gallardo-Saavedra, E. Franco-Mejia, L. Hernández-Callejo, Ó. Duque-Pérez, H. Loaiza-Correa, and E. Alfaro-Mejia. Aerial thermographic inspection of photovoltaic plants: Analysis and selection of the equipment. In *Proceedings of the 2017 Proceedings ISES Solar World Congress*, IEA SHC, Abu Dhabi, UAE, volume 29, 2017.

Sara Gallardo-Saavedra, Luis Hernández-Callejo, and Oscar Duque-Perez. Technological review of the instrumentation used in aerial thermographic inspection of photovoltaic plants. *Renewable and Sustainable Energy Reviews*, 93:566–579, 2018. doi: 10.1016/j.rser.2018.05.027.

Ross Girshick, Jeff Donahue, Trevor Darrell, and Jitendra Malik. Rich feature hierarchies for accurate object detection and semantic segmentation. In *Proceedings of the IEEE Conference on Computer Vision and Pattern Recognition*, pages 580–587, 2014.

Anastasios Golnas. PV system reliability: An operator's perspective. In *2012 IEEE 38th Photovoltaic Specialists Conference (PVSC) PART 2*, pages 1–6. IEEE, 2012.

Fabian Hauf, Basaran Bahadir Kocer, Alan Slatter, Hai-Nguyen Nguyen, Oscar Pang, Ronald Clark, Edward Johns, and Mirko Kovac. Learning tethered perching for aerial robots. In *2023 IEEE International Conference on Robotics and Automation (ICRA)*. IEEE, 2023.

Kaiming He, Xiangyu Zhang, Shaoqing Ren, and Jian Sun. Deep residual learning for image recognition. In *Proceedings of the IEEE Conference on Computer Vision and Pattern Recognition*, pages 770–778, 2016.

Chris Henry, Sahadev Poudel, Sang-Woong Lee, and Heon Jeong. Automatic detection system of deteriorated PV modules using drone with thermal camera. *Applied Sciences*, 10(11):3802, 2020.

Luis Hernández-Callejo, Sara Gallardo-Saavedra, and Víctor Alonso-Gómez. A review of photovoltaic systems: Design, operation and maintenance. *Solar Energy*, 188:426–440, 2019. doi: 10.1016/j.solener.2019.06.017.

Boon Ho, Basaran Bahadir Kocer, and Mirko Kovac. Vision based crown loss estimation for individual trees with remote aerial robots. *ISPRS Journal of Photogrammetry and Remote Sensing*, 188:75–88, 2022.

Ingeborg Høiaas, Katarina Grujic, Anne Gerd Imenes, Ingunn Burud, Espen Olsen, and Nabil Belbachir. Inspection and condition monitoring of large-scale photovoltaic power plants: A review of imaging technologies. *Renewable and Sustainable Energy Reviews*, 161:112353, 2022.

Athar Hussain, Ankit Batra, and Rupendra Pachauri. An experimental study on effect of dust on power loss in solar photovoltaic module. *Renewables: Wind, Water, and Solar*, 4(1):9, 2017. ISSN 2198-994X. doi: 10.1186/s40807-017-0043-y.

Ulrike Jahn, Magnus Herz, David Parlevliet, Marco Paggi, Ioannis Tsanakas, Joshua Stein, Karl Berger, Samuli Ranta, Roger French, Mauricio Richter, et al. *Review on* infrared and electroluminescence imaging for PV field applications. 2018. ISBN 978-3-906042-53-4.

Ahmad Maroof Karimi, Justin S. Fada, Mohammad Akram Hossain, Shuying Yang, Timothy J. Peshek, Jennifer L. Braid, and Roger H. French. Automated pipeline for photovoltaic module electroluminescence image processing and degradation feature classification. *IEEE Journal of Photovoltaics*, 9(5):1324–1335, 2019.

Shehryar Khattak, Christos Papachristos, and Kostas Alexis. Keyframe-based thermal–inertial odometry. *Journal of Field Robotics*, 37(4):552–579, 2020.

Basaran Bahadir Kocer, Tegoeh Tjahjowidodo, Mahardhika Pratama, and Gerald Gim Lee Seet. Inspection-while-flying: An autonomous contact-based nondestructive test using UAV-tools. *Automation in Construction*, 106:102895, 2019. doi: 10.1016/j.autcon.2019.102895.

Basaran Bahadir Kocer, Mohamad Abdul Hady, Harikumar Kandath, Mahardhika Pratama, and Mirko Kovac. Deep neuromorphic controller with dynamic topology for aerial robots. In *2021 IEEE International Conference on Robotics and Automation (ICRA)*, pages 110–116. IEEE, 2021a. doi: 10.1109/ICRA48506. 2021.9561729.

Basaran Bahadir Kocer, Boon Ho, Xuanhao Zhu, Peter Zheng, André Farinha, Feng Xiao, Brett Stephens, Fabian Wiesemüller, Lachlan Orr, and Mirko Kovac. Forest drones for environmental sensing and nature conservation. In *2021 Aerial Robotic Systems Physically Interacting with the Environment (AIRPHARO)*, pages 1–8, 2021b. doi: 10.1109/AIRPHARO52252.2021.9571033.

Basaran Bahadir Kocer, Harvey Stedman, Patryk Kulik, Izaak Caves, Nejra Van Zalk, Vijay M. Pawar, and Mirko Kovac. Immersive view and interface design for teleoperated aerial manipulation. In *2022 IEEE/RSJ International Conference on Intelligent Robots and Systems (IROS)*, pages 4919–4926. IEEE, 2022.

Simon Koch, Thomas Weber, Christian Sobottka, Andreas Fladung, Patrick Clemens, and Juliane Berghold. Outdoor electroluminescence imaging of crystalline photovoltaic modules: Comparative study between manual ground-level inspections and drone-based aerial surveys. In *32nd European Photovoltaic Solar Energy Conference and Exhibition*, pages 1736–1740, 2016.

Alex Krizhevsky, Ilya Sutskever, and Geoffrey E. Hinton. Imagenet classification with deep convolutional neural networks. *Communications of the ACM*, 60(6):84–90, 2017.

Baojie Li, Claude Delpha, Demba Diallo, and A. Migan-Dubois. Application of artificial neural networks to photovoltaic fault detection and diagnosis: A review. *Renewable and Sustainable Energy Reviews*, 138:110512, 2021.

Claire Mantel, Frederik Villebro, Gisele Alves dos Reis Benatto, Harsh Rajesh Parikh, Stefan Wendlandt, Kabir Hossain, Peter Behrensdorff Poulsen, Sergiu Spataru, Dezso Séra, and Søren Forchhammer. Machine learning prediction of defect types for electroluminescence images of photovoltaic panels. page 1, 2019. ISSN 1996756X. doi: 10.1117/12.2528440.

Sachin Mehta, Amar P. Azad, Saneem A. Chemmengath, Vikas Raykar, and Shivkumar Kalyanaraman. DeepSolarEye: Power loss prediction and weakly supervised soiling localization via fully convolutional networks for solar panels. In *2018 IEEE Winter Conference on Applications of Computer Vision (WACV)*, pages 333–342. IEEE, 2018.

Binbin Ni, Pingguo Zou, Qiang Li, and Yabin Chen. Intelligent defect detection method of photovoltaic modules based on deep learning. *Proceedings of the 2018 International Conference on Transportation & Logistics, Information & Communication, Smart City (TLICSC 2018), 161*, pages 167–173, 2018. doi: 10.2991/ tlicsc-18.2018.27.

Lachlan Orr, Brett Stephens, Basaran Bahadir Kocer, and Mirko Kovac. A high payload aerial platform for infrastructure repair and manufacturing. In *2021 Aerial Robotic Systems Physically Interacting with the Environment (AIRPHARO)*, pages 1–6, 2021. doi: 10.1109/AIRPHARO52252.2021.9571052.

R. Pierdicca, E.S. Malinverni, F. Piccinini, M. Paolanti, A. Felicetti, and P. Zingaretti. Deep convolutional neural network for automatic detection of damaged photovoltaic cells. *International Archives of the Photogrammetry, Remote Sensing and Spatial Information Sciences - ISPRS Archives*, 42(2):893–900, 2018. ISSN 16821750. doi: 10.5194/isprs-archives-XLII-2-893-2018.

Dhanup S. Pillai and N. Rajasekar. A comprehensive review on protection challenges and fault diagnosis in PV systems. *Renewable and Sustainable Energy Reviews*, 91:18–40, 2018.

Paolo Bellezza Quater, Francesco Grimaccia, Sonia Leva, Marco Mussetta, and Mohammadreza Aghaei. Light Unmanned Aerial Vehicles (UAVs) for cooperative inspection of PV plants. *IEEE Journal of Photovoltaics*, 4(4):1107–1113, 2014. ISSN 21563381. doi: 10.1109/JPHOTOV.2014.2323714.

Sheikh Aminur Rahaman, Tania Urmee, and David A. Parlevliet. PV system defects identification using remotely piloted aircraft (RPA) based infrared (IR) imaging: A review. *Solar Energy*, 206:579–595, 2020.

Joseph Redmon and Ali Farhadi. YOLOv3: An incremental improvement. *arXiv preprint arXiv:1804.02767*, 2018.

Travis Sarver, Ali Al-Qaraghuli, and Lawrence L. Kazmerski. A comprehensive review of the impact of dust on the use of solar energy: History, investigations, results, literature, and mitigation approaches. *Renewable and Sustainable Energy Reviews*, 22:698–733, 2013. ISSN 13640321. doi: 10.1016/j.rser.2012.12.065.

Connor Shorten and Taghi M. Khoshgoftaar. A survey on image data augmentation for deep learning. *Journal of Big Data*, 6(1):1–48, 2019.

Karen Simonyan and Andrew Zisserman. Very deep convolutional networks for large-scale image recognition. *arXiv preprint arXiv:1409.1556*, 2014.

Marina Sokolova, Nathalie Japkowicz, and Stan Szpakowicz. Beyond accuracy, F-score and ROC: A family of discriminant measures for performance evaluation Marina. *AI 2006: Advances in Artificial Intelligence*, 4304(1), 2006. doi: 10.1007/11941439.

Nitish Srivastava, Geoffrey Hinton, Alex Krizhevsky, Ilya Sutskever, and Ruslan Salakhutdinov. Dropout: A simple way to prevent neural networks from overfitting. *Journal of Machine Learning Research* 15:1929–1958, 2014. doi: 10.1109/ICAEES.2016.7888100.

Harvey Stedman, Basaran Bahadir Kocer, Nejra Hady Van Zalk, Mirko Kovac, and Vijay M. Pawar. Evaluating immersive teleoperation interfaces: Coordinating robot radiation monitoring tasks in nuclear facilities. In *2023 IEEE International Conference on Robotics and Automation (ICRA)*. IEEE, 2023.

Brett Stephens, Lachlan Orr, Basaran Bahadir Kocer, Hai-Nguyen Nguyen, and Mirko Kovac. An aerial parallel manipulator with shared compliance. *IEEE Robotics and Automation Letters*, 7(4):11902–11909, 2022.

Wuqin Tang, Qiang Yang, Kuixiang Xiong, and Wenjun Yan. Deep learning based automatic defect identification of photovoltaic module using electroluminescence images. *Solar Energy*, 201:453–460, 2020. ISSN 0038092X. doi: 10.1016/j.solener.2020.03.049.

Thorsten Trupke. *Photoluminescence and electroluminescence characterization in silicon photovoltaics*, chapter 7.2, pages 322–338. John Wiley & Sons, Ltd., 2017. ISBN 9781118927496. doi: 10.1002/9781118927496.ch30.

Du Ming Tsai, Shih Chieh Wu, and Wei Chen Li. Defect detection of solar cells in electroluminescence images using Fourier image reconstruction. *Solar Energy Materials and Solar Cells*, 99:250–262, 2012. ISSN 09270248. doi: 10.1016/j.solmat.2011.12.007.

John A. Tsanakas, Long Ha, and Claudia Buerhop. Faults and infrared thermographic diagnosis in operating C-SI photovoltaic modules: A review of research and future challenges. *Renewable and Sustainable Energy Reviews*, 62:695–709, 2016.

Din Chang Tseng, Yu Shuo Liu, and Chang Min Chou. Automatic finger interruption detection in electroluminescence images of multicrystalline solar cells. *Mathematical Problems in Engineering*, 2015:879675, 2015. ISSN 15635147. doi: 10.1155/2015/879675.

Jhon Jairo Vega Díaz, Michiel Vlaminck, Dionysios Lefkaditis, Sergio Alejandro Orjuela Vargas, and Hiep Luong. Solar panel detection within complex backgrounds using thermal images acquired by UAVs. *Sensors*, 20(21):6219, 2020.

Ashia C. Wilson, Rebecca Roelofs, Mitchell Stern, Nathan Srebro, and Benjamin Recht. The marginal value of adaptive gradient methods in machine learning. *Advances in Neural Information Processing Systems 30 (NIPS 2017)*, pages 4149–4159, December 2017. ISSN 10495258.

Feng Xiao, Peter Zheng, Julien di Tria, Basaran Bahadir Kocer, and Mirko Kovac. Optic flow-based reactive collision prevention for mavs using the fictitious obstacle hypothesis. *IEEE Robotics and Automation Letters*, 6(2):3144–3151, 2021. doi: 10.1109/LRA.2021.3062317.

Kaichao You, Mingsheng Long, Jianmin Wang, and Michael I. Jordan. How Does Learning Rate Decay Help Modern Neural Networks? 2019.

Ketao Zhang, Pisak Chermprayong, Feng Xiao, Dimos Tzoumanikas, Barrie Dams, Sebastian Kay, Basaran Bahadir Kocer, Alec Burns, Lachlan Orr, Christopher Choi, et al. Aerial additive manufacturing with multiple autonomous robots. *Nature*, 609(7928):709–717, 2022.

Yang Zhao, Ke Zhan, Zhen Wang, and Wenzhong Shen. Deep learning-based automatic detection of multitype defects in photovoltaic modules and application in real production line. *Progress in Photovoltaics: Research and Applications*, 29(4):471–484, 2021. doi: 10.1002/pip.3395.

Peter Zheng, Xinkai Tan, Basaran Bahadir Kocer, Erdeng Yang, and Mirko Kovac. TiltDrone: A fully-actuated tilting quadrotor platform. *IEEE Robotics and Automation Letters*, 5(4):6845–6852, 2020. doi: 10.1109/LRA.2020.3010460.

17

Aerial Repair and Aerial Additive Manufacturing

Yusuf Furkan Kaya[1,2], Lachlan Orr[1,2], Basaran Bahadir Kocer[1,3]*, and Mirko Kovac[1,2]

[1]*Aerial Robotics Laboratory, Imperial College London, London, UK*
[2]*Laboratory of Sustainability Robotics, Swiss Federal Laboratories for Materials Science and Technology, Dübendorf, Switzerland*
[3]*School of Civil, Aerospace and Design Engineering, University of Bristol, Bristol, UK*

17.1 Review of State of the Art in Additive Manufacturing at Architectural Scales

Construction is the core action for human habitation and shelter production. Access to these fundamental needs matures through the essential characteristics of the construction methods, such as pace, suppleness, and cost. These features also challenge the access and build envelope of construction in hard-to-reach areas and extreme environments. In addition, workers' health is at stake considering the dynamic and risky nature of construction sites and these conventional construction tasks. Therefore, research on robotic-based construction systems is experiencing substantial growth because of the promising potential of response to these deficiencies [Tay et al., 2017; Delgado et al., 2019].

The robotic construction field's lead research area is onsite and off-site continuous and discrete (such as brick-laying) additive manufacturing processes that use ground-based static or mobile robotic systems. Hence, it is possible to find the equivalents of these applications in industry as well as in intensive studies in universities. Figure 17.1 demonstrates the main categories of the robotic platforms deployed in these applications. As it can be seen, the first three categories are ground-based gantry robots and robotic arms. And the last category is the novel application of aerial additive manufacturing (aerial AM or AAM), which is still in its infancy and brings a different perspective by deploying aerial vehicles with a

*Email: b.kocer@bristol.ac.uk

Infrastructure Robotics: Methodologies, Robotic Systems and Applications, First Edition.
Edited by Dikai Liu, Carlos Balaguer, Gamini Dissanayake, and Mirko Kovac.

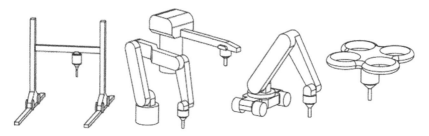

Figure 17.1 Illustrations of the main categories of additive construction platforms currently being employed in research and commercial applications. Gantry, Fixed Arm, Mobile Arm, UAV (From left to right).

robotic manipulator to produce large-scale structures with additive manufacturing methods [Zhang et al., 2022]. This novel production method facilitates multiagent parallel additive manufacturing with an unrestrained build envelope in hard-to-access zones. These characteristics of AAM respond to the deficiencies of ground-based systems and hold enormous potential and promise for the field of robotic construction. However, in most cases, research in aerial construction methods has been preceded by a ground-based demonstration, and these studies have great relevance to directing future work in aerial robotic construction. Therefore, understanding the broader context of additive manufacturing techniques on construction sites helps to evaluate further and discuss AAM methods.

The technology is initially deployed in 1996, and a proliferation has been seen in the last decade. Figure 17.2 represents the technological development timeline of AM. Among these several additive manufacturing processes, continuous robotic manufacturing of concrete is mainly based on two types, known as binder jetting and material extrusion. Contour crafting (CC) [Khoshnevis, 2004], which uses the material extrusion technique, is the pioneering study into the off-site AM of

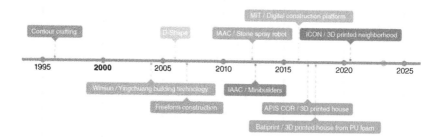

Figure 17.2 The timeline of the technological development of additive manufacturing in construction.

large-scale cementitious construction. The system combines an X, Y, and Z gantry frame with a six-axis nozzle that can generate smooth surfaces for the aimed structure. Approximately one decade later, D-shape [Dshape, 2023] started to use binder jetting to produce architectural scale structures. This system uses an ink binder to create the layers of the structure within the same gantry logic as mentioned with CC. In parallel, the Chinese private company Winsun Decoration Design Engineering [Global, 2023] is known as the inventor of the 3D print nozzle. Even though the company treats these technologies as commercially sensitive and takes privacy very seriously, Ghaffar et al. [2018] state that the cementitious material developed is reinforced with fiberglass, and the production speed of their technology is significantly fast. Similar research to CC is done by the University of Loughborough [Lim et al., 2012] for onsite AM of high-performance concrete in large-scale manufacturing. The produced structure can resist up to 110 MPa of compression, and its dimensions are 2.0, 0.9, and 0.8 meter. These studies were basically seen as proof of concept and gave the first signals of the possibility of robotic production in construction works and have been continuously developed since then.

Since the early 2010s, there has been an incredible leap in the number of private or academic institutions searching and experimenting on different ground-based robotic AM systems. Therefore, a large body of academic review research on additive manufacturing of large concrete structures already exists in the literature [Al Rashid et al., 2020; Gharbia et al., 2019; Buswell et al., 2018; Al Jassmi et al., 2018]. The following section does not attempt to list every 3D concrete printer developed but tries to encompass the design space and highlight relevant novel approaches that can apply to aerial vehicles.

As well as academia, several private companies are trying to do on-site printing of large-scale structures with similar technological approaches and develop robotic systems accordingly. To begin with, the company WASP (World's Advanced Saving Project) developed a couple of large-scale gantry printers for concrete extrusion. The first public demonstration carried out was a print of a clay house using low-cost building materials with a 12 meter gantry containing a delta-robot for end-effector positioning [WASP Delta, 2023]. This concept was later extended into a modular system to allow prints of unlimited footprint size [WASP Crane WASP, 2023]. Another leading 3D concrete printing company is ICON. Their latest project to build an entire neighborhood of 3D printed houses proves how this technology has evolved and become a preferred construction method because of the promised high delivery speed, increased efficiency, and relatively low costs [ICON, 2023]. Finally, a team in IAAC (Institute for Advanced Architecture of Catalonia) realized another small research project of a stone spray robot [Kulik et al., 2012]. This tethered ground-based robotic arm uses sand or soil and glue to manufacture architectural structures additively on diagonal or vertical surfaces.

Mobile robotic systems' research has witnessed the same proliferation as fixed robots. The usage of these systems allows for an arbitrary printing canvas that is less restricted than a fixed gantry robot. Apis Cor, a company based in Boston, uses an easy-to-transport multiaxis robotic arm named "Frank" as an alternative to gantry robots [Cor, 2022]. This system can print a 38 m^2 structure within 24 h. Another mobile system is presented by Helm et al. [2012], capable of fulfilling several construction tasks while fitting through a doorway.

Further work has been done elsewhere on the challenges of multiagent construction [Zhang et al., 2018; Tiryaki et al., 2019; Sustarevas et al., 2019] with the motion planning and control implications being discussed. Similar motion planning for aerial vehicles is more straightforward in theory due to the additional degree of freedom on height. Unmanned aerial vehicles (UAVs) can fly over obstacles and each other to avoid collisions. As individual units are very small compared to the built structure, there are fewer limits on what can be built. Unlike larger fixed arms, they can fly into restricted areas and work. Sustarevas et al. [2018] developed an all-terrain base for mobile construction systems. Dakhli and Lafhaj [2017] developed a similar ground-based system that can perform modular construction tasks and evaluate the efficiency of this system in comparison to traditional manual construction and proved that the robotic system is more efficient and cost saving. A different approach is demonstrated by having a swarm of mobile robots designed by IAAC researchers [Jokic et al., 2014] that can cooperatively produce concrete structures of any volume. These robots, which work as complementary, are relatively small, unlike other onsite robotic production systems. Another mobile robot developed by Keating et al. [2017], known as the "Digital Construction Platform," has several construction capabilities and built a large-scale polyurethane (PU) foam dome structure to demonstrate its real scale application.

Obtaining a large-scale architectural structure is an interdisciplinary study that includes diverse expertise like material science, robotics, and architecture. Therefore, these robotic construction systems cannot be considered without the identification of the specific material to be produced and the unique way of producing it. For that reason, as well as the robotic platform development processes, experimental 3D printing studies continue intensively to realize large-scale robotic construction with novel materials.

Rusenova et al. [2018] have presented a nonregular 3D printing method by jamming bulk materials with string reinforcement to produce reversible structures. Dritsas et al. [2020] developed fungal-like adhesive material (FLAM) to produce large-scale architectural elements using current modes of production like fused deposition modeling (FDM). Another sustainable and vernacular approach to robotic construction is the MUD Frontier [Burry et al., 2020]. Its main goal is to combine domestic vernacular construction knowledge in clay-based material usage with 21st-century technology.

Multiple demonstrations of printing using expanding polyurethane foam have been carried out. As mentioned earlier, digital construction platform by Keating et al. [2017] used PU foam to print a large-scale semi-dome. Batiprint [Batiprint3D, 2023] is a French company developing construction printing methods with polyurethane foam. They have completed a print of a single-storey house by extruding a polyurethane foam mold using a mobile robotic arm, which was later filled with cement [Subrin et al., 2018]. The method outlined appears to have relied heavily on manual intervention with the arm printing sections of the wall and then being manually relocated. Polyurethane foam allows a much larger final print volume to be produced for the same volume of material transported to the construction site. This may make it an attractive candidate for mobile platforms to reduce the frequency with which they need to restock raw material. In the context of aerial construction, this would significantly increase the capacity and printing speed of a single UAV.

Considering all these studies, current large-scale additive manufacturing systems necessitate a horizontal homogeneous plane to construct and a team of experts to manage the construction process. Furthermore, the fixed gantry robots restrict the structure's scale with their predefined canvases. Even though mobile robots have better construction envelopes than these fixed gantry robots and can easily expand them into two dimensions, the third dimension is still constrained with the robot's hardware. On the other hand, aerial robots have an unrestricted build envelope, can access hard to reach zones, and perform construction tasks in harsh environments. Due to these aspects, aerial robotic construction research has grown in the past decade. The following chapter comprehensively explores these studies.

17.2 Review of Demonstrations of Aerial Manufacturing and Repair

Aerial robots are entirely agnostic to terrain conditions. This makes them attractive for performing missions in distant zones or at height without requiring any supporting infrastructure. For that reason, the use of aerial robots for construction tasks in dynamic and hazardous construction sites brings a significant potential to increase safety and efficiency, consequently decreasing costs. Indirectly, this potential causes it to be the subject of intense research. The conceptual representation of aerial repair and AAM with a variety of material in different extreme environment and at height is represented in Figure 17.3.

Figure 17.3 Conceptual representations of aerial repair and aerial additive manufacturing methods. (a) Conceptual representation of AAM in extra-terrestrial environments. (b) Conceptual representation of AAM in polar regions. (c–e) Agility of AAM in hard-to-reach environments. (f) AAM on vertical surfaces. (g) Autonomous inspection and deployment of AAM in repair tasks. (h) AAM of metals.

Lindsey et al. [2011] presented the first instance of discrete aerial construction by assembling specifically designed modules. The rectangular prism units with magnetic joints on both edges were relocated to produce $50\,cm^3$ hollow cubes. The most prominent structure built was a combination of five hollow cubes. However, because the units were particularly designed with magnets, it would be expensive and not structurally viable to construct a large-scale structure with this methodology.

The flight assembled architectural installation by Augugliaro et al. [2014] was a similar representation of modular aerial construction. Two small drones cooperatively carried small rectangular polyurethane prisms for 18 h to build a 6 meter tall tower. This tower, which is assumed to be a small conceptual representation of a giant vertical city, was the first attempt for a large-scale demonstration of aerial construction. The polyurethane bricks were lightweight and easy to handle for small drones but not strong enough to resist external conditions such as wind and rain.

These demonstrations were performed in an indoor environment with infrared tracking cameras for localization. This type of experimental setup is convenient for demonstrating new assembly techniques with small robots but cannot facilitate large-scale building with architectural materials on a construction site. They also relied on building materials optimized for ease of transportation and assembly, rather than their applicability to large-scale buildings.

Another point of view in the research field of modular aerial construction is concentrating on the design of the modular units to ease the process and increase efficiency and precision of the relocation done by Goessens et al. [2018] and Latteur et al. [2015] in two-phase research. The modular units called "dricks" and "droxels" have a shape to decrease and eliminate the possibility of errors caused by drones' low precision rate while locating the units. The aerial robot used in the research is the largest compared to the previous studies. It has a frame size of around 180 cm and a payload of 25 kg.

Wood et al. [2019] took a step further and designed "cyber-physical" modules with onboard sensors and computing that can be reconfigured using a UAV. This expensive approach is challenging to compare with simple block stacking or truss assembly directly; however, it demonstrates how future aerial construction may feedback into how structures in public spaces are designed and used.

In parallel to these discrete AAM processes, Hunt et al. [2014] proposed the first aerial platform to deposit polyurethane foam during flight continuously. This material is deposited from two syringes to a mixing nozzle. To allow the mixed material to have a workable state, it is held 1990s in a 10 ml syringe. Dams et al. [2020] later extended this research to different densities of polyurethane foam to analyze their applicability for AAM. These studies reveal the necessity of high positioning accuracy compared to the earlier discrete assembly methods. This leads to developing a delta manipulator to provide the required precision [Chermprayong et al., 2019]. This technology was used on a repair task of a pipeline.

Nettekoven and Topcu [2021] developed an aerial platform capable of using FDM to 3D print with polylactic acid (PLA). For the clarity and success of the print, the hexacopter, which comes into close contact with the surface, uses surface friction to dampen the vibrations and sudden changes in position caused by the rotors and air movements. Being so close to the surface and decreasing the vibration makes the printing relatively successful. However, this application is only limited to one layer of printing.

Braithwaite et al. [2018], Augugliaro et al. [2013], and Mirjan et al. [2013] demonstrate the use of aerial vehicles with spools of string to construct simple tensile structures between fixed points. A single drone can carry a large amount of lightweight, small-diameter rope and produce a large structure in a single flight so long as solid anchor points are available. This work is expanded in a subsequent paper [Augugliaro et al., 2015] to demonstrate knot tying with a similar system. Tensile construction could be useful in applications where existing solid supports are already in place, such as bridges between trees as buildings [Hauf et al., 2023]. Parallel research by Saldana et al. [2018] presented a temporal modular aerial structure. This formation is created by merging quadrotors, each surrounded by a cuboid skeleton.

17.2.1 Demands and Challenges

To realize the potential of aerial robots in repair, construction, and manufacturing, new tools and methodologies should be developed. The selected challenges in this study can provide insights into the current bottlenecks and challenges. This could further motivate new approaches and solutions in this context.

In situ Material Deposition with Aerial Robots: Deploying aerial robots for in situ applications poses significant challenges due to the unpredictable nature of outdoor conditions. To overcome this obstacle, it is essential to equip the system with a reliable onboard localization and controller system. One of the mitigation strategies can be based on integrating a data-driven wind and disturbance estimation method into the control architecture. By complementing the existing control system with accurate wind and disturbance information, the aerial robot can enhance its robustness and adaptability, thereby improving performance in challenging outdoor environments.

Accurate Trajectories and Precision: A vast number of outstanding applications with aerial robots including remote environmental sensing [Ho et al., 2022] utilize advanced control algorithms including variants of model predictive controllers [Tzoumanikas et al., 2019; Kocer et al., 2019] and learning-based controllers [Kocer et al., 2021]. However, there is still a need for approaches that can capture unexpected changes and disturbances without degrading the controller performance for conducting accurate tracking of the construction or repair trajectories. This problem is also coupled with the quality of the feedback, which is provided by the localization systems. One option is to use real-time kinematic GPS for outside manufacturing. However, it may not be applied indoors and additional sensors including cameras and lidar may be required. The research with visual and lidar odometry also requires further research to handle changes in the environment, illumination, and the places where the features are less [Chen et al., 2019; Chow et al., 2019]. We focused on developing parallel manipulators that stabilize the drone disruptions on defined trajectories. This could be also defined as an active trajectory publisher (e.g. [Stephens et al., 2022]) where it can detect the target geometry to orient itself while setting required setpoints for the flying base.

Distributed and Networked Aerial Robots: Aerial manufacturing in future may include multiple agents conducting construction onsite [Zhang et al., 2022]. For example, in a swarm construction task, each aerial robot can move based on the trajectories of the neighbor robot [Petersen et al., 2019]. This can provide multiple advantages including task partitioning, additional redundancy, scalability, and coverage. However, such a networked system brings additional challenges due to the communication, delay, computation complexity with the increased number of agents, and the complex network structures with links [Eren et al., 2017]. These approaches can include local and relative decision frameworks that could be useful for the emergent cases including collision prevention [Xiao et al., 2021].

Sustainable Energy for Aerial Robots: One of the main bottlenecks of the aerial robot applications is to have limited energy sources that result in a shorter operation times. A longer operation time is a desired expectation, and the initial attempts explored the use of tethered flight [Boukoberine et al., 2019], perching [Nguyen et al., 2019], and in-flight battery swap [Jain and Mueller, 2020]. There is also an interest toward the use of the battery charging stations that can provide continuous mission for the aerial robots. However, an ideal solution could be depending on the battery research focusing on battery chemistry, cell and pack engineering to have longer operations [Viswanathan et al., 2022]. The following areas could be useful to address energy autonomy challenge for complex tasks with aerial robots: (i) energy dense power sources; (ii) efficient power management; (iii) onboard power generation and energy harvesting; (iv) perching and physical contact to extend flight time; (v) efficient task prioritization in multiple drone applications; and (vi) energy efficient material deposition.

Closed Loop Vision: Current robot platforms explore the use of the vision to determine the next actions for the construction [Tish et al., 2020]. For aerial robots, this problem is coupled with the localization where the system needs to rely on its perception sources while conducting constructive actions. This is still a challenge for the aerial robot applications where the motion capture system is not available.

Security and Resilience: Efficient deployment of the aerial robots in the field is subject to the security and reliability threats. In particular, cybersecurity and resilience against potential attacks are major challenges for the real construction applications [Eren et al., 2017]. The impact of such infiltrations could be more catastrophic if the agent of interest is part of a networked system. Therefore, the countermeasures will be more important in future for the manufacturing by aerial robots.

Regulations: The issue of regulations plays a crucial role in the deployment of UAVs and significantly influences where and how they can be flown. To navigate the regulatory landscape and ensure compliant UAV deployment, new developments would be required in the following areas but not limited to: (i) legal framework; (ii) flight restrictions; (iii) operator certifications; (iv) beyond visual line of sight operations; and (v) data privacy and security. Regarding where and how to fly, it is imperative to carefully select deployment locations that comply with regulations, avoid restricted areas, and prioritize safety. Conducting flights in designated UAV-friendly areas, such as approved construction areas or dedicated testing facilities, can help mitigating safety and security issues. Furthermore, considering the operational requirements, flight objectives, and potential impact on surrounding environments and communities is essential in determining the appropriate locations and flight strategies. By adhering to regulations and implementing responsible flight practices, UAV deployments can be conducted effectively and in accordance with legal requirements.

Lightweight Material and Cost of Transportation: For a construction in challenging environments, each additional payload affects the cost and sustainability. Additional bulk volumes of material also affect emission when it is moved from resource to the construction area. To address the limitations of aerial robots with limited payload capabilities and exploit the advantages of aerial robots, we envision a future where hard to reach places could be a better place for aerial material deposition agents. Furthermore, additional research and development efforts can focus on exploring and formulating lightweight additive materials specifically designed for UAV-based additive manufacturing.

17.2.2 Future Prospects

Further development of novel aerial construction methods demonstrates great potential for the industry and the current technological boundaries. In near future, aerial manufacturing would be a more viable candidate when construction is required in an extraterrestrial place where the system and material transportation will be critical.

One of the potential methods can be subtractive aerial manufacturing, which an aerial platform with a manipulator subtracts from a block of materials to construct the aimed structure. A good base for this technology can be the methodology proposed by Duenser et al. [2020], which differentiate itself with other subtractive manufacturing methods by deploying a flexible curvilinear hot wire manipulated by two robotic arm simultaneously. A potential application by a single aerial agent is illustrated in Figure 17.4. This can be also achieved by a multiple drones holding a hot wire from both ends where the systems can be actuated fully [Zheng et al., 2020].

Another aerial construction technology might entail the holographic production of any structure with aerial vehicles, which is illustrated in Figure 17.5. This could be also beneficial in the sensitive areas to construct a digital copy of the intended design. Furthermore, a holographic representation built by drones can be also used as a guidepost during the construction [Jahn et al., 2020]. Another challenge in the

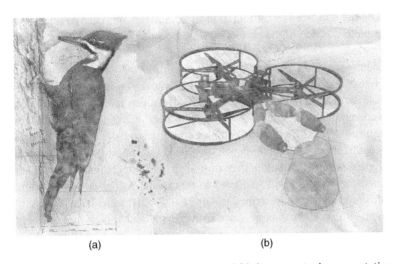

(a)   (b)

Figure 17.4 (a) The Red-head Woodpecker and (b) the conceptual representation of aerial substractive manufacturing with a custom high-payload aerial platform [Orr et al., 2021].

Figure 17.5 The conceptual visualization of aerial additive manufacturing of light (Holodrone).

construction site is to estimate the remaining time, and this approach could project the uncompleted section of the project that can be a reference for the remaining time estimation.

Initial demonstrations of AAM were focused on deposition of polyurethane foam. A system carrying a small volume of material and a lightweight manipulator could be relatively lightweight (2–5 kg). However, to carry construction materials currently in use such as cement, a considerably higher payload system will be required. Not only will more complex and heavy-duty deposition systems be required, but larger manipulators will be needed to support them. To perform SLAM for localization in an outdoors environment, additional sensors and higher performance computation will also need to be carried. A high payload system to meet this requirement is described in Orr et al. [2021]. In its current state of development, it can carry a large delta manipulator and a high-force cement extruder. It also features a 3D lidar for precise spatial localization [Kocer et al., 2022]. Unlike previous systems developed at the Aerial Robotics Laboratory, its manipulator is in a forward-facing orientation for repair and building from vertical surfaces.

17.3 Initial Experimental Evaluations

Our initial investigations present experimental findings on the trajectory and printing precision in AAM. A custom trajectory generator was developed to

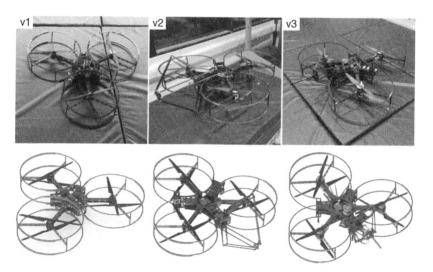

Figure 17.6 Evolution of aerial platforms for aerial repair and manufacturing.

generate circular trajectories for printing tests. The initial experiments focused on achieving two-layered printing using a custom-made aerial platform. The trajectory precision was assessed with a VICON camera system, demonstrating a degree of accuracy within a +/−3 cm range. The printing process exhibited satisfactory results, with the majority of the material reaching the intended surface despite minor spray caused by the propellers' downwash. The use of a FrothPak mini solution proved effective in achieving successful multilayering. These findings contribute to advancing the understanding and capabilities of AAM systems [Grillot, 2018].

These experimental findings demonstrate the successful implementation of circular trajectories and the achievement of multilayered printing using an AAM system. The precision of the trajectory and the material deposition capabilities contribute to expanding the potential applications of UAV-based additive manufacturing. A systematic pipeline of our system can be found in Zhang at et al. [2022] and design iteration for the base aerial platform is illustrated in Figure 17.6. Development stages of the platform can be found in Orr et al. [2021].

17.4 Conclusion and Discussion

In conclusion, AAM has the potential to revolutionize the construction and repair industry by allowing for the creation of new structures and the performance of repairs in challenging or inaccessible locations, including extraterrestrial places. The use of aerial agents enables the easy transportation of materials to these locations, further expanding the potential applications of AAM. While much progress has been made in the development of aerial manufacturing systems, there is still a need for more advanced platforms, localization techniques, and end effectors to achieve the level of precision and stability required for successful aerial building and repair. Our current work on large-scale aerial platforms and the development of AAM techniques aim to address this need and demonstrate the feasibility of AAM for a range of applications.

Bibliography

Hamad Al Jassmi, Fady Al Najjar, and Abdel-Hamid Ismail Mourad. Large-scale 3D printing: The way forward. In *IOP Conference Series: Materials Science and Engineering*, volume 324, page 012088. IOP Publishing, 2018.

Ans Al Rashid, Shoukat Alim Khan, Sami G. Al-Ghamdi, and Muammer Koç. Additive manufacturing: Technology, applications, markets, and opportunities for the built environment. *Automation in Construction*, 118:103268, 2020.

Federico Augugliaro, Ammar Mirjan, Fabio Gramazio, Matthias Kohler, and Raffaello D'Andrea. Building tensile structures with flying machines. In *2013 IEEE/RSJ International Conference on Intelligent Robots and Systems*, pages 3487–3492. IEEE, 2013.

Frederico Augugliaro, Sergei Lupashin, Michael Hamer, Cason Male, Markus Hehn, Mark W. Mueller, Jan Sebastian Willmann, Fabio Gramazio, Matthias Kohler, and Raffaello D'Andrea. The flight assembled architecture installation: Cooperative construction with flying machines. *IEEE Control Systems Magazine*, 34(4):46–64, 2014.

Federico Augugliaro, Emanuele Zarfati, Ammar Mirjan, and Raffaello D'Andrea. Knot-tying with flying machines for aerial construction. In *2015 IEEE/RSJ International Conference on Intelligent Robots and Systems (IROS)*, pages 5917–5922. IEEE, 2015.

Batiprint3D. Batiprint3d shaping tomorrow, 2023. URL https://www.batiprint3d.com. Accessed on 7/7/2023.

Mohamed Nadir Boukoberine, Zhibin Zhou, and Mohamed Benbouzid. Power supply architectures for drones-a review. In *IECON 2019-45th Annual Conference of the IEEE Industrial Electronics Society*, volume 1, pages 5826–5831. IEEE, 2019.

Adam Braithwaite, Talib Alhinai, Maximilian Haas-Heger, Edward McFarlane, and Mirko Kovač. Tensile web construction and perching with nano aerial vehicles. *Robotics Research*, 1:71–88, 2018.

Jane Burry, Jenny E. Sabin, Bob Sheil, and Marilena Skavara. *Fabricate 2020*. UCL Press, 2020.

Richard A. Buswell, W.R. Leal De Silva, Scott Z. Jones, and Justin Dirrenberger. 3D printing using concrete extrusion: A roadmap for research. *Cement and Concrete Research*, 112:37–49, 2018.

Leong Jing Chen, John Henawy, Basaran Bahadir Kocer, and Gerald Gim Lee Seet. Aerial robots on the way to underground: An experimental evaluation of VINS-mono on visual-inertial odometry camera. In *2019 International Conference on Data Mining Workshops (ICDMW)*, pages 91–96. IEEE, 2019.

Pisak Chermprayong, Ketao Zhang, Feng Xiao, and Mirko Kovac. An integrated delta manipulator for aerial repair: A new aerial robotic system. *IEEE Robotics & Automation Magazine*, 26(1):54–66, 2019.

Jiun Fatt Chow, Basaran Bahadir Kocer, John Henawy, Gerald Seet, Zhengguo Li, Wei Yun Yau, and Mahardhika Pratama. Toward underground localization: LiDAR inertial odometry enabled aerial robot navigation. *arXiv preprint arXiv:1910.13085*, 2019.

A. Cor. A revolutionary robotic 3D printer: Apis cor, 2022. URL https://www.apis-cor.com/3dprinter. Accessed on 7/7/2023.

Zakaria Dakhli and Zoubeir Lafhaj. Robotic mechanical design for brick-laying automation. *Cogent Engineering*, 4(1):1361600, 2017.

Barrie Dams, Sina Sareh, Ketao Zhang, Paul Shepherd, Mirko Kovac, and Richard J. Ball. Aerial additive building manufacturing: Three-dimensional printing of polymer structures using drones. *Proceedings of the Institution of Civil Engineers-Construction Materials*, 173(1):3–14, 2020.

Juan Manuel Davila Delgado, Lukumon Oyedele, Anuoluwapo Ajayi, Lukman Akanbi, Olugbenga Akinade, Muhammad Bilal, and Hakeem Owolabi. Robotics and automated systems in construction: Understanding industry-specific challenges for adoption. *Journal of Building Engineering*, 26:100868, 2019.

Stylianos Dritsas, Yadunund Vijay, Samuel Halim, Ryan Teo, Naresh Sanandiya, and Javier G. Fernandez. Cellulosic biocomposites for sustainable manufacturing. *Fabricate*, 2020:74–81, 2020.

Dshape. D-shape project, 2023. URL https://d-shape.com. Accessed on 7/7/2023.

Simon Duenser, Roi Poranne, Bernhard Thomaszewski, and Stelian Coros. RoboCut: Hot-wire cutting with robot-controlled flexible rods. *ACM Transactions on Graphics (TOG)*, 39(4):98:1–98:15, 2020.

Utku Eren, Anna Prach, Başaran Bahadır Koçer, Saša V. Raković, Erdal Kayacan, and Behçet Açıkmeşe. Model predictive control in aerospace systems: Current state and opportunities. *Journal of Guidance, Control, and Dynamics*, 40(7):1541–1566, 2017.

Seyed Hamidreza Ghaffar, Jorge Corker, and Mizi Fan. Additive manufacturing technology and its implementation in construction as an eco-innovative solution. *Automation in Construction*, 93:1–11, 2018.

M. Gharbia, Alice Chang-Richards, and Runyang Zhong. Robotic technologies in concrete building construction: A systematic review. In *International Symposium On Automation and Robotics in Construction*. The International Association for Automation and Robotics in Construction ..., 2019.

W. Global. 3D Printing Architecture's Future 2022, 2023. URL http://www.winsun 3d.com/En. Accessed on 01/10/2023.

Sébastien Goessens, Caitlin Mueller, and Pierre Latteur. Feasibility study for drone-based masonry construction of real-scale structures. *Automation in Construction*, 94:458–480, 2018.

F. Hauf, B.B. Kocer, A. Slatter, H.-N. Nguyen, O. Pang, and R. Clark. Learning tethered perching for aerial robots. In *2023 IEEE International Conference on Robotics and Automation (ICRA)*, pages 5917–5922. IEEE, 2023.

Volker Helm, Selen Ercan, Fabio Gramazio, and Matthias Kohler. Mobile robotic fabrication on construction sites: DimRob. In *2012 IEEE/RSJ International Conference on Intelligent Robots and Systems*, pages 4335–4341. IEEE, 2012.

Boon Ho, Basaran Bahadir Kocer, and Mirko Kovac. Vision based crown loss estimation for individual trees with remote aerial robots. *ISPRS Journal of Photogrammetry and Remote Sensing*, 188:75–88, 2022.

Graham Hunt, Faidon Mitzalis, Talib Alhinai, Paul A. Hooper, and Mirko Kovac. 3D printing with flying robots. In *2014 IEEE International Conference on Robotics and Automation (ICRA)*, pages 4493–4499. IEEE, 2014.

ICON. ICON Icon and Lennar to build largest neighborhood of 3D-printed homes codesigned by BIG-Bjarke Ingels Group ICON build2022, 2023. URL https://www.

iconbuild.com/updates/icon-and-lennar-to-build-largest-neighborhood-of-3d-printed-homes-codesigned. Accessed on 7/7/2023.

Gwyllim Jahn, Cameron Newnham, Nick van den Berg, Melissa Iraheta, and Jackson Wells. Holographic Construction. In *Impact: Design With All Senses: Proceedings of the Design Modelling Symposium*, Berlin 2019, pages 314–324. Springer-Verlag, 2020.

Karan P. Jain and Mark W. Mueller. Flying batteries: In-flight battery switching to increase multirotor flight time. In *2020 IEEE International Conference on Robotics and Automation (ICRA)*, pages 3510–3516. IEEE, 2020.

Sasa Jokic, Petr Novikov, Stuart Maggs, Dori Sadan, Shihui Jin, and Cristina Nan. Robotic positioning device for three-dimensional printing. *arXiv preprint arXiv:1406.3400*, 2014.

Steven J. Keating, Julian C. Leland, Levi Cai, and Neri Oxman. Toward site-specific and self-sufficient robotic fabrication on architectural scales. *Science Robotics*, 2(5):eaam8986, 2017.

Behrokh Khoshnevis. Automated construction by contour crafting–related robotics and information technologies. *Automation in Construction*, 13(1):5–19, 2004.

Basaran Bahadir Kocer, Mehmet Efe Tiryaki, Mahardhika Pratama, Tegoeh Tjahjowidodo, and Gerald Gim Lee Seet. Aerial robot control in close proximity to ceiling: A force estimation-based nonlinear MPC. In *2019 IEEE/RSJ International Conference on Intelligent Robots and Systems (IROS)*, pages 2813–2819. IEEE, 2019.

Basaran Bahadir Kocer, Mohamad Abdul Hady, Harikumar Kandath, Mahardhika Pratama, and Mirko Kovac. Deep neuromorphic controller with dynamic topology for aerial robots. In *2021 IEEE International Conference on Robotics and Automation (ICRA)*, pages 110–116. IEEE, 2021.

Basaran Bahadir Kocer, Lachlan Orr, Brett Stephens, Yusuf Furkan Kaya, Tetiana Buzykina, Asiya Khan, and Mirko Kovac. An intelligent aerial manipulator for wind turbine inspection and repair. In *2022 UKACC 13th International Conference on Control (CONTROL)*, pages 226–227. IEEE, 2022.

Anna Kulik, Inder Shergill, and Petr Novikov. Stone spray robot, 2012. URL https://www.dezeen.com/2012/08/22/stone-spray-robot-by-anna-kulik-inder-shergill-and-petr-novikov/. Accessed on 7/7/2023.

Pierre Latteur, Sébastien Goessens, Jean-Sébastien Breton, Justin Leplat, Zhao Ma, and Caitlin Mueller. Drone-based additive manufacturing of architectural structures. In *Proceedings of IASS Annual Symposia*, pages 1–12. International Association for Shell and Spatial Structures (IASS), 2015.

Sungwoo Lim, Richard A. Buswell, Thanh T. Le, Simon A. Austin, Alistair G.F. Gibb, and Tony Thorpe. Developments in construction-scale additive manufacturing processes. *Automation in Construction*, 21:262–268, 2012.

Quentin Lindsey, Daniel Mellinger, and Vijay Kumar. Construction of cubic structures with quadrotor teams. In *Proceedings of the Robotics: Science & Systems VII 7*, 2011.

Ammar Mirjan, Fabio Gramazio, Matthias Kohler, Federico Augugliaro, and Raffaello D'Andrea. Architectural fabrication of tensile structures with flying machines. *Green Design, Materials and Manufacturing Processes*, 513–518, Taylor & Francis Group, London, 2013, ISBN 978-1-138-00046-9.

Alexander Nettekoven and Ufuk Topcu. A 3D printing hexacopter: Design and demonstration. In *2021 International Conference on Unmanned Aircraft Systems (ICUAS)*, pages 1472–1477. IEEE, 2021.

Hai-Nguyen Nguyen, Robert Siddall, Brett Stephens, Alberto Navarro-Rubio, and Mirko Kovač. A passively adaptive microspine grapple for robust, controllable perching. In *2019 2nd IEEE International Conference on Soft Robotics (RoboSoft)*, pages 80–87. IEEE, 2019.

Lachlan Orr, Brett Stephens, Basaran Bahadir Kocer, and Mirko Kovac. A high payload aerial platform for infrastructure repair and manufacturing. In *2021 Aerial Robotic Systems Physically Interacting with the Environment (AIRPHARO)*, pages 1–6. IEEE, 2021.

Kirstin H. Petersen, Nils Napp, Robert Stuart-Smith, Daniela Rus, and Mirko Kovac. A review of collective robotic construction. *Science Robotics*, 4(28):eaau8479, 2019.

Gergana Rusenova, Falk K. Wittel, Petrus Aejmelaeus-Lindström, Fabio Gramazio, and Matthias Kohler. Load-bearing capacity and deformation of jammed architectural structures. *3D Printing and Additive Manufacturing*, 5(4):257–267, 2018.

David Saldana, Bruno Gabrich, Guanrui Li, Mark Yim, and Vijay Kumar. ModQuad: The flying modular structure that self-assembles in midair. In *2018 IEEE International Conference on Robotics and Automation (ICRA)*, pages 691–698. IEEE, 2018.

Brett Stephens, Lachlan Orr, Basaran Bahadir Kocer, Hai-Nguyen Nguyen, and Mirko Kovac. An aerial parallel manipulator with shared compliance. *IEEE Robotics and Automation Letters*, 7(4):11902–11909, 2022.

Kévin Subrin, Thomas Bressac, Sébastien Garnier, Alexandre Ambiehl, Elodie Paquet, and Benoit Furet. Improvement of the mobile robot location dedicated for habitable house construction by 3D printing. *IFAC-PapersOnLine*, 51(11):716–721, 2018.

Julius Sustarevas, Daniel Butters, Mohammad Hammid, George Dwyer, Robert Stuart-Smith, and Vijay M. Pawar. Map-a mobile agile printer robot for on-site construction. In *2018 IEEE/RSJ International Conference on Intelligent Robots and Systems (IROS)*, pages 2441–2448. IEEE, 2018.

Julius Sustarevas, K.X. Benjamin Tan, David Gerber, Robert Stuart-Smith, and Vijay M. Pawar. YouWasps: Towards autonomous multi-robot mobile deposition for construction. In *2019 IEEE/RSJ International Conference on Intelligent Robots and Systems (IROS)*, pages 2320–2327. IEEE, 2019.

Yi Wei Daniel Tay, Biranchi Panda, Suvash Chandra Paul, Nisar Ahamed Noor Mohamed, Ming Jen Tan, and Kah Fai Leong. 3D printing trends in building and construction industry: A review. *Virtual and Physical Prototyping*, 12(3):261–276, 2017.

Mehmet Efe Tiryaki, Xu Zhang, and Quang-Cuong Pham. Printing-while-moving: A new paradigm for large-scale robotic 3D printing. In *2019 IEEE/RSJ International Conference on Intelligent Robots and Systems (IROS)*, pages 2286–2291. IEEE, 2019.

Daniel Tish, Nathan King, and Nicholas Cote. Highly accessible platform technologies for vision-guided, closed-loop robotic assembly of unitized enclosure systems. *Construction Robotics*, 4:19–29, 2020.

Dimos Tzoumanikas, Wenbin Li, Marius Grimm, Ketao Zhang, Mirko Kovac, and Stefan Leutenegger. Fully autonomous micro air vehicle flight and landing on a moving target using visual–inertial estimation and model-predictive control. *Journal of Field Robotics*, 36(1):49–77, 2019.

Venkatasubramanian Viswanathan, Alan H. Epstein, Yet-Ming Chiang, Esther Takeuchi, Marty Bradley, John Langford, and Michael Winter. The challenges and opportunities of battery-powered flight. *Nature*, 601(7894):519–525, 2022.

WASP Crane WASP. The infinite 3d printer 3d wasp2022, 2023. URL https://www.3dwasp.com/en/3d-printer-house-crane-wasp/. Accessed on 7/7/2023.

WASP Delta. Delta WASP 3MT Concrete, 2023. URL https://www.3dwasp.com/en/concrete-3d-printer-delta-wasp-3mt-concrete/. Accessed on 7/7/2023.

Dylan Wood, Maria Yablonina, Miguel Aflalo, Jingcheng Chen, Behrooz Tahanzadeh, and Achim Menges. Cyber Physical Macro Material as a UAV [re]Configurable Architectural System. In J. Willmann, P. Block, M. Hutter, K. Byrne, and T. Schork, editors, *Robotic Fabrication in Architecture, Art and Design 2018: Foreword by Sigrid Brell-Çokcan and Johannes Braumann, Association for Robots in Architecture*, pages 320–335. Springer, 2019.

Feng Xiao, Peter Zheng, Julien Di Tria, Basaran Bahadir Kocer, and Mirko Kovac. Optic flow-based reactive collision prevention for MAVs using the fictitious obstacle hypothesis. *IEEE Robotics and Automation Letters*, 6(2):3144–3151, 2021.

Xu Zhang, Mingyang Li, Jian Hui Lim, Yiwei Weng, Yi Wei Daniel Tay, Hung Pham, and Quang-Cuong Pham. Large-scale 3D printing by a team of mobile robots. *Automation in Construction*, 95:98–106, 2018.

Ketao Zhang, Pisak Chermprayong, Feng Xiao, Dimos Tzoumanikas, Barrie Dams, Sebastian Kay, Basaran Bahadir Kocer, Alec Burns, Lachlan Orr, Christopher Choi, et al. Aerial additive manufacturing with multiple autonomous robots. *Nature*, 609(7928):709–717, 2022.

Peter Zheng, Xinkai Tan, Basaran Bahadir Kocer, Erdeng Yang, and Mirko Kovac. TiltDrone: A fully-actuated tilting quadrotor platform. *IEEE Robotics and Automation Letters*, 5(4):6845–6852, 2020.

Grillot C., *Automatic Extrusion System for Aerial Additive Manufacturing*, in *Department of Aeronautics*. 2018, Imperial College London.

Index

Note: Page number followed by "*f*" denotes figures and page number followed by "*t*" denotes tables respectively.

Infrastructure Robotics: Methodologies, Robotic Systems and Applications, First Edition.
Edited by Dikai Liu, Carlos Balaguer, Gamini Dissanayake, and Mirko Kovac.
© 2024 The Institute of Electrical and Electronics Engineers, Inc. Published 2024 by John Wiley & Sons, Inc.

IEEE Press Series on Systems Science and Engineering

Editor: **MengChu Zhou**, *New Jersey Institute of Technology*
Co-Editors: **Han-Xiong Li**, *City University of Hong-Kong*
Margot Weijnen, *Delft University of Technology*

The focus of this series is to introduce the advances in theory and applications of systems science and engineering to industrial practitioners, researchers, and students. This series seeks to foster system of systems multidisciplinary theory and tools to satisfy the needs of the industrial and academic areas to model, analyze, design, optimize, and operate increasingly complex man-made systems ranging from control systems, computer systems, discrete event systems, information systems, networked systems, production systems, robotic systems, service systems, and transportation systems to Internet, sensor networks, smart grid, social network, sustainable infrastructure, and systems biology.

Printed and bound by CPI Group (UK) Ltd, Croydon, CR0 4YY

16/04/2025

14658598-0004